Examkrackers MCAT®

BIOLOGY

8TH EDITION

OSOTE
PUBLISHING

ISBN 10: 1-893858-62-6 (Volume 1)
ISBN 13: 978-1-893858-62-6 (5 Volume Set)

8th Edition

To purchase additional copies of this book or the rest of the 5 volume set, call 1-888-572-2536 or fax orders to
1-859-255-0109.

Examkrackers.com

Osote.com

Audioosmosis.com

Cover/inside layout design/illustrations: Examkrackers' staff

Printed and bound in the United States of America.

Acknowledgements

Although I am the author, the hard work and expertise of many individuals contributed to this book. The idea of writing in two voices, a science voice and an MCAT voice, was the creative brainchild of my imaginative friend Jordan Zaretsky. I would like to thank Scott Calvin for lending his exceptional science talent and pedagogic skills to this project. I also must thank thirteen years worth of ExamKrackers students for doggedly questioning every explanation, every sentence, every diagram, and every punctuation mark in the book, and for providing the creative inspiration that helped me find new ways to approach and teach biology. Finally, I wish to thank my wife, Silvia, for her support during the difficult times in the past and those that lie ahead.

I also wish to thank the following individuals:

Contributors

Jennifer Birk-Goldschmidt
Patrick Butler
Dr. Jerry Johnson
Jordie Mann
Graeme McHenry
Timothy Peck
Gibran Shaikh
Vinita Takiar
Sara Thorp
Ruopeng Zhu

Read This Section First!

This manual contains all the biology tested on the MCAT® and more. It contains more biology than is tested on the MCAT® because a deeper understanding of basic scientific principles is often gained through more advanced study. In addition, the MCAT® often presents passages with imposing topics that may intimidate the test-taker. Although the questions don't require knowledge of these topics, some familiarity will increase the confidence of the test-taker.

In order to answer questions quickly and efficiently, it is vital that the test-taker understand what is, and is not, tested directly by the MCAT®. To assist the test-taker in gaining this knowledge, this manual will use the following conventions. Any term or concept which is tested directly by the MCAT® will be written in **red, bold type**. To ensure a perfect score on the MCAT®, you should thoroughly understand all terms and concepts that are in **red, bold type** in this manual. Sometimes it is not necessary to memorize the name of a concept, but it is necessary to understand the concept itself. These concepts will also be in **bold and red**. It is important to note that the converse of the above is not true: just because a topic is not in **bold and red**, does not mean that it is not important.

Any formula that must be memorized will also be written in **red, bold type.**

If a topic is discussed purely as background knowledge, it will be written in *italics*. If a topic is written in italics, it is not likely to be required knowledge for the MCAT® but may be discussed in an MCAT® passage. Do not ignore items in italics, but recognize them as less important than other items. Answers to questions that directly test knowledge of italicized topics are likely to be found in an MCAT® passage.

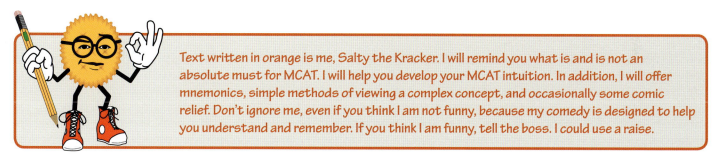

Text written in orange is me, Salty the Kracker. I will remind you what is and is not an absolute must for MCAT. I will help you develop your MCAT intuition. In addition, I will offer mnemonics, simple methods of viewing a complex concept, and occasionally some comic relief. Don't ignore me, even if you think I am not funny, because my comedy is designed to help you understand and remember. If you think I am funny, tell the boss. I could use a raise.

Each chapter in this manual should be read three times: twice before the class lecture, and once immediately following the lecture. During the first reading, you should not write in the book. Instead, read purely for enjoyment. During the second reading, you should both highlight and take notes in the margins. The third reading should be slow and thorough.

The 24 questions in each lecture should be worked during the second reading before coming to class. The in-class exams in the back of the book are to be done in class after the lecture. Do not look at them before class.

Warning: Just attending the class will not raise your score. You must do the work. Not attending class will obstruct dramatic score increases. If you have Audio Osmosis, then listen to the appropriate lecture before and after you read a lecture.

If you are studying independently, read the lecture twice before doing the in-class exam and then once after doing the in-class exam. If you have Examkrackers *MCAT® Audio Osmosis With Jordan and Jon*, listen to that before taking the in-class exam and then as many times as necessary after taking the exam.

A scaled score conversion chart is provided on the answer page. This is not meant to be an accurate representation of your MCAT score. Do not become demoralized by a poor performance on these exams; they are not accurate reflections of your performance on the real MCAT®. The thirty minute exams have been designed to educate. They are similar to an MCAT® but with most of the easy questions removed. We believe that you can answer most of the easy questions without too much help from us, so the best way to raise your score is to focus on the more difficult questions. This method is one of the reasons for the rapid and celebrated success of the Examkrackers prep course and products.

If you find yourself struggling with the science or just needing more practice materials, use the Examkrackers 1001 Questions series. These books are designed specifically to teach the science. If you are already scoring 10s or better, these books are not for you.

You should take advantage of the forums at www.examkrackers.com. The bulletin board allows you to discuss any question in the book with an MCAT® expert at Examkrackers. All discussions are kept on file so you have a bank of discussions to which you can refer to any question in this book.

Although we are very careful to be accurate, errata is an occupational hazard of any science book, especially those that are updated regularly as is this one. We maintain that our books have fewer errata than any other prep book. Most of the time what students are certain are errata is the student's error and not an error in the book. So that you can be certain, any errata in this book will be listed as it is discovered at www.examkrackers.com on the bulletin board. Check this site initially and periodically. If you discover what you believe to be errata, please post it on this board and we will verify it promptly. We understand that this system calls attention to the very few errata that may be in our books, but we feel that this is the best system to ensure that you have accurate information for your exam. Again, we stress that we have fewer errata than any other prep book on the market. The difference is that we provide a public list of our errata for your benefit.

Study diligently, trust this book to guide you, and you will reach your MCAT® goals.

Table of Contents

BIOLOGICAL SCIENCES

DIRECTIONS. Most questions in the Biological Sciences test are organized into groups, each preceded by a descriptive passage. After studying the passage, select the one best answer to each question in the group. Some questions are not based on a descriptive passage and are also independent of each other. You must also select the one best answer to these questions. If you are not certain of an answer, eliminate the alternatives that you know to be incorrect and then select an answer from the remaining alternatives. A periodic table is provided for your use. You may consult it whenever you wish.

PERIODIC TABLE OF THE ELEMENTS

1 H 1.0																	2 He 4.0
3 Li 6.9	4 Be 9.0											5 B 10.8	6 C 12.0	7 N 14.0	8 O 16.0	9 F 19.0	10 Ne 20.2
11 Na 23.0	12 Mg 24.3											13 Al 27.0	14 Si 28.1	15 P 31.0	16 S 32.1	17 Cl 35.5	18 Ar 39.9
19 K 39.1	20 Ca 40.1	21 Sc 45.0	22 Ti 47.9	23 V 50.9	24 Cr 52.0	25 Mn 54.9	26 Fe 55.8	27 Co 58.9	28 Ni 58.7	29 Cu 63.5	30 Zn 65.4	31 Ga 69.7	32 Ge 72.6	33 As 74.9	34 Se 79.0	35 Br 79.9	36 Kr 83.8
37 Rb 85.5	38 Sr 87.6	39 Y 88.9	40 Zr 91.2	41 Nb 92.9	42 Mo 95.9	43 Tc (98)	44 Ru 101.1	45 Rh 102.9	46 Pd 106.4	47 Ag 107.9	48 Cd 112.4	49 In 114.8	50 Sn 118.7	51 Sb 121.8	52 Te 127.6	53 I 126.9	54 Xe 131.3
55 Cs 132.9	56 Ba 137.3	57 La* 138.9	72 Hf 178.5	73 Ta 180.9	74 W 183.9	75 Re 186.2	76 Os 190.2	77 Ir 192.2	78 Pt 195.1	79 Au 197.0	80 Hg 200.6	81 Tl 204.4	82 Pb 207.2	83 Bi 209.0	84 Po (209)	85 At (210)	86 Rn (222)
87 Fr (223)	88 Ra 226.0	89 Ac⁼ 227.0	104 Unq (261)	105 Unp (262)	106 Unh (263)	107 Uns (262)	108 Uno (265)	109 Une (267)									

*	58 Ce 140.1	59 Pr 140.9	60 Nd 144.2	61 Pm (145)	62 Sm 150.4	63 Eu 152.0	64 Gd 157.3	65 Tb 158.9	66 Dy 162.5	67 Ho 164.9	68 Er 167.3	69 Tm 168.9	70 Yb 173.0	71 Lu 175.0
⁼	90 Th 232.0	91 Pa (231)	92 U 238.0	93 Np (237)	94 Pu (244)	95 Am (243)	96 Cm (247)	97 Bk (247)	98 Cf (251)	99 Es (252)	100 Fm (257)	101 Md (258)	102 No (259)	103 Lr (260)

Molecular Biology; Cellular Respiration

LECTURE 1

1.1 Introduction

This lecture discusses the structure and basic functions of the major chemical components of living cells and their surroundings. Although most of the details of this biochemistry are not required on the MCAT, this knowledge does create a strong base from which to understand the rest of the manual.

Most biological molecules can be classified as lipids, proteins, carbohydrates or nucleotide derivatives. Each of these types of molecules possesses a carbon skeleton. Together with water and minerals, they form living cells and their environment.

Cellular respiration provides the energy needed for muscles to do work.

Water

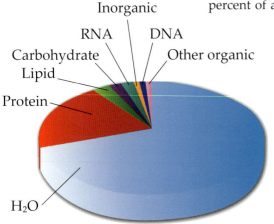

Inorganic
RNA   DNA
Carbohydrate   Other organic
Lipid
Protein
H₂O

Water is the solvent in which the chemical reactions of living cells take place. 70 to 80 percent of a cell's mass is due to water (chart on left). Water is a small polar molecule that can **hydrogen bond**. Most compounds as light as water would exist as a gas at typical cell temperatures. The ability of water to hydrogen bond allows it to maintain its liquid state in the cellular environment. Hydrogen bonding also provides strong cohesive forces between water molecules. These cohesive forces "squeeze" **hydrophobic** (Greek: hydros → water, phobos → fear) molecules away from water, and cause them to aggregate. **Hydrophilic** (Greek: philos → love) molecules dissolve easily in water because their negatively charged ends attract the positively charged hydrogens of water, and their positively charged ends attract the negatively charged oxygen of water (Figure 1.1). Thus, water molecules surround (solvate) a hydrophilic molecule separating it from the group.

Hydrogen bonding between individual water molecules creates cohesive forces that are strong enough to support this Raft Spider.

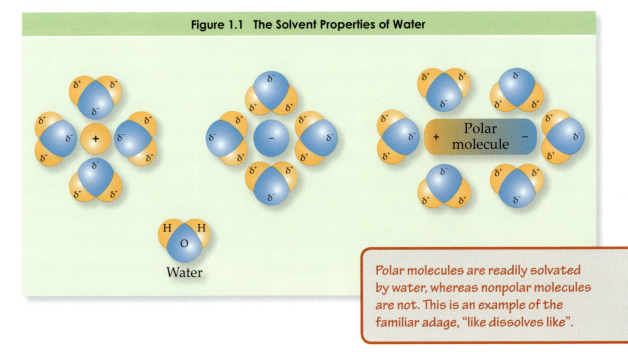

Figure 1.1   The Solvent Properties of Water

H   H
O
Water

Polar molecule

Polar molecules are readily solvated by water, whereas nonpolar molecules are not. This is an example of the familiar adage, "like dissolves like".

Besides acting as a solvent, water often acts as a reactant or product. Most macromolecules of living cells are broken apart via **hydrolysis** (Greek: lysis → separation), and are formed via **dehydration** synthesis. (Hydrolysis and dehydration are discussed in Biology Lecture 7, and in the Organic Chemistry manual.)

1.3 Lipids

A **lipid** is any biological molecule that has low solubility in water and high solubility in nonpolar organic solvents. Because they are hydrophobic, they make excellent barriers separating aqueous environments. Six major groups of lipids are: fatty acids, triacylglycerols, phospholipids, glycolipids, steroids, and terpenes (Figure 1.2).

Besides being lipids themselves, **fatty acids** are the building blocks for most, but not all, complex lipids. They are long chains of carbons truncated at one end by a carboxylic acid. There is usually an even number of carbons, with the maximum number of carbons in humans being 24. Fatty acids can be saturated or unsaturated. **Saturated fatty acids** possess only single carbon-carbon bonds. **Unsaturated fatty acids** contain one or more carbon-carbon double bonds. Oxidation of fatty acids liberates large amounts of chemical energy for a cell. Most fats reach the cell in the form of fatty acids, and not as triacylglycerols.

Triacylglycerols, phospholipids, and glycolipids are sometimes referred to as fatty acids.

Omega-3 fatty acids are lipids. Omega-3 fatty acids are found in fish and other seafood including algae and krill, some plants, and nut oils.

Triacylglycerols (Latin: tri → three), commonly called **triglycerides** or simply **fats** and **oils,** are constructed from a three carbon backbone called **glycerol**, which is attached to three fatty acids (Figure 1.2). Their function in a cell is to store energy. They may also function to provide thermal insulation and padding to an organism. **Adipocytes** (Latin: adips → fat, Greek: kytos → cell), also called fat cells, are specialized cells whose cytoplasm contains almost nothing but triglycerides.

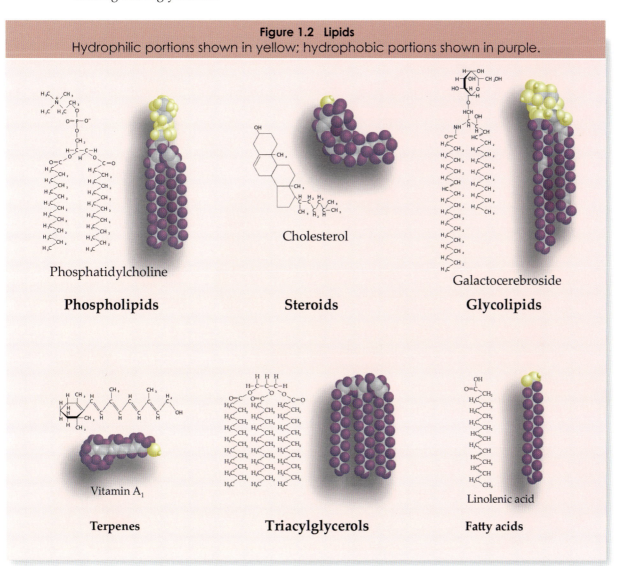

Figure 1.2 Lipids
Hydrophilic portions shown in yellow; hydrophobic portions shown in purple.

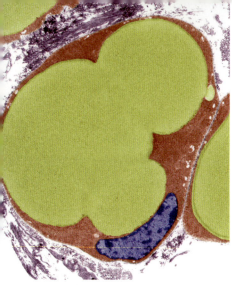

Each fat cell or adipocyte consists of a large central lipid droplet (yellow) surrounded by a thin layer of cytoplasm (red) containing the nucleus (blue). Fat cells store energy as an insulating layer of fat under the skin.

Phospholipids are built from a glycerol backbone as well, but a polar phosphate group replaces one of the fatty acids. The phosphate group lies on the opposite side of the glycerol from the fatty acids making the phospholipid polar at the phosphate end and nonpolar at the fatty acid end. This condition is called **amphipathic** (Latin: ambo → both) and makes phospholipids especially well suited as the major component of membranes. The polar end of the phospholipid to the right is in pink.

Glycolipids (Greek: glucus → sweet) are similar to phospholipids, except that glycolipids have one or more carbohydrates attached to the three-carbon glycerol backbone instead of the phosphate group. Glycolipids are also amphipathic. They are found in abundance in the membranes of myelinated cells composing the human nervous system.

Steroids are four ringed structures. They include some hormones, *vitamin D*, and cholesterol, an important membrane component.

Terpenes are a sixth class of lipids which include *vitamin A*, a vitamin important for vision.

Another class of lipids (not shown in Figure 1.2 but often listed as a fatty acid) is the 20 carbon *eicosanoids* (Greek: eikosi → twenty). Eicosanoids include prostaglandins, thromboxanes, and leukotrienes. Eicosanoids are released from cell membranes as local hormones that regulate, among other things, blood pressure, body temperature, and smooth muscle contraction. (See Paracrine System in Lecture 4 for more on local hormones.) Aspirin is a commonly used inhibitor of the synthesis of prostaglandins.

Since lipids are insoluble in aqueous solution, they are transported in the blood via *lipoproteins*. A lipoprotein contains a lipid core surrounded by phospholipids and *apoproteins* (apoproteins are discussed below). Thus the lipoprotein is able to dissolve lipids in its hydrophobic core, and then move freely through the aqueous solution due to its hydrophilic shell. Lipoproteins are classified by their density. The greater the ratio of lipid to protein, the lower the density. The major classes of lipoproteins in humans are chylomicrons, *very low density lipoproteins (VLDL)*, *low density lipoproteins (LDL)*, and *high density lipoproteins (HDL)*. (For more on lipoproteins, see Biology Lecture 6.11 – Fats.)

Lipoprotein

Know these major functions of lipids:

1. phospholipids serve as a structural component of membranes;

2. triacylglycerols store metabolic energy, provide thermal insulation and padding;

3. steroids regulate metabolic activities; and

4. some fatty acids (eicosanoids) even serve as local hormones.

1.4 Proteins

Peanuts are a source of protein.

Proteins are built from a chain of **amino acids** linked together by **peptide bonds** (Peptide bonds are discussed in Organic Chemistry Lecture 4). Thus proteins are sometimes referred to as **polypeptides** (Greek: polys → many). Nearly all proteins in all species are built from the same 20 α-amino acids. They are called alpha amino acids because the amine is attached to the carbon in the alpha position to the carbonyl. In humans, ten of the amino acids are **essential**. In other words the body cannot manufacture these 10, so they must be ingested directly. Each amino acid in a polypeptide chain is referred to as a *residue*; very small polypeptides are sometimes referred to as peptides. The amino acids typically differ from each other only in their **side chains**, often designated as the R group. The side chain is also attached to the α-carbon. Digested proteins reach the cells of the human body as single amino acids. The 20 amino acids are shown in Figure 1.3.

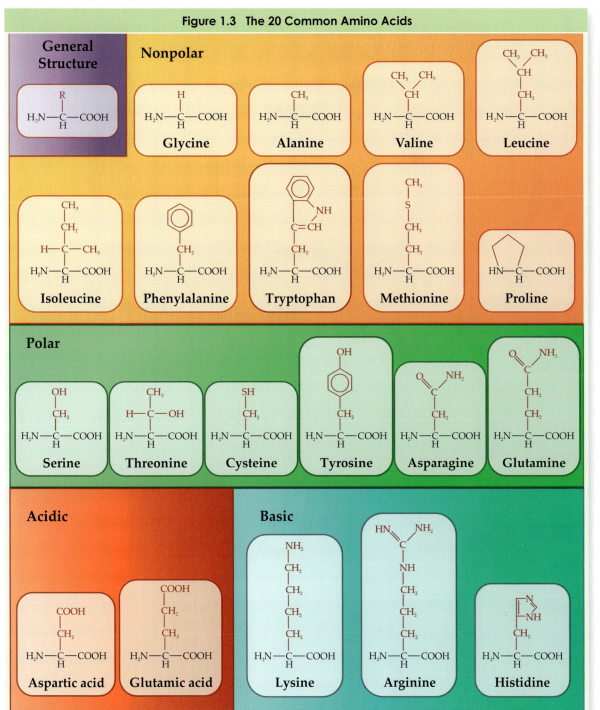

Figure 1.3 The 20 Common Amino Acids

The amino acid stuctures shown in Figure 1.3 are artificial. Amino acids in solution will always carry one or more charges. The position and nature of the charges will depend upon the pH of the solution.

Protein is found in nearly all unprocessed foods.

The number and sequence of amino acids in a polypeptide is called the **primary structure**. Once the primary structure is formed, the single chain can twist into an **α-helix**, or lie along side itself and form a **β-pleated sheet**. With β-pleated sheets, the connecting segments of the two strands of the sheet can lie in the same direction (*parallel*) or in opposite directions (*antiparallel*). Both α-helices and β-pleated sheets are reinforced by hydrogen bonds between the carbonyl oxygen and the hydrogen on the amino group. A single protein usually contains both structures at various locations along its chain. The α-helix and the β-pleated sheets are the **secondary structure** and contribute to the *conformation* of the protein. All proteins have a primary structure and most have a secondary structure. Larger proteins (globular, fibrous/structural, etc...) can have a tertiary and quaternary structure. The **tertiary structure** refers to the three dimensional shape formed when the peptide chain curls and folds. Five forces create the tertiary structure: 1.) covalent **disulfide bonds** between two cysteine amino acids on different parts of the chain, 2.) electrostatic (ionic) interactions mostly between acidic and basic side chains 3.) hydrogen bonds, 4.) van der Waals forces, 5.) hydrophobic side chains pushed away from water toward the center of the protein (see Figure 1.4).

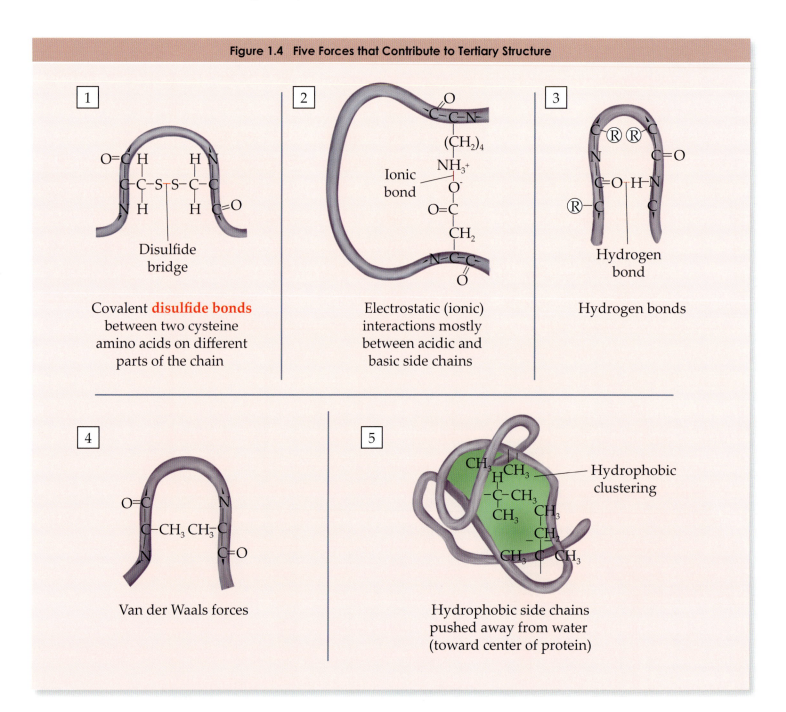

Figure 1.4   Five Forces that Contribute to Tertiary Structure

Covalent **disulfide bonds** between two cysteine amino acids on different parts of the chain

Electrostatic (ionic) interactions mostly between acidic and basic side chains

Hydrogen bonds

Van der Waals forces

Hydrophobic side chains pushed away from water (toward center of protein)

In addition to these forces, the amino acid proline induces turns in the polypeptide that will disrupt both α-helix and β-pleated sheet formation. When two or more polypeptide chains bind together, they form the **quaternary structure** of the protein. The same five forces at work in the tertiary structure can also act to form the quaternary structure.

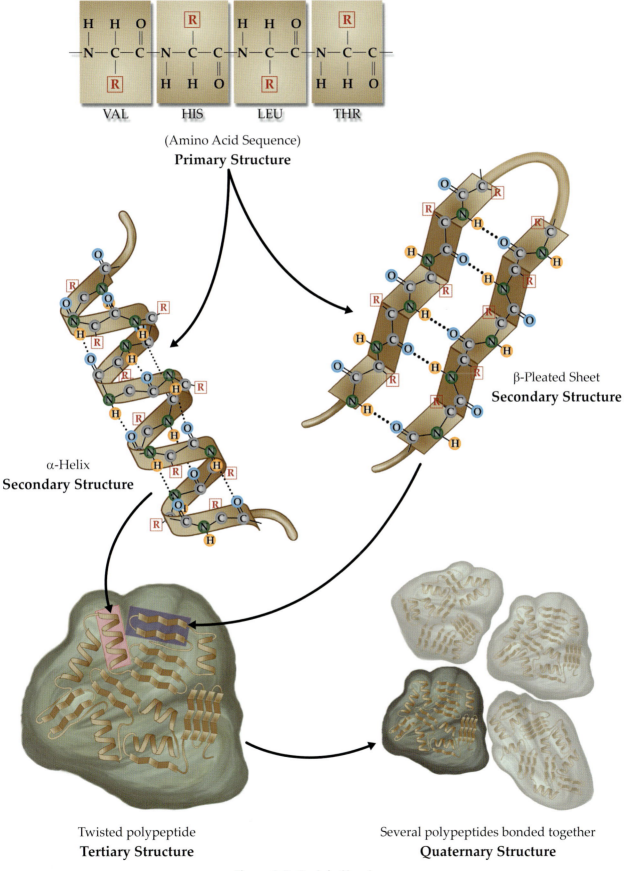

Figure 1.5 Protein Structure

When the conformation is disrupted, the protein is said to be **denatured**. A denatured protein has lost most of its secondary, tertiary, and quaternary structure. Some denaturing agents and the forces that they disrupt are given in Table 1.1. Very often, once the denaturing agent is removed, the protein will spontaneously refold to its original conformation. This suggests that the amino acid sequence plays a key role in the conformation of a protein.

Properly folded protein

Denaturation

Denatured protein

Table 1.1   Denaturing Agents

Denaturing Agents	Forces Disrupted
Urea	Hydrogen bonds
Salt or Change in pH	Electrostatic bonds
Mercaptoethanol	Disulfide bonds
Organic solvents	Hydrophobic forces
Heat	All forces

As alluded from the previous page, there are two types of proteins: *globular* and *structural*. There are more types of globular proteins than types of structural proteins. Globular proteins function as enzymes (i.e. pepsin), hormones (i.e. insulin), membrane pumps and channels (i.e. Na^+/K^+ pump and voltage gated sodium channels), membrane receptors (i.e. nicotinic receptors on a post-synaptic neuron), intercellular and intracellular transport and storage (i.e. hemoglobin and myoglobin), osmotic regulators (i.e. albumin), in the immune response (i.e. antibodies), and more.

Structural proteins are made from long polymers. They maintain and add strength to cellular and matrix structure. *Collagen* (Figure 1.6), a structural protein made from a unique type of helix, is the most abundant protein in the body. Collagen fibers add great strength to, among others, skin, tendons, ligaments, and bone. Microtubules, which make up eukaryotic flagella and cilia, are made from globular tubulin, which polymerizes under the right conditions to become a structural protein.

Glycoproteins are proteins with carbohydrate groups attached. These are a component of cellular plasma membranes. *Proteoglycans* are also a mixture of proteins and carbohydrates, but they generally consist of more than 50% carbohydrates. Proteoglycans are the major component of the extracellular matrix as discussed in Biology Lecture 4 – Cellular Matrix.

Heat, salt, and changes in pH can cause a protein to lose its higher-level conformations. Notice, for example, that the denatured form of the protein does not contain any of the α-helices that the properly folded protein has. Denaturing agents rarely affect the primary structure of a protein, which contains the essential information for conformation. Thus, mildly denatured proteins can often spontaneously return to their original conformation.

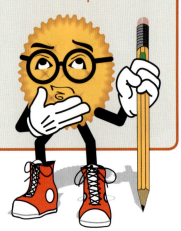

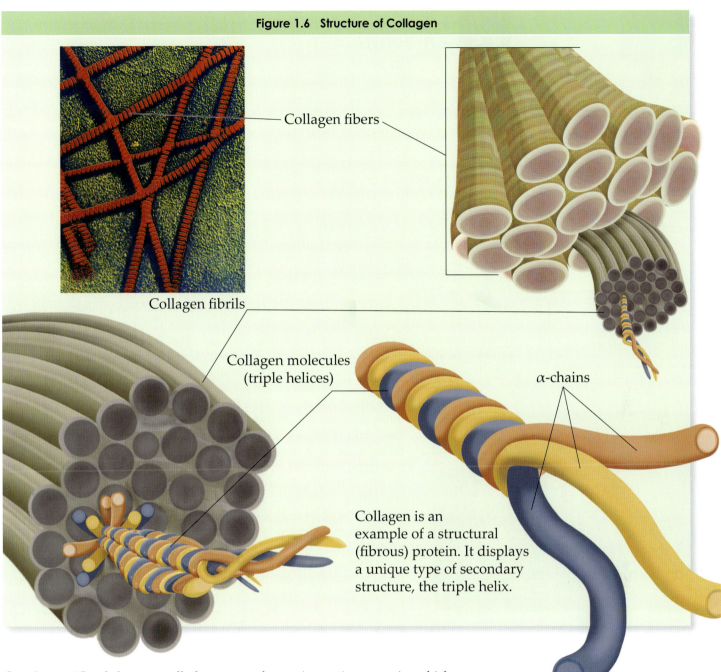

Figure 1.6 Structure of Collagen

Collagen fibers

Collagen fibrils

Collagen molecules
(triple helices)

α-chains

Collagen is an example of a structural (fibrous) protein. It displays a unique type of secondary structure, the triple helix.

Cytochromes (Greek: kytos → cell, chroma → color or pigment) are proteins which require a *prosthetic* (nonproteinaceous) *heme* (Greek: haima → blood) group in order to function. Cytochromes get their name from the color that they add to the cell. Examples of cytochromes are hemoglobin and the cytochromes of the electron transport chain in the inner-membrane of mitochondria. Proteins containing nonproteinaceous components are called *conjugated proteins*.

Proteins are important. Understand the different structures, 1°, 2°, 3°, and 4°, and the bonding involved. Know what denaturation means. The rest is just good background knowledge to help you read MCAT passages. You don't have to memorize the structures of each amino acid, but recognize the basic structure of a generic amino acid. Although nucleic acids, some lipids, and even some carbohydrates contain nitrogen, when you see nitrogen on the MCAT, think protein.

Cytochrome proteins carry out electron transport via oxidation and reduction of the heme group.

Diabetics monitor how much glucose is in their blood.

> In the absence of insulin, only the brain and the liver continue to absorb glucose.

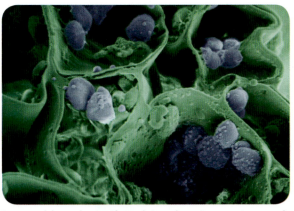

Large chloroplasts (found in plants) contain starch granules made by photosynthesis.

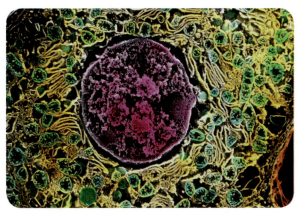

Liver cells contain large amounts of glycogen. This helps the liver regulate blood glucose levels.

1.5 Carbohydrates

As implied by the name, **carbohydrates** (also called sugars or saccharides) are made from carbon and water. They have the empirical formula $C(H_2O)$. Five and six carbon carbohydrates (pentoses and hexoses) are the most common in nature. The six carbon carbohydrate called **glucose** (Greek: glucus → sweet) is the most commonly occurring six carbon carbohydrate. Glucose normally accounts for 80% of the carbohydrates absorbed by humans. Essentially all digested carbohydrates reaching body cells have been converted to glucose by the liver or enterocytes. Glucose exists in aqueous solution in an unequal equilibrium heavily favoring the ring form over the chain form. The ring form has two **anomers**. In the first anomer, α-glucose, the hydroxyl group on the anomeric carbon (carbon number one) and the methoxy group (carbon number six) are on opposite sides of the carbon ring. In β-glucose the hydroxyl group and the methoxy group are on the same side of the carbon ring. The cell can oxidize glucose transferring its chemical energy to a more readily useable form, ATP. If the cell has sufficient ATP, glucose is polymerized to the polysaccharide, **glycogen** or converted to fat. As shown in Figure 1.7, glycogen is a branched glucose polymer with alpha linkages. Glycogen is found in all animal cells, but especially large amounts are found in muscle and liver cells. The liver regulates the blood glucose level, so liver cells are one of the few cell types capable of reforming glucose from glycogen and releasing it back into the blood stream. Only certain epithelial cells in the digestive tract and the proximal tubule of the kidney are capable of absorbing glucose against a concentration gradient. This is done via a secondary active transport mechanism down the concentration gradient of sodium. All other cells absorb glucose via facilitated diffusion. Insulin increases the rate of facilitated diffusion for glucose and other monosaccharides. In the absence of insulin, only neural and hepatic cells are capable of absorbing sufficient amounts of glucose via the facilitated transport system. Plants form **starch** and **cellulose** from glucose. Starch comes in two forms: *amylose* and *amylopectin*. Amylose is an isomer of cellulose that may be branched or unbranched and has the same alpha linkages as glycogen. Amylopectin resembles glycogen but has a different branching structure. Cellulose has beta linkages. Most animals have the enzymes to digest the alpha linkages of starch and glycogen but not the beta linkages of cellulose. Some animals such as cows have bacteria in their digestive systems that release an enzyme to digest the beta linkages in cellulose. Recent research suggests that certain insects do produce an enzyme to digest the beta linkages of cellulose.

> Not bad, but I think I prefer alpha linkages in my salad.

Figure 1.7   Glucose and Glucose Polymers

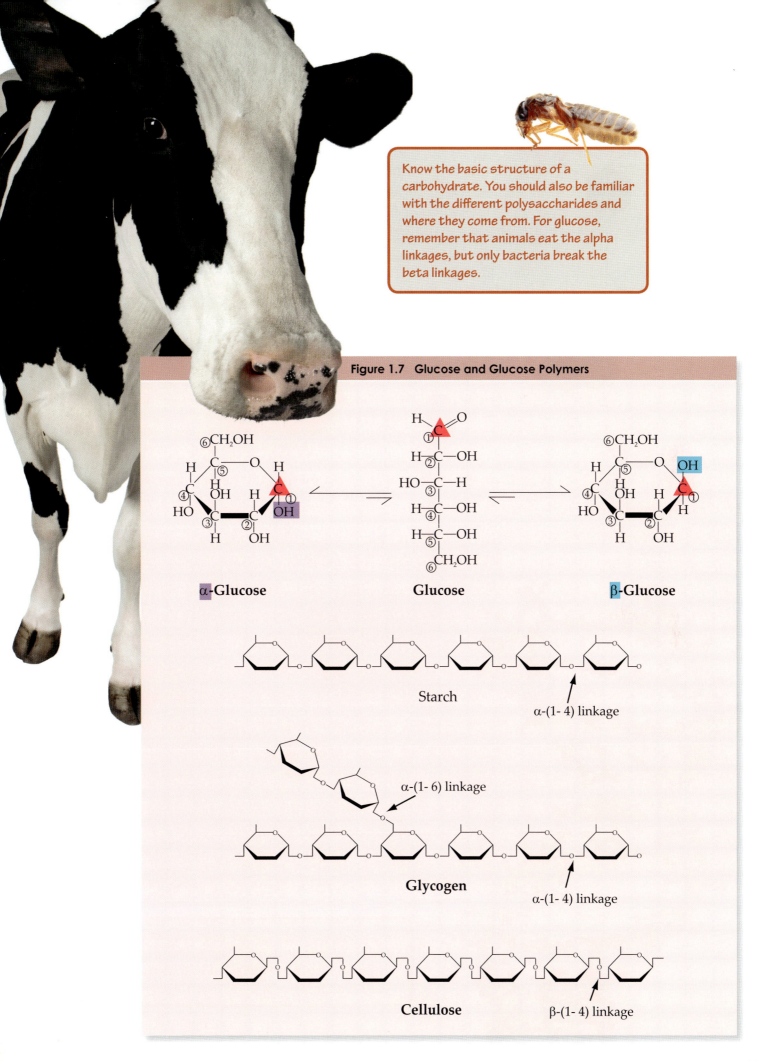

α-Glucose   Glucose   β-Glucose

Starch

α-(1- 4) linkage

α-(1- 6) linkage

Glycogen

α-(1- 4) linkage

Cellulose   β-(1- 4) linkage

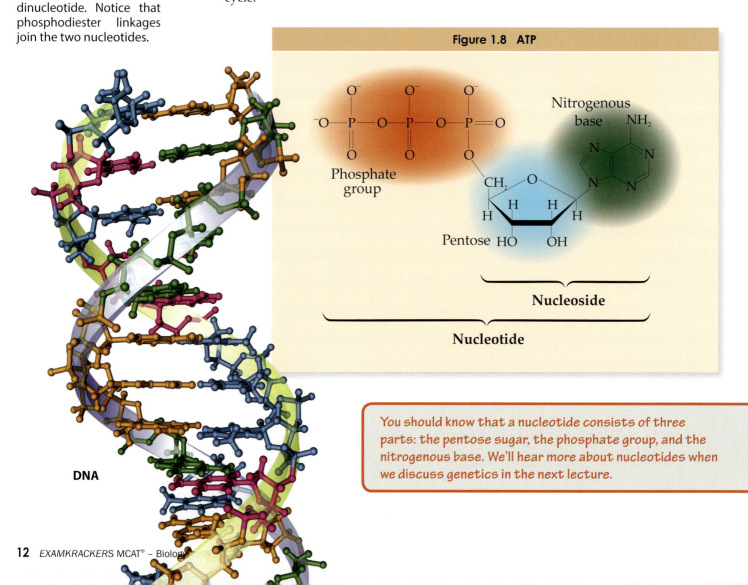

NADH is an example of a dinucleotide. Notice that phosphodiester linkages join the two nucleotides.

Nucleotides are composed of three components (Figure 1.8):

1. a five carbon sugar;

2. a nitrogenous base;

3. a phosphate group.

The most common nitrogenous bases in nucleotides are adenine, guanine, cytosine, thymine, and uracil. Nucleotides form polymers to create the **nucleic acids**, **DNA** and **RNA**. In nucleic acids, nucleotides are joined together by **phosphodiester bonds** between the phosphate group of one nucleotide and the 3rd carbon of the pentose of the other nucleotide forming long strands. By convention, a strand of nucleotides in a nucleic acid is written as a list of its nitrogenous bases. A nucleotide attached to the number 3 carbon (3') of its neighbor, follows that neighbor in the list. In other words, nucleotides are written 5'→ 3'. In typical DNA, two strands are joined by hydrogen bonds to make the structure called a **double helix**. Adenine and thymine form two hydrogen bonds, while cytosine and guanine form three. By convention, DNA is written so that the top strand runs 5'→ 3' and the bottom runs 3'→ 5'. In typical RNA there is only one strand and no helix is formed; also uracil replaces thymine. (More will be said about nucleic acids in Biology Lecture 2.)

Other important nucleotides include **ATP** (adenosine triphosphate: Figure 1.8), the source of readily available energy for the cell; also **cyclic AMP**, an important component in many second messenger systems; **NADH** and **FADH$_2$**, the coenzymes involved in the Krebs cycle.

Figure 1.8 ATP

Phosphate group

Nitrogenous base

Pentose

Nucleoside

Nucleotide

> You should know that a nucleotide consists of three parts: the pentose sugar, the phosphate group, and the nitrogenous base. We'll hear more about nucleotides when we discuss genetics in the next lecture.

DNA

1.7 Minerals

Minerals are the dissolved inorganic ions inside and outside the cell. By creating electrochemical gradients across membranes, they assist in the transport of substances entering and exiting the cell. They can combine and solidify to give strength to a matrix, such as *hydroxyapatite* in bone. Minerals also act as cofactors (discussed later in this lecture) assisting enzyme or protein function. For instance, iron is a mineral found in *heme*, the prosthetic group of *cytochromes*.

This fractured surface of a bone consists of an organic matrix of collagen fibers and mineral-based molecules such as hydroxyapatite and chondroitin sulfate.

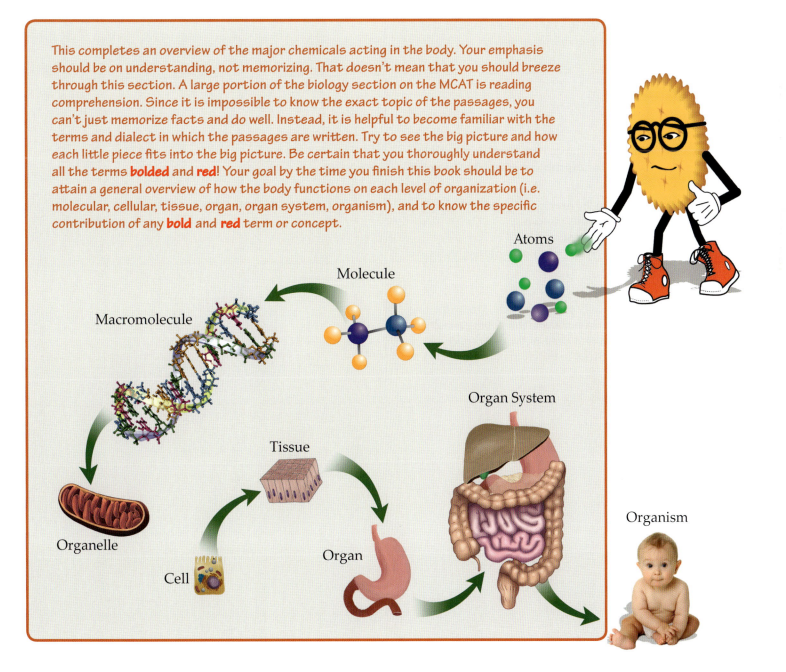

This completes an overview of the major chemicals acting in the body. Your emphasis should be on understanding, not memorizing. That doesn't mean that you should breeze through this section. A large portion of the biology section on the MCAT is reading comprehension. Since it is impossible to know the exact topic of the passages, you can't just memorize facts and do well. Instead, it is helpful to become familiar with the terms and dialect in which the passages are written. Try to see the big picture and how each little piece fits into the big picture. Be certain that you thoroughly understand all the terms **bolded** and **red**! Your goal by the time you finish this book should be to attain a general overview of how the body functions on each level of organization (i.e. molecular, cellular, tissue, organ, organ system, organism), and to know the specific contribution of any **bold** and **red** term or concept.

Atoms

Molecule

Macromolecule

Organelle

Cell

Tissue

Organ

Organ System

Organism

1. The most common catabolic reaction in the human body is:

    A. dehydration.
    B. hydrolysis.
    C. condensation.
    D. elimination.

2. A molecule of DNA contains all of the following EXCEPT:

    A. deoxyribose sugars.
    B. polypeptide bonds.
    C. phosphodiester bonds.
    D. nitrogenous bases.

3. Which of the following is a carbohydrate polymer that is stored in plants and digestible by animals?

    A. starch
    B. glycogen
    C. cellulose
    D. glucose

4. Excessive amounts of nitrogen are found in the urine of an individual who has experienced a period of extended fasting. This is most likely due to:

    A. glycogenolysis in the liver.
    B. the breakdown of body proteins.
    C. lipolysis in adipose tissue.
    D. a tumor on the posterior pituitary causing excessive ADH secretion.

5. Proline is not technically an α-amino acid. Due to the ring structure of proline, it cannot conform to the geometry of the α-helix and creates a bend in the polypeptide chain. This phenomenon assists in the creation of what level of protein structure?

    A. primary
    B. secondary
    C. tertiary
    D. quaternary

6. Metabolism of carbohydrate and fat spare protein tissue. All of the following are true of fats EXCEPT:

    A. Fats may be used in cell structure.
    B. Fats may be used as hormones.
    C. Fats are a more efficient form of energy storage than proteins.
    D. Fats are a less efficient form of energy storage than carbohydrates.

7. Which of the following is found in the RNA but not the DNA of a living cell?

    A. thymine
    B. a double helix
    C. an additional hydroxyl group
    D. hydrogen bonds

8. Like cellulose, chitin is a polysaccharide that cannot be digested by animals. Chitin differs from cellulose by possessing an acetyl-amino group at the second carbon. What molecule is a reactant in the breaking of the β-1,4-glycoside linkages of cellulose and chitin?

    A. water
    B. oxygen
    C. α-1,4-glucosidase
    D. β-1,4-glucosidase

1.8 Enzymes

Virtually all biological reactions are governed by enzymes. Although there are a few nucleic acids that act as enzymes, typically **enzymes** are globular proteins. The function of any enzyme is to act as a **catalyst**, lowering the energy of activation for a biological reaction and increasing the rate of that reaction. Enzymes increase reaction rates by magnitudes of as much as thousands of trillions. This is a much greater increase than typical lab catalysts. Such extreme control over reaction rates gives enzymes the ability to pick and choose which reactions will or will not occur inside a cell. Enzymes, like any catalysts, are not consumed nor permanently altered by the reactions which they catalyze. Only a small amount of catalyst is required for any reaction. Like any catalyst, enzymes do not alter the **equilibrium** of a reaction. (See Chemistry Lecture 2 for more on equilibrium and catalysts.)

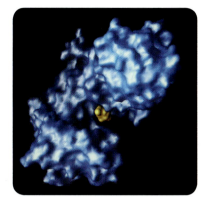

Hexokinase lowers the activation energy for the phosphorylation of glucose.

The reactant or reactants upon which an enzyme works are called the **substrates**. Substrates are generally smaller than the enzyme. The position on the enzyme to where the substrate binds, usually with numerous noncovalent bonds, is called the **active site**. The enzyme bound to the substrate is called the **enzyme-substrate complex**.

Normally, enzymes are designed to work only on a specific substrate or group of closely related substrates. This is called **enzyme specificity**. The **lock and key theory** is an example of enzyme specificity. In this theory, the active site of the enzyme has a specific shape like a lock that only fits a specific substrate, the key. The lock and key model explains some but not all enzymes. In a second theory called the **induced fit** model, the shape of both the enzyme and the substrate are altered upon binding. Besides increasing specificity, the alteration actually helps the reaction to proceed. In reactions with more than one substrate, the enzyme may also orient the substrates relative to each other, creating optimal conditions for a reaction to take place.

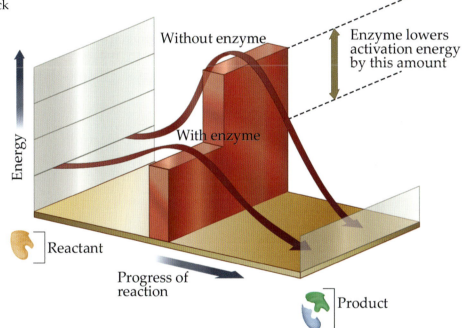

Without enzyme

With enzyme

Enzyme lowers activation energy by this amount

Energy

Reactant

Progress of reaction

Product

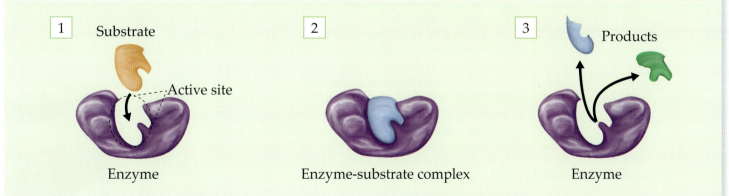

Figure 1.9   Enzymatic Reaction

1   Substrate

Active site

Enzyme

2

Enzyme-substrate complex

3   Products

Enzyme

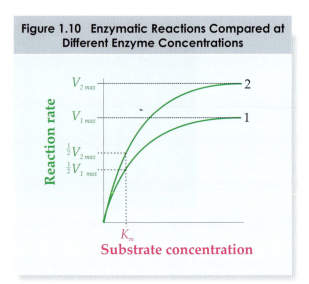

Enzymes exhibit **saturation kinetics**; as the relative concentration of substrate increases, the rate of the reaction also increases, but to a lesser and lesser degree until a maximum rate (V_{max}) has been achieved (Figure 1.10). This occurs because as more substrate is added, individual substrates must begin to wait in line for an unoccupied enzyme. Thus, V_{max} is proportional to enzyme concentration. *Turnover number* is the number of substrate molecules one enzyme active site can convert to product in a given unit of time when an enzyme solution is saturated with substrate. Related to V_{max} is the *Michaelis constant* (K_m). K_m is the substrate concentration at which the reaction rate is equal to $\frac{1}{2}V_{max}$. Unlike V_{max}, K_m does not vary when the enzyme concentration is changed. Under certain conditions, K_m is therefore a good indicator of an enzyme's affinity for its substrate.

> Don't fret over V_{max} and K_m, they're not on the MCAT. If you are able to understand them, however, they provide a deeper understanding of enzyme kinetics, which <u>is</u> on the MCAT. We'll hear more about V_{max} and K_m later, but remember, the stuff in italics is not tested on the MCAT directly.

Temperature and pH also affect enzymatic reactions. At first, as the temperature increases, the reaction rate goes up, but at some point, the enzyme denatures and the rate of the reaction drops off precipitously. For enzymes in the human body, the optimal temperature is most often around 37° C. Enzymes also function within specific pH ranges. The optimal pH varies depending upon the enzyme. For instance, pepsin, active in the stomach, prefers a pH below 2, while trypsin, active in the small intestine, works best at a pH between 6 and 7.

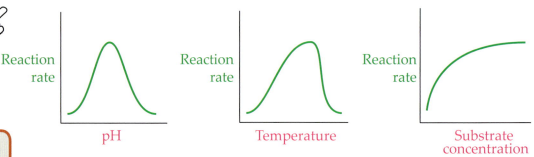

> The important relationships between enzymes and their environment are represented by these three graphs. Memorize these three graphs and understand them.

In order to reach their optimal activity, many enzymes require a non-protein component called a **cofactor** (Latin: co- → with or together). Cofactors can be coenzymes or metal ions. **Coenzymes** are divided into two types: *cosubstrates* and *prosthetic groups*. Both types are organic molecules. Cosubstrates reversibly bind to a specific enzyme, and transfer some chemical group to another substrate. The cosubstrate is then reverted to its original form by another enzymatic reaction. This reversion to original form is what distinguishes a cosubstrate from normal substrates. **ATP** is an example of a cosubstrate type of coenzyme. Prosthetic groups, on the other hand, remain covalently bound to the enzyme throughout the reaction, and, like the enzyme, emerge from the reaction unchanged. Coenzymes are often **vitamins** or vitamin derivatives. (Vitamins are essential [cannot be produced by the body] organic molecules.) As mentioned before, *heme* is a prosthetic group. Heme binds with *catalase* in peroxisomes to degrade hydrogen peroxide. Metal ions are the second type of cofactor. Metal ions can act alone or with a prosthetic group. Typical metal ions that function as cofactors in the human body are iron, copper, manganese, magnesium, calcium, and zinc. An enzyme without its cofactor is called an *apoenzyme* (Greek: apo- → away from) and is completely nonfunctional. An enzyme with its cofactor is called a *holoenzyme* (Greek: holos → whole, entire, complete).

> Just know that some enzymes need cofactors to function, and that cofactors are either minerals or coenzymes. Also remember that many coenzymes are vitamins or their derivatives.

1.9 Enzyme Inhibition

Enzyme activity can be inhibited. Enzyme inhibitors can be classified according to three different mechanisms: irreversible inhibitors, competitive inhibitors, and noncompetitive inhibitors (Figure 1.11). Agents which bind covalently to enzymes and disrupt their function are **irreversible inhibitors**. A few irreversible inhibitors bind noncovalently. Irreversible inhibitors tend to be highly toxic. For example, *penicillin* is an irreversible inhibitor that binds to a bacterial enzyme that assists in the manufacturing of peptidoglycan cell walls.

> Irreversible inhibitors bond to enzymes and disrupt their function. Irreversible inhibitors tend to be highly toxic.

Competitive inhibitors compete with the substrate by binding reversibly with noncovalent bonds to the active site. Since, typically, they bind directly to the active site for only a fraction of a second, they block the substrate from binding during that time. Of course, the reverse is also true; if the substrate binds first, it blocks the inhibitor from binding. Thus, competitive inhibitors raise the apparent K_m but do not change V_{max}. In other words, in the presence of a competitive inhibitor, the rate of the reaction can be increased to the original, uninhibited V_{max} by increasing the concentration of the substrate. Overcoming inhibition by increasing substrate concentration is the classic indication of a competitive inhibitor. Competitive inhibitors often resemble the substrate. *Sulfanilamide* is an antibiotic which competitively inhibits a bacterial enzyme that manufactures folic acid leading to the death of bacterial cells. Although humans require folic acid, sulfanilamide does not harm humans because we use a different enzymatic pathway to manufacture folic acid.

Noncompetitive inhibitors bind noncovalently to an enzyme at a spot other than the active site and change the conformation of the enzyme. Noncompetitive inhibitors do not prevent the substrate from binding, and they bind just as readily to enzymes that have a substrate as to those that don't. Noncompetitive inhibitors do not resemble the substrate, so they commonly act on more than one enzyme. Unlike competitive inhibitors, they cannot be overcome by excess substrate, and they lower V_{max}. They do not, however, lower the enzyme affinity for the substrate, so K_m remains the same.

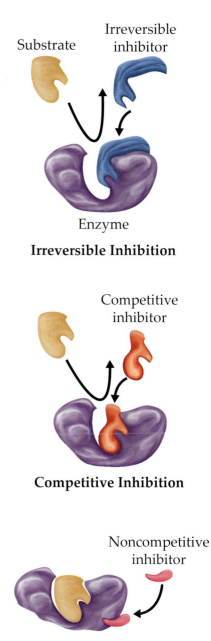

Irreversible Inhibition

Competitive Inhibition

Noncompetitive Inhibition

Figure 1.11 Enzyme Inhibitors

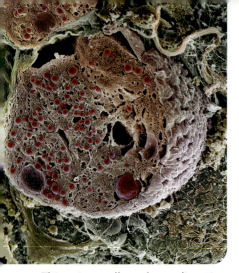

This acinar cell produces digestive enzymes in zymogen granules (purple). These enzymes are excreted into the pancreatic ducts and carried to the small intestine, where they are activated and aid in the breakdown of carbohydrates, fats, and proteins.

1.10 Enzyme Regulation

Enzymes select which reactions take place within a cell, so the cell must regulate enzyme activity. Enzymes are regulated by four primary means:

1. *Proteolytic cleavage (irreversible covalent modification)*—Many enzymes are released into their environment in an inactive form called a **zymogen** or **proenzyme** (Greek: pro → before). When specific peptide bonds on zymogens are cleaved, the zymogens become irreversibly activated. Activation of zymogens may be instigated by other enzymes, or by a change in environment. For instance, pepsinogen (notice the "–ogen" at the end indicating zymogen status) is the zymogen of pepsin and is activated by low pH.

2. *Reversible covalent modification*—Some enzymes are activated or deactivated by phosphorylation or the addition of some other modifier such as AMP. The removal of the modifier is almost always accomplished by hydrolysis. Phosphorylation typically occurs in the presence of a *protein kinase*.

3. *Control proteins*—Control proteins are protein subunits that associate with certain enzymes to activate or inhibit their activity. *Calmodulin* or *G-proteins* are typical examples of control proteins.

4. **Allosteric interactions**—Allosteric regulation is the modification of the enzyme configuration resulting from the binding of an activator or inhibitor at a specific binding site on the enzyme. (Allosteric regulation is discussed below.)

Normally, an enzyme governs just one reaction in a series of reactions. If one of the products downstream in a reaction series comes back and inhibits the enzymatic activity of an earier reaction, this phenomenon is called **negative feedback** or **feedback inhibition**. Negative feedback provides a shut down mechanism for a series of enzymatic reactions when that series has produced a sufficient amount of product. Most enzymes work within some type of negative feedback cycle. **Positive feedback** also occurs, where the product returns to activate the enzyme. Positive feedback mechanisms occur less often than negative feedback.

Negative feedback inhibition is typical in many amino acid synthesis pathways. It is wasteful and unnecessary to synthesize amino acids that are readily available in the environment. Therefore, upstream enzymes involved in a particular synthetic metabolic pathway typically have allosteric inhibitory sites that bind the final amino acid product. If the final product is present in the environment, further synthesis will shut down.

Figure 1.12   Enzyme Regulation

Substrate

Enzyme 1

Intermediate A

Feedback inhibition

Enzyme 2

Intermediate B

Enzyme 3

End-product

Feedback inhibitors do not resemble the substrates of the enzymes that they inhibit. Instead, they bind to the enzyme and cause a conformational change. This is called **allosteric regulation** (Greek: allos → different or other, stereos → solid). There exist both **allosteric inhibitors** and **allosteric activators**. All allosteric inhibitors and activators are not necessarily noncompetitive inhibitors, because many alter K_m without affecting V_{max}. Allosteric enzymes do not exhibit typical kinetics because they normally have several binding sites for different inhibitors, activators, and even substrates. At low substrate concentrations, small increases in substrate concentration increase enzyme efficiency as well as reaction rate. The first substrate changes the shape of the enzyme allowing other substrates to bind more easily. This phenomenon is called **positive cooperativity**. **Negative cooperativity** occurs as well. It is cooperativity in the presence of the allosteric inhibitor *2,3BPG* that gives the oxygen dissociation curve of hemoglobin its sigmoidal shape.

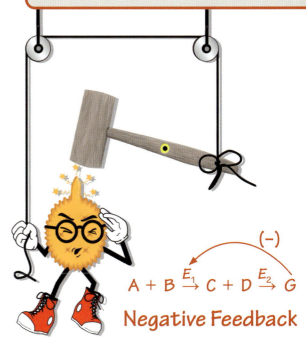

Negative Feedback

1.11 Enzyme Classification

Enzymes are named according to the reactions that they catalyze. Very often, the suffix "-ase" is simply added to the end of the substrate upon which the enzyme acts. For instance, acetylcholinesterase acts upon the ester group in acetylcholine.

Enzymes are classified into six categories:

1. *oxidoreductases*
2. *transferases;*
3. *hydrolases;*
4. *lyases;*
5. *isomerases;*
6. *ligases.*

The only distinction between classifications that might be of interest to an MCAT taker is between lyases and ligases. A lyase that catalyzes addition of one substrate to a double bond of a second substrate is sometimes called a *synthase*. ATP synthase is an example of a lyase. A ligase also governs an addition reaction, but requires energy from ATP or some other nucleotide. Ligases are sometimes called *synthetases*.

Kinases and *phosphatases* may also come up on the MCAT. A kinase is an enzyme which phosphorylates something, while a phosphatase is an enzyme which dephosphorylates something. Often times a kinase phosphorylates another enzyme in order to activate or deactivate it. *Hexokinase* is the enzyme which phosphorylates glucose as soon as it enters a cell.

9. Enzymes are required by all living things because enzymes:

   A. raise the free energy of chemical reactions.
   B. properly orient reactants and lower activation energy.
   C. increase the temperature of reacting molecules.
   D. increase the number of reacting molecules.

10. All of the following must change the rate of an enzyme-catalyzed reaction EXCEPT:

    A. changing the pH.
    B. lowering the temperature.
    C. decreasing the concentration of substrate.
    D. adding a noncompetitive inhibitor.

11. Since an increase in temperature increases the reaction rate, why isn't the elevation of temperature a method normally used to accelerate enzyme-catalyzed reactions?

    A. Raising the temperature causes the reaction to occur too quickly.
    B. Raising the temperature does not sufficiently surmount the activation energy barrier.
    C. Heat changes the configuration of proteins.
    D. Heat does not increase the probability of molecular collision.

12. Which of the following is (are) true concerning feedback inhibition?

    I. It often acts by inhibiting enzyme activity.
    II. It works to prevent a build up of excess nutrients.
    III. It only acts through enzymes.

    A. I only
    B. II only
    C. I and II only
    D. I, II, and III

13. One mechanism of enzyme inhibition is to inhibit an enzyme without blocking the active site, but by altering the shape of the enzyme molecule. This mechanism is called:

    A. competitive inhibition.
    B. noncompetitive inhibition.
    C. feedback inhibition.
    D. positive inhibition.

14. The continued production of progesterone caused by the release of HCG from the growing embryo is an example of:

    A. positive feedback.
    B. negative feedback.
    C. feedback inhibition.
    D. feedback enhancement.

15. Peptidases that function in the stomach most likely:

    A. *increase* their function in the small intestine due to *increased* hydrogen ion concentration.
    B. *decrease* their function in the small intestine due to *increased* hydrogen ion concentration.
    C. *increase* their function in the small intestine due to *decreased* hydrogen ion concentration.
    D. *decrease* their function in the small intestine due to *decreased* hydrogen ion concentration.

16. The rate of a reaction slows when the reaction is exposed to a competitive inhibitor. Which of the following might overcome the effects of the inhibitor?

    A. decreasing enzyme concentration
    B. increasing temperature
    C. increasing substrate concentration
    D. The effects of competitive inhibition cannot be overcome.

1.12 Cellular Metabolism

Metabolism is all cellular chemical reactions. It consists of *anabolism* (Greek: ana → up, ballein → to throw), molecular synthesis, and *catabolism* (Greek: kata → down), molecular degradation. There are three basic stages of *catabolic* metabolism:

1) Macromolecules (polysaccharides, proteins, and lipids) are broken down into their constituent parts (monosaccharides, amino acids, and fatty acids and glycerol) releasing little or no energy.

2) Constituent parts are oxidized to acetyl CoA, pyruvate or other metabolites forming some ATP and reduced coenzymes (NADH and $FADH_2$) in a process that does not directly utilize oxygen.

3) If oxygen is available and the cell is capable of using oxygen, these metabolites go into the citric acid cycle to capture large amounts of energy (more NADH, $FADH_2$, or ATP); otherwise the coenzyme NAD^+ and other byproducts are either recycled or expelled as waste. The second and third stages, the energy acquiring stages, are called **respiration**. If oxygen is used, the respiration is aerobic; if oxygen is not used, the respiration is anaerobic.

Thought Provoker

Which of the following are anabolic? Catabolic?

1. Beta oxidation of fats

2. Cholesterol synthesis

3. Glucose synthesis (gluconeogenesis)

4. Glycogen degregation (glycogenolysis)

5. Photosynthesis

Answer: See page 28

Notes:

1.13 Glycolysis

Anaerobic respiration (Latin: an → not or without, aer → air, respirare → to breath) is respiration in which oxygen is not required. **Glycolysis** is the first stage of anaerobic and aerobic respiration. Glycolysis (Figure 1.13) is the series of reactions that breaks a 6-carbon glucose molecule into two 3-carbon molecules of **pyruvate.** (Pyruvate is just the conjugate base of **pyruvic acid**). Other important products from glycolysis are two molecules of ATP each from ADP, inorganic phosphate and water, and two molecules of NADH each from the reduction of NAD⁺. All living cells and organisms are capable of breaking down glucose to pyruvate; the most common chemical pathway for this is glycolysis. Glycolysis will operate in both the presence and absence of oxygen; it neither requires oxygen, nor is poisoned by it. The reactions of glycolysis occur in the **cytosol** (fluid portion) of living cells.

The first step of glycolysis occurs upon the entry of glucose into any human cell. *Hexokinase* phosphorylates glucose to *glucose 6-phosphate* with a phosphate group from ATP. (The liver and pancreas use an *isozyme* [an enzyme with the same function] of hexokinase called *glucokinase*.) Under normal cellular conditions the phosphorylation of glucose is irreversible, and assists the facilitated diffusion mechanism which transports glucose into the cell. (The liver, which must make glucose from glycogen and export it, possesses a special enzyme, *glucose 6-phosphatase*, which dephosphorylates glucose 6-phosphate to reform glucose. Glucose 6-phosphatase is also found in kidney cells.) Phosphorylated molecules cannot diffuse through the membrane. Although this is the first step in glycolysis, the process does not necessarily continue. Glucose 6-phosphate may be converted to *glucose 1-phosphate* and then to glycogen. If glucose 6-phosphate follows the glycolytic pathway, it goes to *fructose 6-phosphate* in the second step of glycolysis. In the third step of glycolysis, a second phosphate group is added at the expense of one more ATP. This step is irreversible and commits the molecule to the glycolytic pathway. Now, the six carbon *fructose 1,6-bisphosphate* is broken into Glyceraldehyde 3-phosphate (PGAL) and dihydroxyacetone phosphate. Dihydroxyacetone phosphate is subsequently converted to PGAL. Up to this point, no energy has been captured from the breakdown of glucose, but two ATPs have been spent.

Next, each 3-carbon molecule is phosphorylated while reducing one NAD⁺ to NADH. The resulting 3-carbon molecules each transfer one of their phosphate groups to an ADP to form one ATP each in **substrate level phosphorylation**. (Substrate level phosphorylation is the formation of ATP from ADP and inorganic phosphate using the energy released from the decay of high energy phosphorylated compounds as opposed to using the energy from diffusion of ions down their concentration gradient, as with oxidative phosphorylation.) The remaining 3-carbon molecules go through three more steps before donating their phosphate group to ADP to yield ATP and pyruvate. Altogether, 2 ATPs are spent and 4 ATPs are produced. The two pyruvate molecules and the two NADH molecules that are left are still relatively high energy molecules.

The products of carbohydrate digestion in the alimentary tract are approximately 80% glucose, and 20% fructose and galactose. *Fructose* and *galactose* are monosaccharides. Much of the fructose and galactose ingested by humans is converted into glucose in the liver enterocytes; however, fructose can enter glycolysis as fructose 6-phosphate or *glyceraldehyde 3-phosphate*, and galactose can be converted to glucose 6-phosphate to enter glycolysis. Simple table sugar is a disaccharide made from glucose and fructose. *Lactose* is a disaccharide found in milk, and is broken down into glucose and galactose in the small intestine. 95% of the monosaccharides in the blood are glucose.

Streptococcus spp., a bacterium taken from a human mouth, uses glycolysis to capture energy from glucose.

Notice that glycolysis has two stages: a six carbon stage and a three carbon stage. The six carbon stage expends two ATPs to phosphorylate the molecule; kind of like "priming a pump". The three carbon stage synthesizes two ATP with each three carbon molecule. Also recognize pyruvate and NADH (the names, not the structures), but don't worry too much about the names of the other chemicals. Just recognize them as part of glycolysis. Know the products of glycolysis, especially the net production of 2 ATPs.

Glycolysis

Figure 1.13

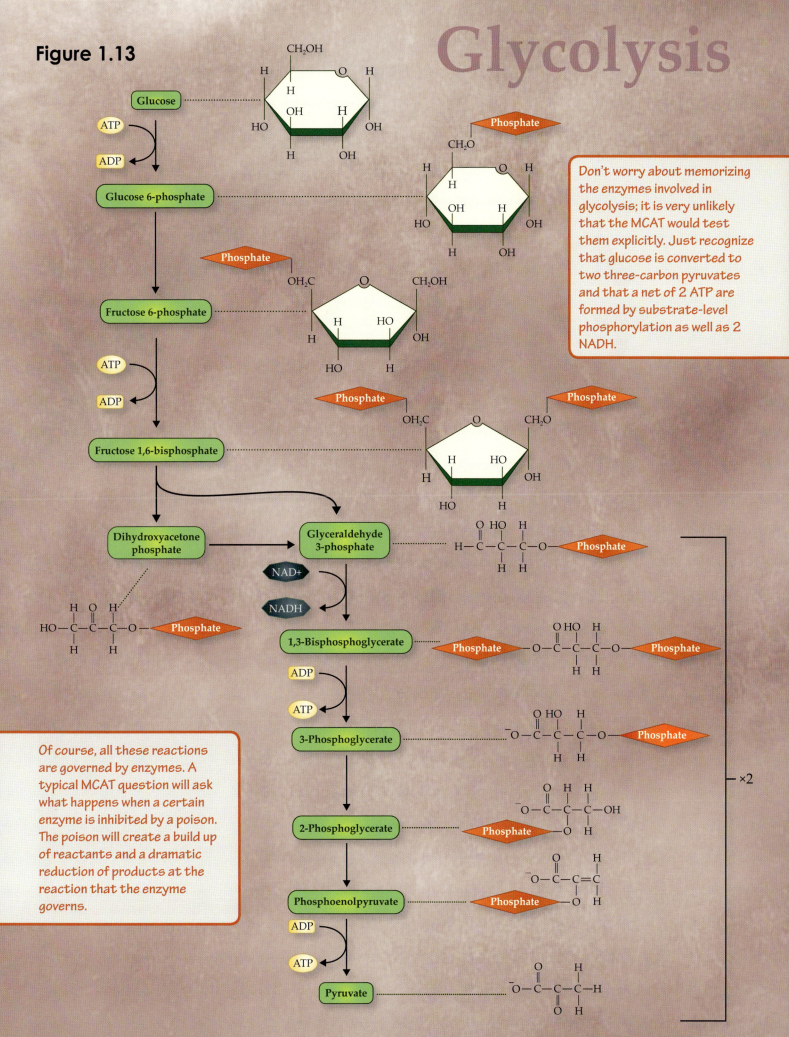

Don't worry about memorizing the enzymes involved in glycolysis; it is very unlikely that the MCAT would test them explicitly. Just recognize that glucose is converted to two three-carbon pyruvates and that a net of 2 ATP are formed by substrate-level phosphorylation as well as 2 NADH.

Of course, all these reactions are governed by enzymes. A typical MCAT question will ask what happens when a certain enzyme is inhibited by a poison. The poison will create a build up of reactants and a dramatic reduction of products at the reaction that the enzyme governs.

1.14 Fermentation

Fermentation is anaerobic respiration. It includes the process of glycolysis, the reduction of pyruvate to ethanol or lactic acid, and the oxidation of the NADH back to the NAD$^+$. Yeast and some microorganisms produce ethanol, while human muscle cells and other microorganisms produce lactic acid. Fermentation takes place when a cell or organism is either unable to assimilate the energy from NADH and pyruvate, or has no oxygen available to do so. In fermentation, the NAD$^+$ is restored for use in its role in glycolysis as a coenzyme, and the lactic acid or ethanol with carbon dioxide is expelled from the cell as a waste product.

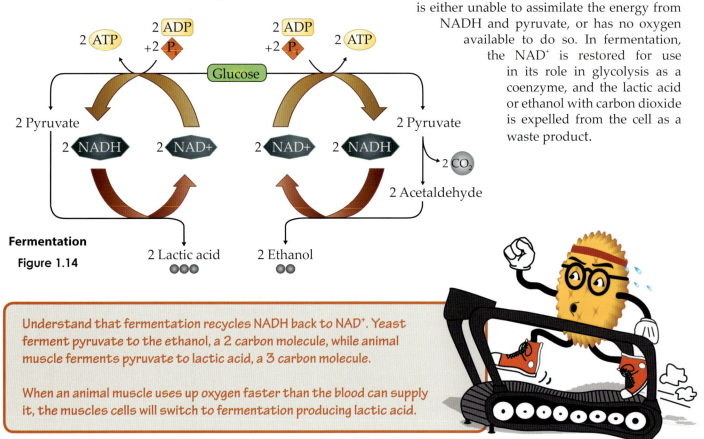

Fermentation

Figure 1.14

2 ATP   2 ADP   +2 P$_i$ Glucose   2 ADP +2 P$_i$ 2 ATP

2 Pyruvate   2 NADH   2 NAD+   2 NAD+   2 NADH   2 Pyruvate

2 CO$_2$

2 Acetaldehyde

2 Lactic acid   2 Ethanol

Understand that fermentation recycles NADH back to NAD$^+$. Yeast ferment pyruvate to the ethanol, a 2 carbon molecule, while animal muscle ferments pyruvate to lactic acid, a 3 carbon molecule.

When an animal muscle uses up oxygen faster than the blood can supply it, the muscles cells will switch to fermentation producing lactic acid.

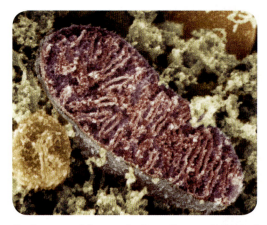

1.15 Aerobic Respiration

Aerobic respiration requires oxygen (Figure 1.15). If oxygen is present in a cell that is capable of aerobic respiration, the products of glycolysis (pyruvate and NADH) will move into the **matrix of a mitochondrion**. The outer membrane of a mitochondrion is permeable to small molecules, and both pyruvate and NADH pass via facilitated diffusion through a large membrane protein called *porin*. The **inner mitochondrial membrane**, however, is less permeable. Although pyruvate moves into the matrix via facilitated diffusion, each NADH (depending upon the mechanism used for transport) may or may not require the hydrolysis of ATP. Once inside the matrix, pyruvate is converted to **acetyl CoA** in a reaction that produces NADH and CO$_2$.

During aerobic respiration the majority of ATPs are produced inside the mitochondrion.

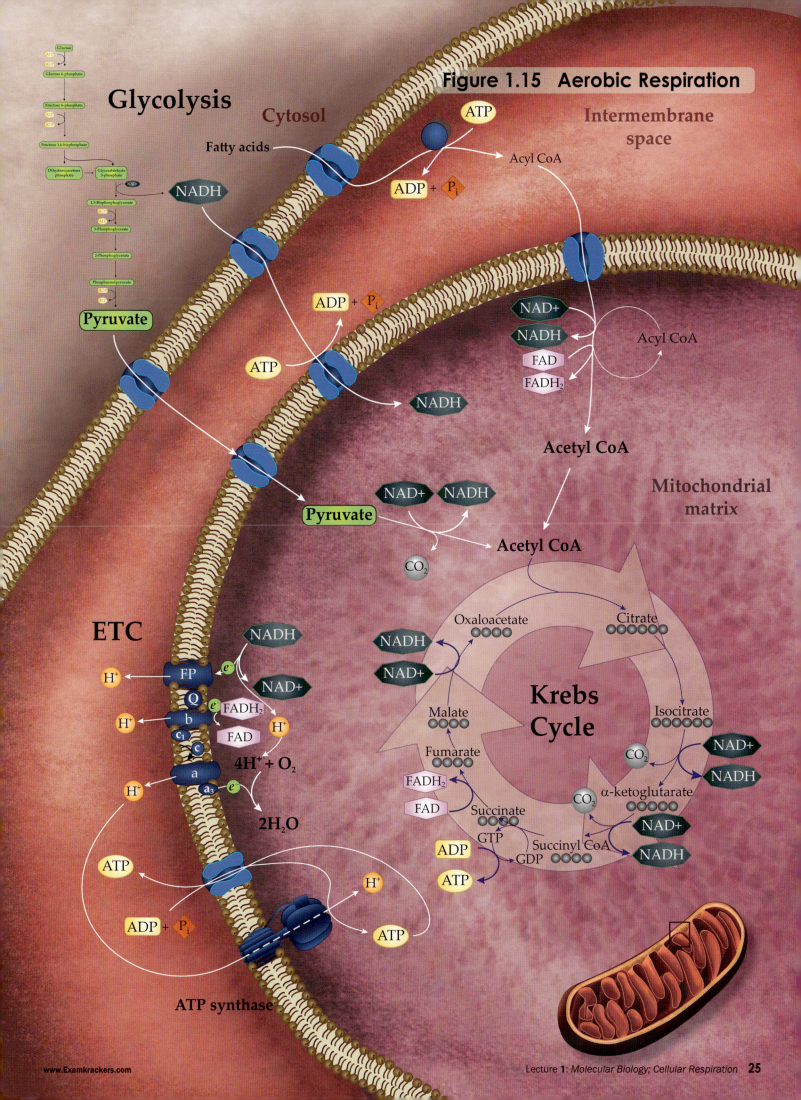

Figure 1.15 Aerobic Respiration

Glycolysis

Cytosol

Intermembrane space

Mitochondrial matrix

Pyruvate

NADH

ATP

Fatty acids

ADP + P$_i$

Acyl CoA

NAD+

NADH

FAD

FADH$_2$

Acyl CoA

Acetyl CoA

ADP + P$_i$

ATP

NADH

Pyruvate

NAD+   NADH

CO$_2$

Acetyl CoA

ETC

NADH

NAD+

H$^+$

FP

Q

b

c$_1$

c

a

a$_3$

e

e

FADH$_2$

FAD

H$^+$

4H$^+$ + O$_2$

2H$_2$O

H$^+$

ATP

ADP + P$_i$

H$^+$

ATP

ATP synthase

Oxaloacetate

Citrate

NADH

NAD+

Malate

Isocitrate

Fumarate

FADH$_2$

FAD

Succinate

GTP

GDP

Succinyl CoA

ADP

ATP

α-ketoglutarate

NAD+

NADH

CO$_2$

CO$_2$

NAD+

NADH

Krebs Cycle

Acetyl CoA is a coenzyme which transfers two carbons (two carbons from pyruvate) to the 4-carbon oxaloacetic acid to begin the **Krebs cycle** (also called the **citric acid cycle**). Each turn of the Krebs cycle (Figure 1.16) produces 1 ATP, 3 NADH, and 1 $FADH_2$. The process of ATP production in the Krebs cycle is called **substrate-level phosphorylation**. During the cycle, two carbons are lost as CO_2, and oxaloacetic acid is reproduced to begin the cycle over again.

Sir Hans Krebs (1900-81), German-British biochemist and Nobel Laureate

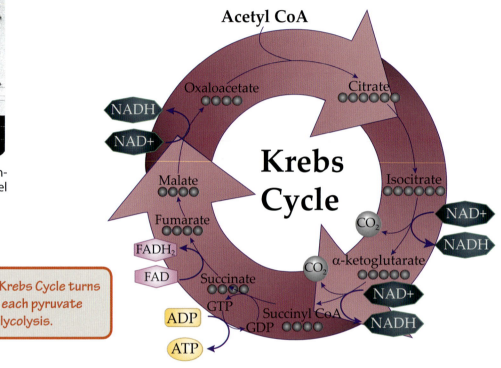

Note that the Krebs Cycle turns twice; once for each pyruvate generated by glycolysis.

Figure 1.16   Kreb's Cycle

Triglycerides can also be catabolized for ATP. Fatty acids are converted into acyl CoA along the outer membrane of the mitochondrion and endoplasmic reticulum at the expense of 1 ATP. They are then brought into the matrix, and two carbons at a time are cleaved from the acyl CoA to make acetyl CoA. This reaction also produces $FADH_2$ and NADH for every two carbons taken from the original fatty acid. Acetyl CoA then enters into the Krebs cycle as usual. The glycerol backbone is converted to PGAL.

Amino acids are deaminated in the liver. The deaminated product is either chemically converted to pyruvic acid or acetyl CoA, or it may enter the Krebs cycle at various stages depending upon which amino acid was deaminated.

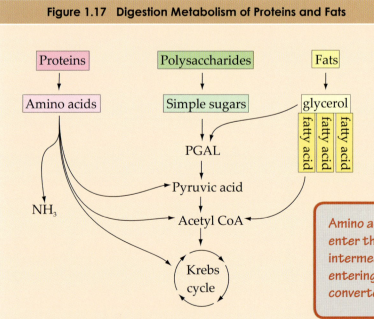

Figure 1.17   Digestion Metabolism of Proteins and Fats

Amino acids must first be deaminated, after which they can either enter the Krebs Cycle as pyruvate or as one of the Krebs Cycle intermediates. Nucleotides must also be deaminated before entering the Krebs Cycle as one of its intermediates. Fats are converted to Acetyl-CoA, which can then enter the Krebs Cycle.

The **electron transport chain (ETC)** (Figure 1.18) is a series of proteins, including cytochromes with heme, in the inner membrane of the mitochondrion. The first protein complex in the series oxidizes NADH by accepting its high energy electrons. Electrons are then passed down the protein series and ultimately accepted by oxygen to form water. As electrons are passed along, protons are pumped into the intermembrane space for each NADH. This establishes a proton gradient called the **proton-motive force** which propels protons through **ATP synthase** to manufacture ATP. Production of ATP in this fashion is called **oxidative phosphorylation**. From 2 to 3 ATPs are manufactured for each NADH. $FADH_2$ works in a similar fashion to NADH, except $FADH_2$ reduces a protein further along in the ETC series, and thus only produces about 2 ATPs.

You should know that aerobic respiration produces about 36 net ATPs (includes glycolysis). You should also know that 1 NADH brings back 2 to 3 ATPs and that 1 $FADH_2$ brings back about 2 ATPs. Know how many NADHs, $FADH_2$s, and ATPs are produced in each turn of the Krebs cycle, and that one glucose produces two turns. Don't worry too much about the fatty acids and amino acids. Just realize that they can be catabolized for energy via the Krebs cycle.

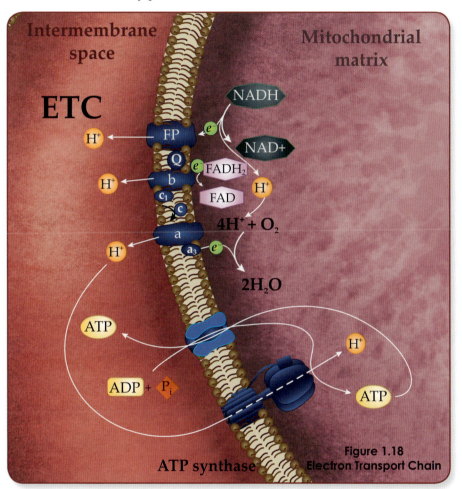

Figure 1.18
Electron Transport Chain

Notice that the environment in the intermembrane space has a lower pH than the matrix.

You don't have to memorize the names of all the chemicals; however, you should recognize them as being in the Krebs cycle. You should pay special attention to acetyl CoA and ATP synthase. Also know the difference between oxidative phosphorylation and substrate level phosphorylation. You should definitely know the products and reactants for respiration:

$$Glucose + O_2 \rightarrow CO_2 + H_2O$$
(This reaction is not balanced)

This is a combustion reaction. Finally, be sure that you remember that the final electron acceptor is oxygen. This is why oxygen is necessary for aerobic respiration.

1.18 Terms You Need To Know

Terms		
α-helix	FADH$_2$	Phosphodiester Bonds
β-pleated Sheet	Fats	Phospholipids
Active Site	Fatty Acids	Polypeptides
Adipocytes	Feedback Inhibition	Positive Feedback
Aerobic Respiration	Fermentation	Primary Structure
Allosteric Regulation	Glucose	Proenzyme
Amino Acids	Glycerol	Proteins
Amphipathic	Glycogen	Proton-Motive Force
Anaerobic Respiration	Glycolysis	Pyruvate
ATP	Hydrogen Bond	Quaternary Structure
ATP Synthase	Hydrolysis	RNA
Carbohydrates	Hydrophilic	Saturation Kinetics
Catalyst	Hydrophobic	Saturated Fatty Acids
Cellular Respiration	Induced-Fit Model	Secondary Structure
Cellulose	Inner Mitochondrial	Side Chains
Citric Acid Cycle	Membrane	Starch
Coenzymes	Irreversible Inhibitors	Steroids
Cofactor	Krebs Cycle	Substrate-Level
Competitive Inhibitors	Lipid	Phosphorylation
Cooperativity	Lock and Key Model	Substrates
Cyclic AMP	Mitochondrial Matrix	Tertiary Structure
Disulfide Bonds	Metabolism	Triglycerides
Dehydration	Minerals	Unsaturated Fatty
Denatured	NADH	Acids
DNA	Negative Feedback	Vitamins
Double Helix	Noncompetitive	Water
Electron Transport	Inhibitors	Zymogen
Chain (ETC)	Nucleic Acids	
Enzymes	Nucleotides	
Enzyme Specificity	Oils	
Enzyme-Substrate	Oxidative	
Complex	Phosphorylation	
Essential Amino Acid	Peptide Bonds	

17. As electrons are passed from one protein complex to another, the final electron acceptor of the electron-transport chain is:

   A. ATP.
   B. H_2O.
   C. NADH.
   D. O_2.

18. In a human renal cortical cell, the Krebs cycle occurs in the:

   A. cytosol.
   B. mitochondrial matrix.
   C. inner mitochondrial membrane.
   D. intermembrane space.

19. As electrons move within the electron transport chain, each intermediate carrier molecule is:

   A. oxidized by the preceding molecules and reduced by the following molecule.
   B. reduced by the preceding molecule and oxidized by the following molecule.
   C. reduced by both the preceding and the following molecules.
   D. oxidized by both the preceding and the following molecules.

20. In aerobic respiration, the energy from the oxidation of NADH:

   A. directly synthesizes ATP.
   B. passively diffuses protons from the intermembrane space into the matrix.
   C. establishes a proton gradient between the intermembrane space and the mitochondrial matrix.
   D. pumps protons through ATP synthase.

21. Which of the following processes occurs under both aerobic and anaerobic conditions?

   A. fermentation
   B. Krebs cycle
   C. glycolysis
   D. oxidative phosphorylation

22. Glycolysis takes place in the cytoplasm of an animal cell. Which of the following is NOT a product or reactant in glycolysis?

   A. glucose
   B. pyruvate
   C. ATP
   D. O_2

23. What is the net ATP production from fermentation?

   A. 0 ATP
   B. 2 ATP
   C. 4 ATP
   D. 8 ATP

24. Heart and liver cells can produce more ATP for each molecule of glucose than other cells in the body. This most likely results from:

   A. a more efficient ATP synthase on the outer mitochondrial membrane.
   B. an additional turn of the Kreb's cycle for each glucose molecule.
   C. a more efficient mechanism for moving NADH produced in glycolysis into the mitochondrial matrix.
   D. production of additional NADH by the citric acid cycle.

———————————————————

Notes:

GENES

I look good in genes...

2.1 The Gene

A **gene** is a sequence of DNA nucleotides that codes for rRNA, tRNA, or a single polypeptide via an mRNA intermediate. (In the case of a virus, a gene may be an RNA sequence.) It is the gene (i.e. DNA sequence), and not the trait (i.e. eye color) that is inherited. Eukaryotes have more than one copy of some genes, while **prokaryotes** have only one copy of each gene. Genes are often referred to as *unique sequence DNA*, while regions of non-coding DNA (found only in eukaryotes) are called *repetitive sequence DNA*. In eukaryotes, unique sequence DNA dominates. Eukaryotic genes that are being actively transcribed by a cell are associated with regions of DNA called euchromatin, while genes not being actively transcribed are associated with tightly packed regions of DNA called heterochromatin.

> Generally speaking: one gene; one polypeptide. One exception is postranscriptional processing of RNA.

There are between 20,000 and 25,000 genes in the human genome. The entire DNA sequence of an organism is called the **genome**. Only a little over 1% of the human genome actually codes for protein. Variation of the nucleotide sequence among humans is small; human DNA differs from individual to individual at approximately 1 nucleotide out of every 1200 or about 0.08%. The variation between humans and chimpanzees is about 2%.

> A small variation in a genome can make a big difference.

The Central Dogma of gene expression is that DNA is transcribed to RNA, which is translated to amino acids forming a protein (Figure 2.1). All living organisms use this same method to express their genes. Retroviruses (not a living organism) store their information as RNA and must first convert their RNA to DNA in order to express their genes.

> DNA ⇨ RNA ⇨ Protein

Figure 2.1   Central Dogma of Molecular Biology

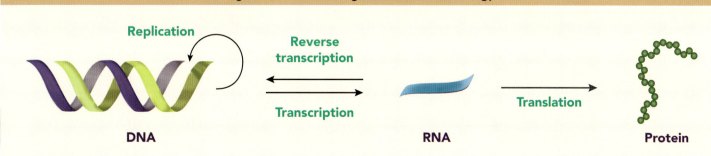

Replication

Reverse transcription

Transcription

Translation

DNA   RNA   Protein

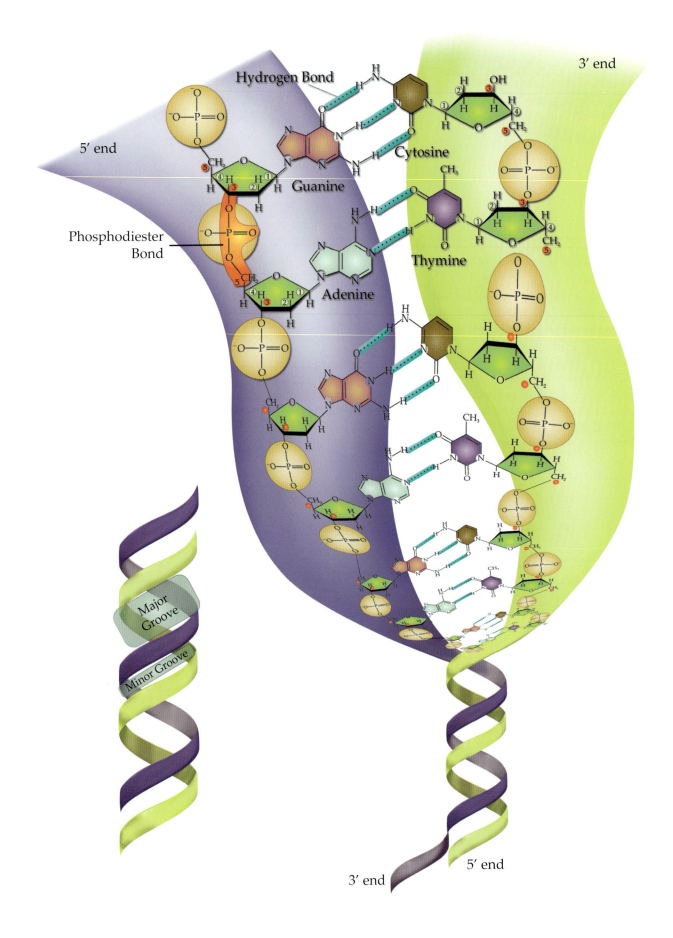

Figure 2.2 DNA Double Helix

2.2 DNA

DNA *(deoxyribonucleic acid)* is a polymer of nucleotides. DNA nucleotides differ from each other only in their nitrogenous base. Four nitrogenous bases exist in DNA: **adenine (A), guanine (G), cytosine (C), thymine (T)**. Adenine and guanine are two ring structures called **purines**, while cytosine and thymine are single ring structures called **pyrimidines** (Figure 2.3). The DNA nucleotide with the base adenine is called *adenosine phosphate*; however, it is common to refer to the nucleotides by their base name only. Each nucleotide is bound to the next by a **phosphodiester bond** between the third carbon of one deoxyribose and the fifth carbon of the other creating the sugar-phosphate backbone of a single strand of DNA with a **5′→ 3′ directionality**. The 5′ and 3′ indicate the carbon numbers on the sugar (Figure 2.2). The end 3′ carbon is attached to an –OH group and the end 5′ arbon is attached to a phosphate group. In a living organism, two DNA strands lie side by side in opposite 3′ → 5′ directions (**antiparallel**) bound together by hydrogen bonds between nitrogenous bases to form a **double stranded** structure. This hydrogen bonding is commonly referred to as **base-pairing**. The length of a DNA strand is measured in base-pairs **(bp)**. Under normal circumstances, the hydrogen bonds form only between specific purine-pyrimidine pairs; adenine forms 2 hydrogen bonds with thymine, and guanine forms 3 hydrogen bonds with cytosine. Therefore, in order for two strands to bind together, their bases must match up in the correct order. Two strands that match in such a fashion are called **complementary strands**. When complementary strands bind together, they curl into a **double helix** (Figure 2.2). The double helix contains two distinct grooves called the major groove and the minor groove. Each groove spirals once around the double helix for every ten base-pairs. The diameter of the double helix is about 2 nanometers or 13 times the diameter of a carbon atom.

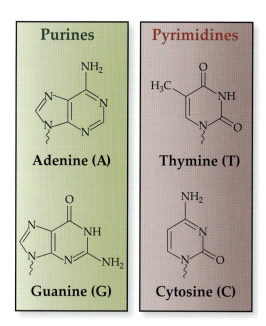

Figure 2.3 Nucleotide Bases in DNA

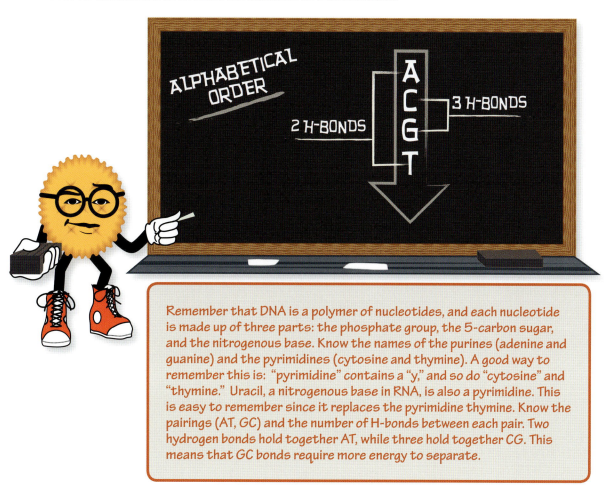

Remember that DNA is a polymer of nucleotides, and each nucleotide is made up of three parts: the phosphate group, the 5-carbon sugar, and the nitrogenous base. Know the names of the purines (adenine and guanine) and the pyrimidines (cytosine and thymine). A good way to remember this is: "pyrimidine" contains a "y," and so do "cytosine" and "thymine." Uracil, a nitrogenous base in RNA, is also a pyrimidine. This is easy to remember since it replaces the pyrimidine thymine. Know the pairings (AT, GC) and the number of H-bonds between each pair. Two hydrogen bonds hold together AT, while three hold together CG. This means that GC bonds require more energy to separate.

One time in each life cycle, a cell replicates its DNA. **DNA replication** is **semiconservative**. This means that when a new double strand is created, it contains one strand from the original DNA, and one newly synthesized strand.

The process of DNA replication (Figure 2.4) is governed by a group of proteins called a *replisome*. Replication does not begin at the end of a chromosome, but toward the middle at a site called the *origin of replication*. A single eukaryotic chromosome contains multiple origins on each chromosome, while replication in prokaryotes usually takes place from a single origin on the circular chromosome. From the origin, two replisomes proceed in opposite directions along the chromosome making replication a **bidirectional** process. The point where a replisome is attached to the chromosome is called the *replication fork*. Each chromosome of eukaryotic DNA is replicated in many discrete segments called *replication units* or *replicons*.

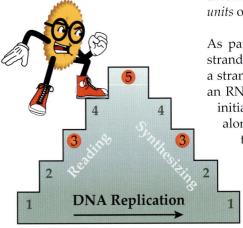

As part of the replisome, *DNA helicase* unwinds the double helix separating the two strands. **DNA polymerase**, the enzyme that builds the new DNA strand, cannot initiate a strand from two nucleotides; it can only add nucleotides to an existing strand. *Primase*, an RNA polymerase, creates an RNA **primer** approximately 10 ribonucleotides long to initiate the strand. DNA polymerase adds deoxynucleotides to the primer and moves along each DNA strand creating a new complementary strand. DNA polymerase reads the parental strand in the $3' \rightarrow 5'$ direction, creating the new complementary strand in the $5' \rightarrow 3'$ direction. (By convention, the nucleotide sequence in DNA is written $5' \rightarrow 3'$ as well. This direction is sometimes referred to as downstream and the $3' \rightarrow 5'$ direction as upstream.) Each nucleotide added to the new strand requires the removal of a *pyrophosphate group* (two phosphates bonded together) from a deoxynucleotide triphosphate. Some of the energy derived from the hydrolysis of the pyrophosphate is used to drive replication. For instance, the DNA nucleotide containing adenine is made by cleaving the second phosphate bond in the deoxy-version of ATP. (The 'deoxy-version of ATP' just means that the hydroxyl group on the 2' carbon has been replaced with a hydrogen.)

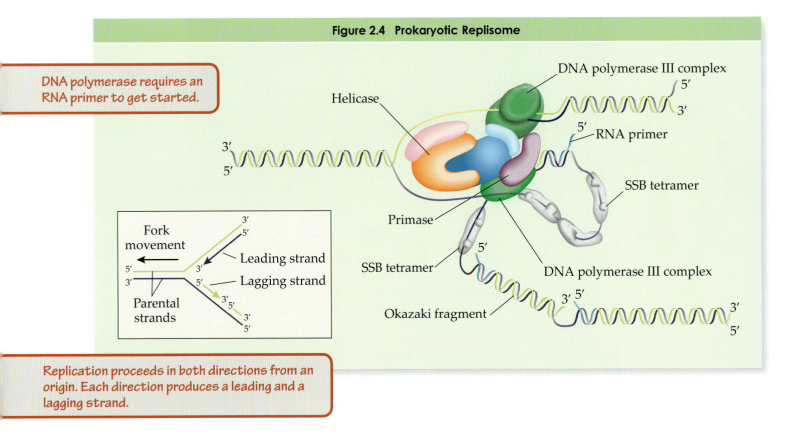

Figure 2.4 Prokaryotic Replisome

DNA polymerase requires an RNA primer to get started.

DNA polymerase III complex

Helicase

RNA primer

SSB tetramer

Primase

Fork movement

Leading strand

Lagging strand

Parental strands

SSB tetramer

DNA polymerase III complex

Okazaki fragment

Replication proceeds in both directions from an origin. Each direction produces a leading and a lagging strand.

Since DNA polymerase reads in only one direction, one strand of DNA is looped around the replisome giving it the same orientation as the other. The single strand in the loop is prevented from folding back onto itself by the *SSB tetramer* proteins (also called *helix destabilizer proteins*). As is shown in Figure 2.4, the polymerization of the new strand is continuously interrupted and restarted with a new RNA primer. This interrupted strand is called the **lagging strand**; the continuous new strand is called the **leading strand**. The lagging strand is made from a series of disconnected strands called **Okazaki fragments**. Okazaki fragments are about 100 to 200 nucleotides long in eukaryotes and about 1000 to 2000 nucleotides long in prokaryotes. **DNA ligase** (Latin: ligare → to fasten or bind) moves along the lagging strand and ties the Okazaki fragments together to complete the polymer. Since the formation of one strand is continuous and the other fragmented, the process of replication is said to be **semidiscontinuous**.

Besides being a polymerase, one of the subunits in DNA polymerase is an exonuclease (it removes nucleotides from the strand). This enzyme automatically proofreads each new strand, and makes repairs when it discovers any mismatched nucleotides, such as thymine matched with guanine. DNA replication in eukaryotes is extremely accurate. Only one base in 10^9–10^{11} is incorrectly incorporated.

In order to complete the copy of an entire genome, replication must be fast. The DNA polymerase shown in Figure 2.4 moves at over 500 nucleotides per second. DNA polymerase in humans moves much more slowly at around 50 nucleotides per second. However, multiple origins of replication allow the over 6 billion base pairs that make up the 46 human chromosomes to be replicated quite quickly. Replication in a human cell requires about 8 hours.

The ends of eukaryotic chromosomal DNA possess telomeres. **Telomeres** are repeated six nucleotide units from 100 to 1,000 units long that protect the chromosomes from being eroded through repeated rounds of replication. *Telomerase* catalyzes the lengthening of telomeres.

Although there are some differences, replication in eukaryotes and prokaryotes is very similar. Except where specified, the process described above is accurate for both.

Replication has five steps:
1. Helicase unzips the double helix;
2. RNA Polymerase builds a primer;
3. DNA Polymerase assembles the leading and lagging strands;
4. the primers are removed;
5. Okazaki fragments are joined.

DNA replication is fast and accurate.

2.4 *RNA*

RNA (ribonucleic acid) is identical to DNA in structure except that:

1) carbon number 2 on the pentose is not "deoxygenated" (it has a hydroxyl group attached);

2) RNA is **single stranded**; and

3) RNA contains the pyrimidine **uracil** (shown in Figure 2.5) instead of thymine.

Unlike DNA, RNA can move through the nuclear pores and is not confined to the nucleus. Three important types of RNA are mRNA, rRNA, and tRNA. **mRNA (messenger RNA)** delivers the DNA code for amino acids to the cytosol where the proteins are manufactured. **rRNA (ribosomal RNA)** combines with proteins to form **ribosomes**, the intracellular complexes that direct the synthesis of proteins. rRNA is synthesized in the **nucleolus**. **tRNA (transfer RNA)** collects amino acids in the cytosol, and transfers them to the ribosomes for incorporation into a protein. Notice the similarity between uracil and thymine. This is a common cause of mutations in DNA.

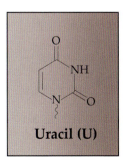

Figure 2.5 Uracil

You should know these differences between DNA and RNA. DNA is made from deoxyribose; RNA is made from ribose. DNA is double stranded; RNA is single stranded. DNA has thymine; RNA has uracil. DNA is produced by replication; RNA is produced by transcription. In eukaryotes, DNA is only in the nucleus and mitochondrial matrix only, while RNA is also in the cytosol.

2.5 Transcription

All RNA is manufactured from a DNA template in a process called **transcription**. Since DNA cannot leave the nucleus or the mitochondrial matrix, eukaryotic transcription must take place only in these two places. The beginning of transcription is called **initiation**. In initiation, a group of proteins called *initiation factors* finds a promoter on the DNA strand, and assembles a *transcription initiation complex*, which includes **RNA polymerase**. Prokaryotes have one type of RNA polymerase, whereas eukaryotes (other than plants) have three: one for rRNA; one for mRNA and some snRNAs; and one for tRNA and other RNAs. A **promoter** is a sequence of DNA nucleotides that designates a beginning point for transcription. The promoter in prokaryotes is located at the beginning of the gene (said to be upstream). The transcription start point is part of the promoter. The first base-pair located at the transcription start point is designated +1; base-pairs located before the start point such as those in the promoter are designated by negative numbers. The most commonly found nucleotide sequence of a promoter recognized by the RNA polymerase of a given species is called the *consensus sequence*. Variation from the consensus sequence causes RNA polymerase to bond less tightly and less often to a given promoter, which leads to those genes being transcribed less frequently.

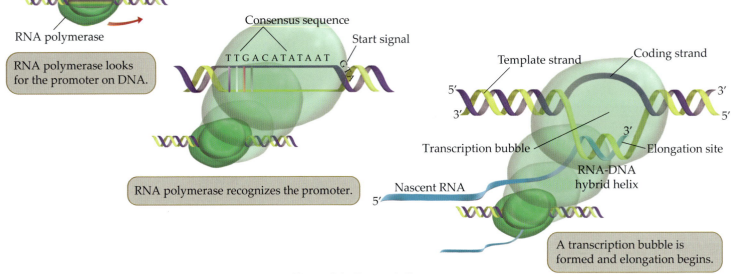

RNA polymerase looks for the promoter on DNA.

RNA polymerase recognizes the promoter.

A transcription bubble is formed and elongation begins.

Figure 2.6 Transcription

After binding to the promoter, RNA polymerase unzips the DNA double helix creating a *transcription bubble*. Next the complex switches to **elongation** mode. In elongation, RNA polymerase transcribes only one strand of the DNA nucleotide sequence into a complementary RNA nucleotide sequence. Only one strand in a molecule of double stranded DNA is transcribed. This strand is called the *template strand* or (−) *antisense strand*. The other strand, called the *coding strand* or (+) *sense strand* protects its partner against degradation. Like DNA polymerase, RNA polymerase moves along the DNA strand in the $3' \rightarrow 5'$ direction building the new RNA strand in the $5' \rightarrow 3'$. Transcription proceeds ten times more slowly than DNA replication. In addition, RNA polymerase does not contain a proofreading mechanism, and the rate of errors for transcription is higher than for replication. (Errors in RNA are not called mutations.) Since the errors are created in RNA, they are not transmitted to progeny. Most genes are transcribed many times in a cell life cycle, so the problems arising from errors in transcription are not generally harmful.

The end of transcription is called **termination**, and requires a special *termination sequence* and special proteins to dissociate RNA polymerase from DNA.

Replication makes no distinction between genes. Instead, genes are activated or deactivated at the level of transcription. For all cells, most regulation of gene expression occurs at the level of transcription via proteins called **activators** and **repressors**. Activators and repressors bind to DNA close to the promoter, and either activate or repress the activity of RNA polymerase. Activators and repressors are often allosterically regulated by small molecules such as cAMP.

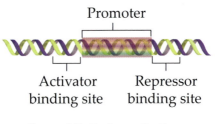

Figure 2.7 Prokaryotic Gene

The primary function of gene regulation in prokaryotes is to respond to the environmental changes. Changes in gene activity are a response to the concentration of specific nutrients in and around the cell. In contrast, lack of change or *homeostasis* of the intracellular and extracellular compartments is the hallmark of multicellular organisms. The primary function of gene regulation in multicellular organisms is to control the intra- and extracellular environments of the cell.

Prokaryotic mRNA typically includes several genes in a single transcript (*polycistronic*), whereas eukaryotic mRNA includes only one gene per transcript (*monocistronic*). The genetic unit usually consisting of the operator, promoter, and genes that contribute to a single prokaryotic mRNA is called the **operon**. A commonly used example of an operon is the *lac* operon. The lac operon codes for enzymes that allow *E. coli* to import and metabolize lactose when glucose is not present in sufficient quantities. Low glucose levels lead to high cAMP levels. cAMP binds to and activates a *catabolite activator protein (CAP)*. The activated CAP protein binds to a *CAP site* located adjacent and upstream to the promoter on the *lac* operon. The promoter is now activated allowing the formation of an initiation complex and the subsequent transcription and translation of three proteins. A second regulatory site on the *lac* operon, called the *operator*, is located adjacent and downstream to the promoter. The operator provides a binding site for a *lac* repressor protein. The *lac* repressor protein is inactivated by the presence of lactose in the cell. The *lac* repressor protein will bind to the operator unless lactose binds to the lac repressor protein.

The binding of the *lac* repressor to the operator in the absence of lactose prevents the transcription of the *lac* genes. Lactose, then, can *induce* the transcription of the *lac* operon only when glucose is not present. The promoter and gene for the *lac* repressor is located adjacent and upstream to the CAP binding site.

Most genetic regulation occurs at transcription when regulatory proteins bind DNA and activate or inhibit its transcription. In other words, the amount of a given type of protein within a cell is likely to be related to how much of its mRNA is transcribed. One reason for this is that mRNA has a short half-life in the cytosol, so soon after transcription is completed, the mRNA is degraded and its protein is no longer translated. A second reason is that many proteins can be transcribed from a single mRNA, so there is an amplifying effect.

An operon is a sequence of bacterial DNA containing an operator, a promoter, and related genes. The genes of an operon are transcribed on one mRNA. Genes outside the operon may code for activators and repressors.

Figure 2.8 Structure of Lac Operon

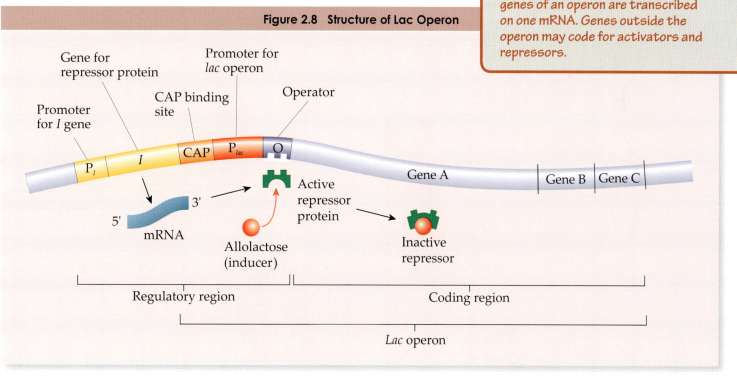

Gene regulation in eukaryotes is more complicated involving the interaction of many genes. Thus more room is required than is available near the promoter. *Enhancers* are short, non-coding regions of DNA found in eukaryotes. Their function is similar to activators and repressors but they act at a much greater distance from the promoter.

You don't need to memorize anything about the lac operon.
Just understand how it works.

Post-transcriptional Processing

Post-transcriptional processing of RNA occurs in both eukaryotic and prokaryotic cells. In prokaryotes, rRNA and tRNA go through posttranscriptional processing, but almost all mRNA is directly translated to protein. In eukaryotes, each type of RNA undergoes posttranscriptional processing and posttranscriptional processing allows for additional gene regulation.

The initial mRNA nucleotide sequence arrived at through transcription is called the **primary transcript** (also called *pre-mRNA*, or *heterogeneous nuclear RNA [hnRNA]*). The primary transcript is processed in three ways: 1) addition of nucleotides; 2) deletion of nucleotides; 3) modification of nitrogenous bases. Even before the eukaryotic mRNA is completely transcribed, its 5′ end is capped in a process using GTP. **The 5′** cap serves as an attachment site in protein synthesis and as a protection against degradation by *exonucleases*. The 3′ end is *polyadenylated* with a **poly A tail**, also to protect it from exonucleases.

The primary transcript is much longer than the mRNA that will be translated into a protein. Before leaving the nucleus, the primary transcript is cleaved into **introns** and **exons**. Enzyme-RNA complexes called small nuclear ribonucleoproteins (**snRNPs** "snurps") recognize nucleotides sequences at the ends of the introns. Several snRNPs associate with proteins to form a complex called a *spliceosome*. Inside the spliceosome, the introns are looped bringing the exons together. The introns are then excised by the spliceosomes and the exons are spliced together to form the single mRNA strand that ultimately codes for a polypeptide. Introns do not code for protein and are degraded within the nucleus. In certain cases, alternative splicing patterns can incorporate different exons into the mature mRNA. Although there are only an estimated 20,000-25,000 protein-coding genes in the human genome, there are about 120,000 proteins made possible by differential splicing of exons. Generally, intron sequences are much longer than exon sequences. Introns represent about 24 % of the genome, while exons represent about 1.1%. The average number of exons per gene is seven. The sequences of DNA that code for introns and exons are also called introns and exons.

> Post-transcriptional modification only occurs in the nucleus in eukaryotes.

> Remember that introns remain in the nucleus, and EXons EXit the nucleus to be translated.

> One reason there are more proteins than genes is that different splicing patterns of the same gene can create different polypeptides.

> Most of a typical gene consists of introns removed by snRNPs in the nucleus.

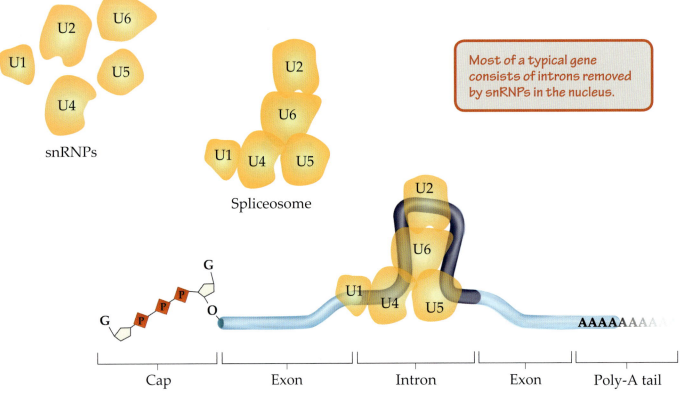

snRNPs

Spliceosome

| Cap | Exon | Intron | Exon | Poly-A tail |

Figure 2.9 Introns

When heated or immersed in high concentration salt solution or high pH solution, the hydrogen bonds connecting the two strands in a double stranded DNA molecule are disrupted, and the strands separate; the DNA molecule is said to be **denatured** or *melted*. The temperature needed to separate DNA strands is called the *melting temperature* (T_m). Since guanine and cytosine make three hydrogen bonds while thymine and adenine make only two, DNA with more G-C base pairs has a greater T_m. Heating to 95°C (just below the boiling point of water) is generally sufficient to denature any DNA sequence. Denatured DNA is less viscous, denser, and more able to absorb UV light.

Separated strands will spontaneously associate with their original partner or any other complementary nucleotide sequence. Thus, the following double stranded combinations can be formed through **nucleic acid hybridization**: DNA-DNA, DNA-RNA, and RNA-RNA. Hybridization techniques enable scientists to identify nucleotide sequences by binding a known sequence with an unknown sequence.

One method bacteria use to defend themselves from viruses is to cut the viral DNA into fragments with restriction enzymes. The bacteria protect their own DNA from these enzymes by *methylation* (adding a –CH$_3$). Methylation is usually, but not always, associated with inactivated genes. **Restriction enzymes** (also called restriction endonucleases) *digest* (cut) nucleic acid only at certain nucleotide sequences along the chain (see Figure 2.10). Such a sequence is called a *restriction site* or *recognition sequence*. Typically, a restriction site will be a **palindromic** sequence four to six nucleotides long. (Palindromic means that it reads the same backwards as forwards.) Most restriction endonucleases cleave the DNA strand unevenly, leaving complementary single stranded ends. These ends can reconnect through hybridization and are termed *sticky ends*. Once paired, the phosphodiester bonds of the fragments can be joined by DNA ligase. There are hundreds of restriction endonucleases known, each attacking a different restriction site. A given sample of DNA of significant size (number of base=pairs) is likely to contain a recognition sequence for at least one restriction endonuclease. Two DNA fragments cleaved by the same endonuclease can be joined together regardless of the origin of the DNA. Such DNA is called **recombinant DNA**; it has been artificially recombined.

> To denature DNA means to separate the two strands of the double helix.

> DNA prefers to be double stranded and will look for a complementary partner.

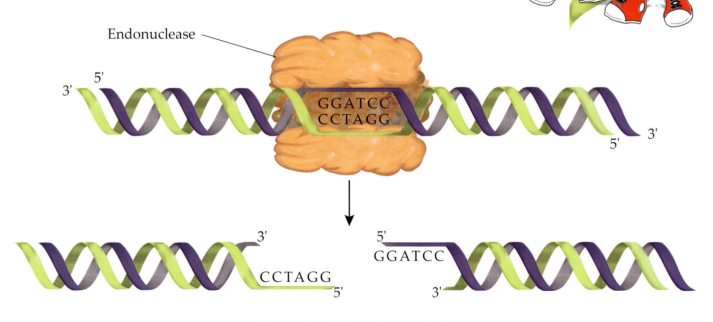

Figure 2.10 Polypeptide Synthesis

Recombinant DNA can be made long enough for bacteria to replicate and then placed within the bacteria using a **vector**, typically a **plasmid** or sometimes an infective virus. The bacteria can then be grown in large quantity forming a **clone** of cells containing the vector with the recombinant DNA fragment. The clones can be saved separately producing a **clone library**.

> To make a DNA library, take your DNA fragment, use a vector to insert it into a bacterium, and reproduce that bacterium like crazy. Now you have clones of bacteria with your DNA fragment.

Because not all bacteria take up the vector and not all vectors take up the DNA fragment, a library may contain some clones that do not contain vectors or contain vectors that do not contain the recombinant DNA fragment. By including in the original vector a gene for resistance to a certain antibiotic and the *lacZ* gene, which enables the bacteria to metabolize the sugar X-gal, libraries can later be screened for the appropriate clones. When an antibiotic is added to a library of clones, clones without resistance will be eliminated. Clones without resistance must not have taken up the vector.

In order to screen out clones that contain the original vector and not the DNA fragment, an endonuclease with a recognition site that cuts the *lacZ* gene in two should be used to place the DNA fragment into the vector. Since the endonuclease cleaves the *lacZ* gene, the *lacZ* gene will not work when the DNA fragment is placed in the vector. Clones with an active *lacZ* gene turn blue in the presence of X-gal. Clones with the cleaved form of the gene do not turn blue. Clones with the DNA fragment then will not turn blue when placed on a medium with X-gal.

Part of the screening process involves actually finding the desired DNA sequence from a library. One technique to find a particular gene in a library is hybridization. The radioactively labeled complementary sequence of the desired DNA fragment (called a **probe**) is used to search the library. The radiolabeled clones are identified by laying them over photographic film which they expose and non-radiolabeled clones do not.

Eukaryotic DNA contains introns. Since bacteria have no mechanism for removing introns, it is useful to clone DNA with no introns. In order to do this, the mRNA produced by the DNA is reverse transcribed using reverse transcriptase. The DNA product is called **complementary DNA** or **cDNA**. Adding DNA polymerase to cDNA produces a double strand of the desired DNA fragment.

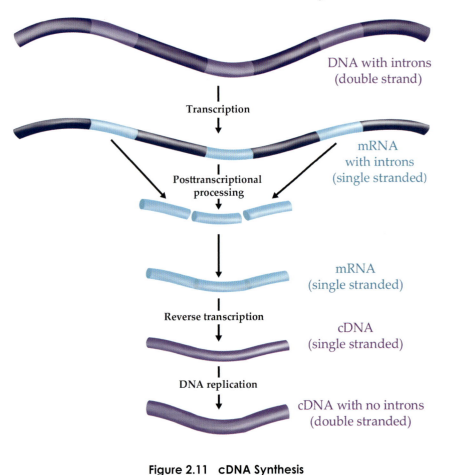

DNA with introns
(double strand)

Transcription

mRNA
with introns
(single stranded)

Posttranscriptional
processing

mRNA
(single stranded)

Reverse transcription

cDNA
(single stranded)

DNA replication

cDNA with no introns
(double stranded)

Figure 2.11 cDNA Synthesis

> cDNA is just DNA reverse transcribed from mRNA. The great thing about cDNA is that it lacks the introns that would normally be found in eukaryotic DNA.

Figure 2.12 Clone Library

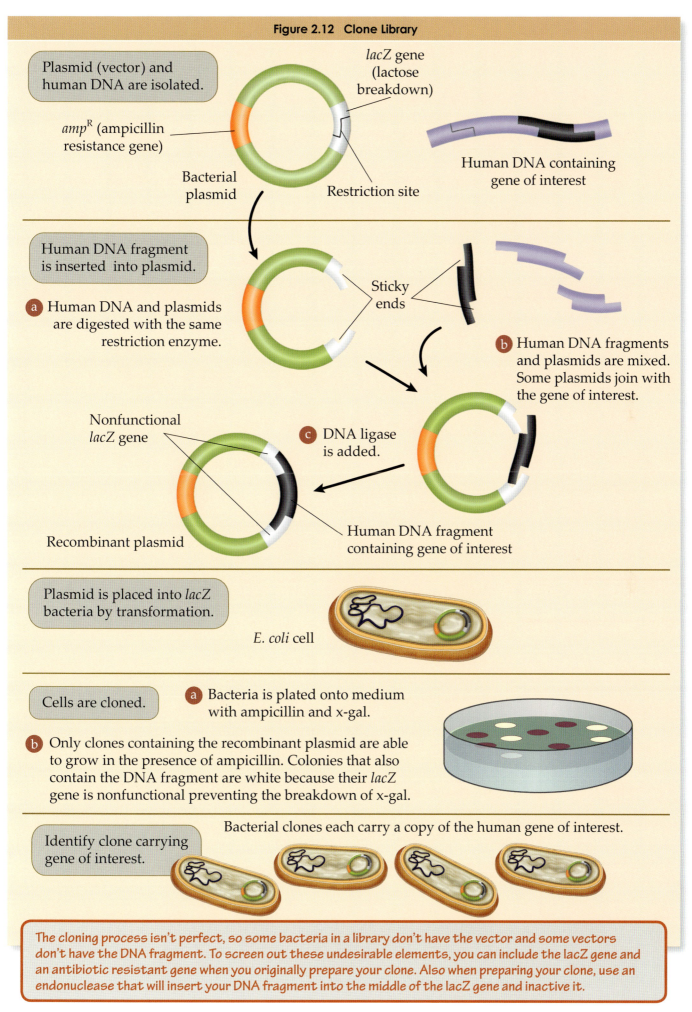

Plasmid (vector) and human DNA are isolated.

amp^R (ampicillin resistance gene)

Bacterial plasmid

lacZ gene (lactose breakdown)

Restriction site

Human DNA containing gene of interest

Human DNA fragment is inserted into plasmid.

a Human DNA and plasmids are digested with the same restriction enzyme.

Sticky ends

b Human DNA fragments and plasmids are mixed. Some plasmids join with the gene of interest.

Nonfunctional *lacZ* gene

c DNA ligase is added.

Recombinant plasmid

Human DNA fragment containing gene of interest

Plasmid is placed into *lacZ* bacteria by transformation.

E. coli cell

Cells are cloned.

a Bacteria is plated onto medium with ampicillin and x-gal.

b Only clones containing the recombinant plasmid are able to grow in the presence of ampicillin. Colonies that also contain the DNA fragment are white because their *lacZ* gene is nonfunctional preventing the breakdown of x-gal.

Identify clone carrying gene of interest.

Bacterial clones each carry a copy of the human gene of interest.

The cloning process isn't perfect, so some bacteria in a library don't have the vector and some vectors don't have the DNA fragment. To screen out these undesirable elements, you can include the lacZ gene and an antibiotic resistant gene when you originally prepare your clone. Also when preparing your clone, use an endonuclease that will insert your DNA fragment into the middle of the lacZ gene and inactive it.

A much faster method of "cloning" called **polymerase chain reaction (PCR)** has been developed using a specialized polymerase enzyme found in a species of bacterium adapted to life in nearly boiling waters. In PCR, the double strand of DNA to be "cloned" or (speaking more precisely) *amplified* is placed in a mixture with many copies of two DNA primers, one for each strand. The mixture is heated to 95°C to denature the DNA. When the mixture is cooled to 60°C, the primers hybridize (or **anneal**) to their complementary ends of the DNA strands. Next, the heat resistant polymerase is added with a supply of nucleotides, and the mixture is heated to 72°C to activate the polymerase. The polymerase amplifies the complementary strands doubling the amount of DNA. The procedure can be repeated many times without adding more polymerase because it is heat resistant. The result is an exponential increase in the amount of DNA. Starting with a single fragment, 20 cycles produces over one million copies (2^{20}). What used to require days with recombinant DNA techniques, can now be done in hours with PCR. Another advantage of PCR is that minute DNA samples can be amplified. PCR requires that the base sequence flanking the ends of the DNA fragment be known, so that the complementary primers can be chosen.

Figure 2.13   PCR

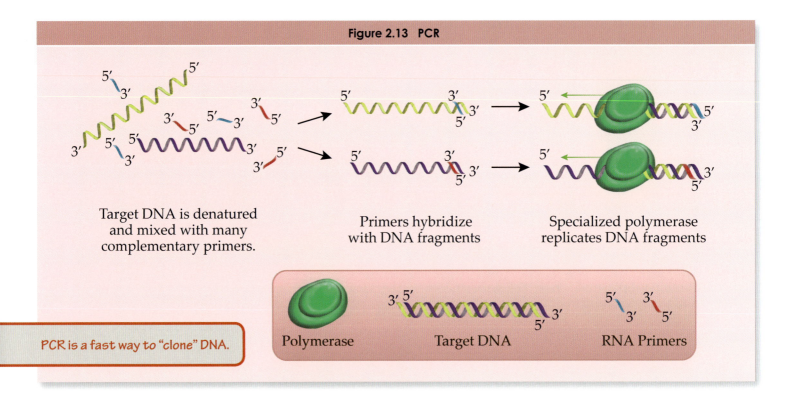

Target DNA is denatured and mixed with many complementary primers.

Primers hybridize with DNA fragments

Specialized polymerase replicates DNA fragments

Polymerase   Target DNA   RNA Primers

PCR is a fast way to "clone" DNA.

The recipe for a Southern blot is:

1. Chop up some DNA;

2. Use an electric field to spread out pieces according to size;

3. Blot it onto a membrane;

4. Add a radioactive probe made from DNA or RNA;

5. Visualize with radiographic film.

Southern blotting is a technique used to identify target fragments of known DNA sequence in a large population of DNA. In a southern blot, the DNA to be identified is cleaved into restriction fragments. The fragments are resolved (separated) according to size by gel electrophoresis. Large fragments move more slowly through the gel than small fragments. Next, the gel is made alkaline to denature the DNA fragments. A membrane, such as a sheet of *nitrocellulose*, is used to blot the gel which transfers the resolved single stranded DNA fragments onto the membrane. A radio-labeled probe with a nucleotide sequence complementary to the target fragment is added to the membrane. The probe hybridizes with and marks the target fragment. The membrane is exposed to radiographic film which reveals the location of the probe and the target fragment.

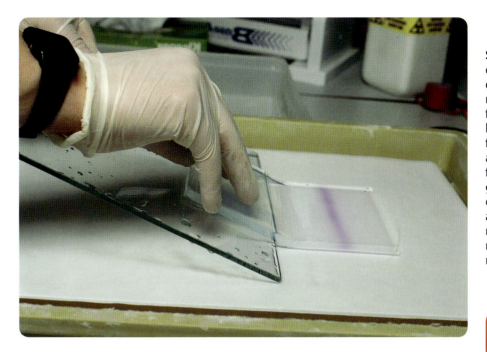

Southern blotting: A researcher transfers an electrophoresis gel containing DNA into a tray of salt solution prior to blotting the DNA onto nitrocellulose paper. The purple dye shows how far electrophoresis has progressed. Southern blotting is a technique to reveal specific fragments of DNA (deoxyribonucleic acid) in a complex mixture. The total DNA is cut into fragments and separated according to size by gel electrophoresis. The DNA is then blotted onto a sheet of nitrocellulose placed on top, and a radioactive DNA probe added to the nitrocellulose. The resulting banding pattern, revealed by autoradiography, can be used to map the structure of particular genes.

A Southern blot identifies specific sequences of DNA by nucleic acid hybridization, and a Northern blot uses the same techniques to identify specific sequences of RNA.

A **Northern blot** is just like a Southern blot, but it identifies RNA fragments, not DNA fragments.

A **Western blot** can detect a particular protein in a mixture of proteins. First a mixture of proteins are resolved by size using electrophoresis. Next they are blotted onto a nitrocellulose membrane. An antibody (the *primary antibody*) specific to the protein in question is then added and binds to that protein. Next, a *secondary antibody-enzyme* conjugate is added. The secondary antibody recognizes and binds to the primary antibody and marks it with the enzyme for subsequent visualization. The reaction catalyzed by the enzyme attached to the secondary antibody can produce a colored, fluorescent or radioactive reaction product which can be visualized or detected with xray film.

If Western blot shows up on the MCAT, it is likely to be described thoroughly in a passage. Just recognize that this is the one that detects a <u>protein</u> with <u>antibodies</u>.

Restriction fragment length polymorphisms (**RFLP**) analysis identifies individuals as opposed to identifying specific genes. The DNA of different individuals possesses different restriction sites and varying distances between restriction sites. The population of humans is *polymorphic* for their restriction sites. After fragmenting the DNA sample with endonucleases, a band pattern unique to an individual is revealed on radiographic film via Southern blotting techniques. RFLPs (pronounced "riflips") are the DNA fingerprints used to identify criminals in court cases.

The genome of one human differs from the genome of another at about one nucleotide in every 1000. These differences have been called *single nucleotide polymorphisms* (SNPs). Like RFLPs, SNPs can serve as a fingerprint to an individuals genome.

25. Which of the following is always true concerning the base composition of DNA?

   A. In each single strand, the number of adenine residues equals the number thymine residues.
   B. In each single strand, the number of adenine residues equals the number of guanine residues.
   C. In a molecule of double stranded DNA, the ratio of adenine residues to thymine residues equals the ratio of cytosine residues to guanine residues.
   D. In a molecule of double stranded DNA, the number of adenine residues plus thymine residues equals the number of cytosine residues plus guanine residues.

26. An mRNA molecule being translated at the rough endoplasmic reticulum is typically shorter than the gene from which it was transcribed because:

   A. the primary transcript was cut as it crossed the nuclear membrane.
   B. normally multiple copies of the mRNA are produced and spliced.
   C. introns in the primary transcript are excised.
   D. several expressed regions of the primary transcript have equal numbers of base pairs.

27. In PCR amplification, a primer is hybridized to the end of a DNA fragment and acts as the initiation site of replication for a specialized DNA polymerase. The DNA fragment to be amplified is shown below. Assuming that the primer attaches exactly to the end of the fragment, which of the following is most likely the primer? (Note: The N stands for any nucleotide.)

   5′-ATGNNNNNNNNNNNNNNNGCT-3′
   DNA fragment

   A. 5′-GCT-3′
   B. 5′-TAC-3′
   C. 5′-TCG-3′
   D. 5′-AGC-3′

28. Which of the following is NOT true concerning DNA replication?

   A. DNA ligase links the Okazaki fragments.
   B. Helicase unwinds the DNA double helix.
   C. Only the sense strand is replicated.
   D. DNA strands are synthesized in the 5′ to 3′ direction.

29. The gene for triose phosphate isomerase from maize (a corn plant) spans over 3400 base pairs of DNA and contains eight introns and nine exons. Which of the following would most likely represent the number of nucleotides found in the mature mRNA after posttranscriptional processing?

   A. 1050
   B. 3400
   C. 6800
   D. 13,600

30. Complementary strands of DNA are held together by:

   A. phosphodiester bonds.
   B. covalent bonds.
   C. hydrophobic interactions.
   D. hydrogen bonds.

31. Eukaryotic mRNA production occurs in the following sequence:

   A. transcription from DNA in the cytoplasm followed by post transcriptional processing on the ribosome.
   B. transcription from DNA in the nucleus followed by post transcriptional processing in the nucleus.
   C. translation from DNA in the nucleus followed by post-transcriptional processing in the nucleus.
   D. translation from DNA in the cytoplasm followed by post-transcriptional processing on the ribosome.

32. In Southern blotting, DNA fragments are separated based upon size during electrophoresis. Which of the following is true of this process?

   A. Positively charged DNA fragments move toward the cathode.
   B. Positively charged DNA fragments move toward the anode.
   C. Negatively charged DNA fragments move toward the cathode.
   D. Negatively charged DNA fragments move toward the anode.

2.8 The Genetic Code

mRNA nucleotides are strung together to form a **genetic code** which translates the DNA nucleotide sequence into an amino acid sequence and ultimately into a protein. There are four different nucleotides in RNA that together must form an unambiguous code for the 20 common amino acids. The number of possible combinations of a row of two nucleotides, where each nucleotide might contain any one of the four nitrogenous bases, is $4^2 = 16$, not enough to code for 20 amino acids (see Table 2.1). Therefore, the code must be a combination of three nucleotides. However, any three nucleotides gives $4^3 = 64$ possible combinations. These are more possibilities than there are amino acids. Thus more than one series of three nucleotides may code for any amino acid; the code is **degenerative**. But any single series of three nucleotides will code for one and only one amino acid; the code is **unambiguous**. In addition, the code is **almost universal**; nearly every living organism uses the same code.

Table 2.1

	A	C	G	T
A	AA	CA	GA	TA
C	AC	CC	GC	TC
G	AG	CG	GG	TG
T	AT	CT	GT	TT

Table 2.2

First position 5′ end ↓	Second position				Third position 3′ end ↓
	U	**C**	**A**	**G**	
U	Phe	Ser	Tyr	Cys	U
	Phe	Ser	Tyr	Cys	C
	Leu	Ser	STOP	STOP	A
	Leu	Ser	STOP	Trp	G
C	Leu	Pro	His	Arg	U
	Leu	Pro	His	Arg	C
	Leu	Pro	Gln	Arg	A
	Leu	Pro	Gln	Arg	G
A	Ile	Thr	Asn	Ser	U
	Ile	Thr	Asn	Ser	C
	Ile	Thr	Lys	Arg	A
	Met	Thr	Lys	Arg	G
G	Val	Ala	Asp	Gly	U
	Val	Ala	Asp	Gly	C
	Val	Ala	Glu	Gly	A
	Val	Ala	Glu	Gly	G

Memorize the start codon, AUG, and the stop codons UAA, UAG, and UGA. It is necessary to understand that a single codon (such as GUC) always codes for only one amino acid, in this case valine; but that there are other codons (GUU, GUA, GUG) that also code for valine. The first part of this means that the code is unambiguous, and the second part means that the code is degenerative. Finally, you must understand probabilities. For instance, you must be able to figure out how many possible codons exist. As discussed above, four possible nucleotides can be placed in each of 3 positions giving $4^3 = 64$.

Try this: A polypeptide contains 100 amino acids. How many possible amino acid sequences are there for this polypeptide?

Answer: See page 62

Three consecutive nucleotides on a strand of mRNA represent a **codon**. All but three possible codons code for amino acids. The remaining codons, UAA, UGA, and UAG, are **stop codons** (also called **termination codons**). Stop codons signal an end to protein synthesis. The start codon, AUG, also acts as a codon for the amino acid methionine.

The genetic code is given in Table 2.2. Be certain that you can read this table. For instance, the codons for lysine (Lys) are AAA and AAG. By convention, a sequence of RNA nucleotides is written 5′ → 3′.

2.9 Translation

Translation (Figure 2.14) is the process of protein synthesis directed by mRNA. Each of the three major types of RNA plays a unique role in translation. mRNA is the template which carries the genetic code from the nucleus to the cytosol in the form of codons. tRNA contains a set of nucleotides that is complementary to the codon, called the **anticodon**. tRNA sequesters the amino acid that corresponds to its anticodon. rRNA with protein makes up the **ribosome**, which provides the site for translation to take place. rRNA actively participates in the translation process.

The ribosome is composed of a **small subunit** and a **large subunit** made from rRNA and many separate proteins. The ribosome and its subunits are measured in terms of *sedimentation coefficients* given in *Svedberg units (S)*. The sedimentation coefficient gives the speed of a particle in a centrifuge, and is proportional to mass and related to shape and density. Prokaryotic ribosomes are smaller than eukaryotic ribosomes. Prokaryotic ribosomes are made from a 30S and a 50S subunit and have a combined sedimentary coefficient of 70S. Eukaryotic ribosomes are made of 40S and 60S subunits and have a combined sedimentary coefficient of 80S. The complex structure of ribosomes requires a special organelle called the **nucleolus** in which to manufacture them. (Prokaryotes do not possess a nucleolus, but synthesis of prokaryotic ribosomes is similar to that of eukaryotic ribosomes.) Although the ribosome is assembled in the nucleolus, the small and large subunits are exported separately to the cytoplasm.

> Notice that the sedimentation coefficients don't add up: 40 + 60 ≠ 80.

After posttranscriptional processing in a eukaryote, mRNA leaves the nucleus through the nuclear pores and enters the cytosol. With the help of *initiation factors* (proteins), The 5′ end attaches to the small subunit of a ribosome. A tRNA possessing the 5′-CAU-3′ anticodon sequesters the amino acid methionine and settles in at the **P site** (peptidyl site). This is the signal for the large subunit to join and form the **initiation complex**. This process is termed **initiation**.

Now **elongation** of the polypeptide begins. A tRNA with its corresponding amino acid attaches to the **A site** (aminoacyl site) at the expense of two GTPs. The C-terminus (carboxyl end) of methionine attaches to the *N-terminus* (amine end) of the amino acid at the A site in a dehydration reaction catalyzed by *peptidyl transferase*, an enzyme possessed by the ribosome. In an elongation step called **translocation**, the ribosome shifts 3 nucleotides along the mRNA toward the 3′ end. The tRNA that carried methionine moves to the **E site** where it can exit the ribosome. The tRNA carrying the nascent (newly formed) dipeptide moves to the P site, clearing the A site for the next tRNA. Translocation requires the expenditure of another GTP. The elongation process is repeated until a stop codon reaches the P site.

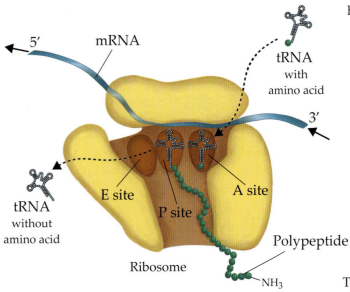

Figure 2.14 Translation

Translation ends when a stop codon is reached in a step called **termination**. When a stop (or *nonsense*) codon reaches the A site, proteins known as *release factors* bind to the A site allowing a water molecule to add to the end of the polypeptide chain. The polypeptide is freed from the tRNA and ribosome, and the ribosome breaks up into its subunits to be used again for another round of protein synthesis later.

> Know the process of translation: initiation, elongation, and termination. Know the role of each type of RNA.

Even as the polypeptide is being translated, it begins folding. The amino acid sequence determines the folding conformation and the folding process is assisted by proteins called *chaperones*. In **post-translational modifications**, sugars, lipids, or phosphate groups may be added to amino acids. The polypeptide may be cleaved in one or more places. Separate polypeptides may join to form the quaternary structure of a protein.

Translation may take place on a free floating ribosome in the cytosol producing proteins that function in the cytosol, or a ribosome may attach itself to the rough ER during translation and inject proteins into the ER lumen. Proteins injected into the ER lumen are destined to become membrane bound proteins of the nuclear envelope, ER, Gogli, lysosomes, plasma membrane, or to be secreted from the cell. Free floating ribosomes are identical in structure to ribosomes that attach to the ER. The growing polypeptide itself may or may not cause the ribosome to attach to the ER depending upon the polypeptide. A 20 amino acid sequence called a **signal peptide** near the front of the polypeptide is recognized by protein-RNA **signal-recognition particle** (SRP) that carries the entire ribosome complex to a receptor protein on the ER. There the protein grows across the membrane where it is either released into the lumen or remains partially attached to the ER. The signal peptide is usually removed by an enzyme. Signal peptides may also be attached to polypeptides to target them to mitochondria, the nucleus, or other organelles.

> Translation begins on a free floating ribosome. A signal peptide at the beginning of the translated polypeptide may direct the ribosome to attach to the ER, in which case the polypeptide is injected into the lumen. Polypeptides injected into the lumen may be secreted from the cell via the Golgi or may remain partially attached to the membrane.

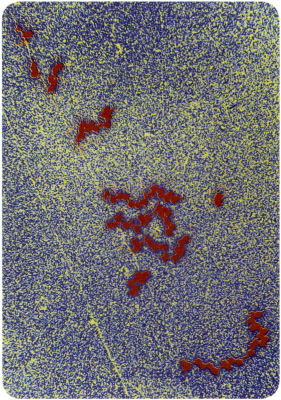

This false-color transmission electron micrograph (TEM) of a structural gene from the bacterium *Eschericia coli* shows the coupled transcription of DNA into mRNA and the simultaneous translation into protein molecules. The DNA fiber runs down the image (in yellow) from top left, with numerous ribosomes (in red) attached to each mRNA chain. The longer chains (called polysomes) are furthest from the point of gene origin.

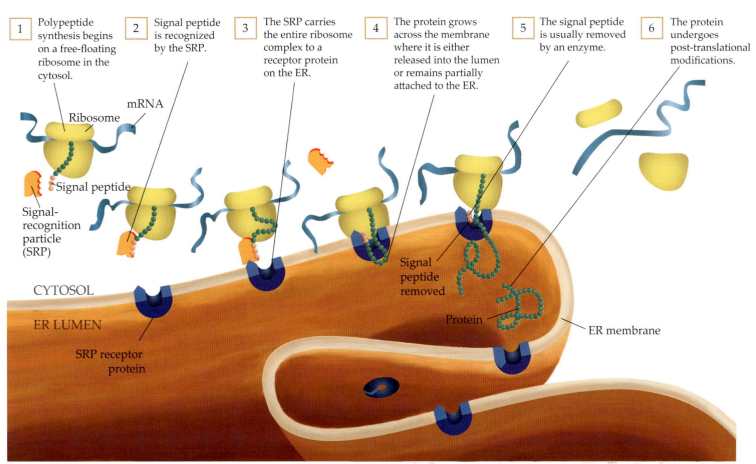

1 Polypeptide synthesis begins on a free-floating ribosome in the cytosol.

2 Signal peptide is recognized by the SRP.

3 The SRP carries the entire ribosome complex to a receptor protein on the ER.

4 The protein grows across the membrane where it is either released into the lumen or remains partially attached to the ER.

5 The signal peptide is usually removed by an enzyme.

6 The protein undergoes post-translational modifications.

Figure 2.15 Polypeptide Synthesis

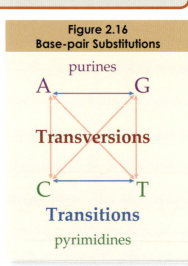

**Figure 2.16
Base-pair Substitutions**

purines

A ↔ G

Transversions

C ↔ T

Transitions

pyrimidines

2.10 | *Mutations*

Any alteration in the genome that is not genetic recombination is called a **mutation**. (The genome is the totality of DNA.) Mutations may occur at the chromosomal level, or the nucleotide level. A **gene mutation** is the alteration in the sequence of DNA nucleotides in a single gene. A **chromosomal mutation** occurs when the structure of a chromosome is changed. In multicellular organisms, a mutation in a somatic cell (Greek: soma → body) is called a *somatic mutation*. A somatic mutation of a single cell may have very little effect on an organism with millions of cells. A mutation in a germ cell, from which all other cells arise, can be very serious for the offspring. Only about one out of every million gametes will carry a mutation for a given gene.

Mutations can be *spontaneous* (occurring due to random errors in the natural process of replication and genetic recombination) or *induced* (occurring due to physical or chemical agents called **mutagens**). The effects on the cell are the same in either case. A mutagen is any physical or chemical agent that increases the frequency of mutation above the frequency of spontaneous mutations. If a mutation changes a single **base-pair** of nucleotides in a double strand of DNA, that mutation is called a **point mutation**. One type of point mutation called a **base-pair substitution mutation** results when one base-pair is replaced by another. A base pair substitution exchanging one purine for the other purine (A ↔ G) or one pyrimidine for the other pyrimidine (C ↔ T) is called a *transition mutation*. A base-pair substitution exchanging a purine for a pyrimidine or a pyrimidine for a purine is called a *transversion mutation*. A **missense mutation** is a base-pair mutation that occurs in the amino acid coding sequence of a gene. A missense mutation may or may not alter the amino acid sequence of a protein, and an alteration of a single amino acid may or may not have serious effects on the function of the protein. (Sickle cell anemia, for instance, is a disease caused by a single amino acid difference in hemoglobin.) If there is no change in protein function, the mutation is called a *neutral mutation*, and if the amino acid is not changed, it is called a *silent mutation*. Even a silent mutation may be significant because it may change the rate of transcription.

Figure 2.17   Example of a Missense Mutation

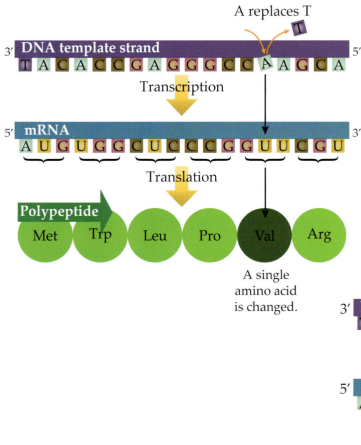

Figure 2.18   Example of a Silent Mutation

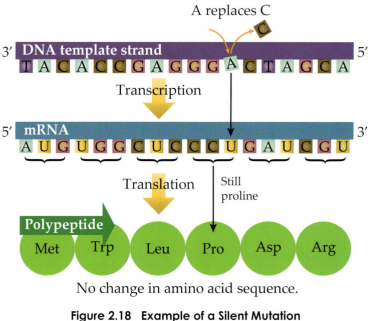

An **insertion or deletion** of a base-pair, may result in a **frameshift mutation**. A frameshift mutation results when the deletions or insertions occur in multiples other than three. This is because the genetic code is read in groups of three nucleotides, and the entire sequence after the mutation will be shifted so that the three base sequences will be grouped incorrectly. For instance, if a single T nucleotide were inserted into the series: AAA|GGG|CCC|AAA, so that it reads AAT|AGG|GCC|CAA|A, each 3-nucleotide sequence downstream from the mutation would be changed because the entire series would be shifted one to the right. On the other hand, if three T nucleotides were inserted randomly, the downstream sequence would not be shifted, and only one or a few 3-nucleotide sequences would be changed. AAT|TAG|GTG|CCC|AAA This is a nonframeshift mutation. Frameshift mutations often result in a completely nonfunctional protein, whereas nonframeshift mutations may still result in a partially or even completely active protein.

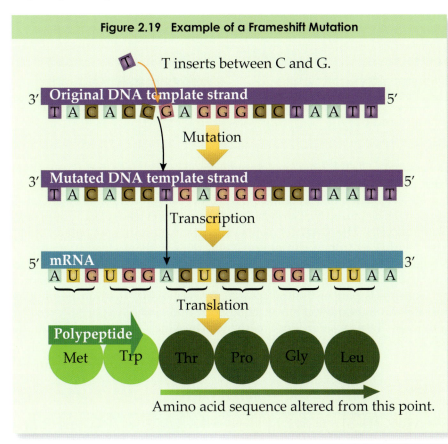

Figure 2.19   Example of a Frameshift Mutation

T inserts between C and G.

3′ **Original DNA template strand** 5′
T A C A C C G A G G G C C T A A T T

Mutation

3′ **Mutated DNA template strand** 5′
T A C A C C T G A G G G C C T A A T T

Transcription

5′ **mRNA** 3′
A U G U G G A C U C C C G G A U U A A

Translation

Polypeptide
Met Trp Thr Pro Gly Leu

Amino acid sequence altered from this point.

Male white lions are found only in wildlife reserves in South Africa, where they are selectively bred. It is not an albino, but a leucistic. Unlike albinism, which is a reduction in just melanin, leucism is caused by a mutation that reduces all types of skin or hair pigment.

If a base-pair substitution or an insertion or deletion mutation creates a stop codon, a **nonsense mutation** results. Nonsense mutations are usually very serious for the cell because they prevent the translation of a functional protein entirely.

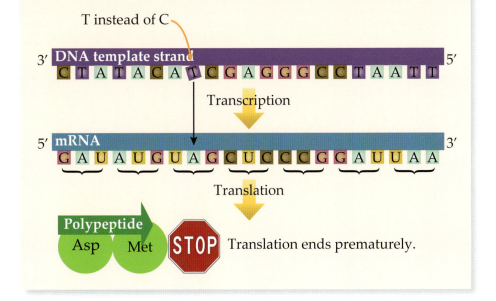

Figure 2.20   Example of a Nonsense Mutation

T instead of C

3′ **DNA template strand** 5′
C T A T A C A T C G A G G G C C T A A T T

Transcription

5′ **mRNA** 3′
G A U A U G U A G C U C C C G G A U U A A

Translation

Polypeptide
Asp Met **STOP** Translation ends prematurely.

Structural changes may occur to a chromosome in the form of deletions, duplications, translocations, and inversions. Chromosomal **deletions** occur when a portion of the chromosome breaks off, or when a portion of the chromosome is lost during homologous recombination and/or crossing over events. **Duplications** occur when a DNA fragment breaks free of one chromosome and incorporates into a homologous chromosome. Deletion or duplication can occur with entire chromosomes (*aneuploidy*) or even entire sets of chromosomes (*polyploidy*).

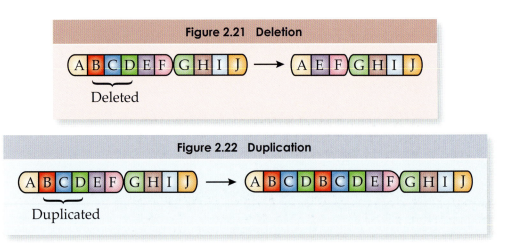

Figure 2.21 Deletion

Figure 2.22 Duplication

When a segment of DNA from one chromosome is exchanged for a segment of DNA on another chromosome, the resulting mutation is called a reciprocal **translocation**. In **inversion** the orientation of a section of DNA is reversed on a chromosome. Translocation and inversion can be caused by transposition. Transposition takes place in both prokaryotic and eukaryotic cells. The DNA segments called **transposable elements** or **transposons** can excise themselves from a chromosome and reinsert themselves at another location. Transposons can contain one gene, several genes, or just a control element. A transposon within a chromosome will be flanked by identical nucleotide sequences. A portion of the flanking sequence is part of the transposon. When moving, the transposon may excise itself from the chromosome and move; it may copy itself and move, or copy itself and stay, moving the copy. Transposition is one mechanism by which a somatic cell of a multicellular organism can alter its genetic makeup without meiosis.

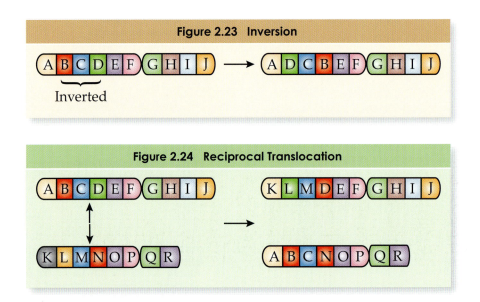

Figure 2.23 Inversion

Figure 2.24 Reciprocal Translocation

A mutation can be a **forward mutation** or a **backward mutation**. These terms refer to an already mutated organism that is mutated again. The mutation can be forward, tending to change the organism even more from its original state, or backward, tending to revert the organism back to its original state. The original state is called the **wild type**. For example, you may be working in the lab with bacteria that normally produce histidine, and you mutate a sample so that the bacteria in that sample no longer produce histidine. Now you have the wild type, his+, and the mutants, his-. If you back mutated the his-, you would produce the wild type, his+. If you forward mutated the his-, they may lose the ability to produce some other amino acid in addition to histidine.

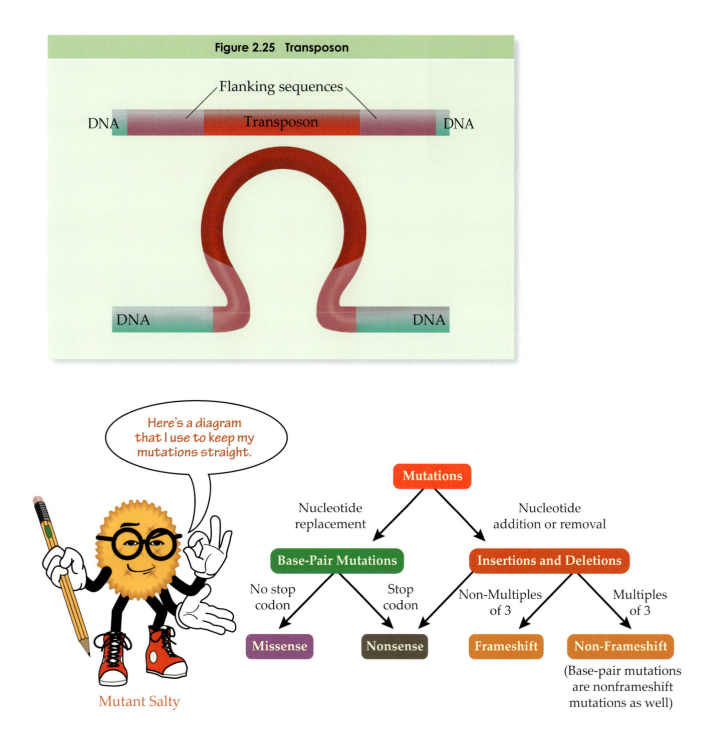

Figure 2.25   Transposon

Flanking sequences

DNA   Transposon   DNA

DNA   DNA

Here's a diagram that I use to keep my mutations straight.

Mutant Salty

Mutations

Nucleotide replacement

Nucleotide addition or removal

Base-Pair Mutations

Insertions and Deletions

No stop codon

Stop codon

Non-Multiples of 3

Multiples of 3

Missense

Nonsense

Frameshift

Non-Frameshift

(Base-pair mutations are nonframeshift mutations as well)

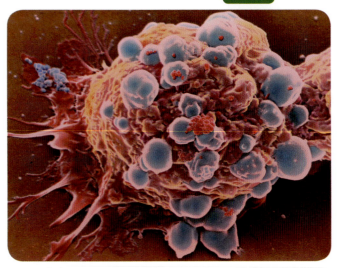

A single breast cancer cell has an uneven surface with blebs (blue) and cytoplasmic projections (red, at left). Clumps of cancerous (malignant) cells form tumors, which possess the ability to invade and destroy surrounding tissues and travel to distant parts of the body to seed secondary tumors. Malignant cells proliferate and grow in a chaotic manner, with defective cell division retained within each new generation of cells. Variations also occur in size and structure of the cancer cell from its original type.

Cancer is the unrestrained and uncontrolled growth of cells. Normal cells divide 20 to 50 times before they stop dividing and die, but cancer cells continue to grow and divide indefinitely. A mass of cancer cells is called a *tumor*. A tumor is benign if it is localized in a small lump. When an individual has a tumor invasive enough to impair function of an organ, the tumor is said to be *malignant* and the individual is said to have cancer. Cancer cells may separate from the tumor and enter the bodies circulatory systems and establish tumors in other parts of the body. This process is called metastasis.

Certain genes that stimulate normal growth in human cells are called *proto-oncogenes*. Proto-oncogenes can be converted to **oncogenes**, genes that cause cancer, by mutagens such as UV radiation, chemicals, or simply by random mutations. Mutagens that can cause cancer are called **carcinogens**.

Notes:

33. If each of the following mRNA nucleotide sequences contains three codons, which one contains a start codon?

   A. 3′-AGGCCGUAG-5′
   B. 3′-GUACCGAAC-5′
   C. 5′-AAUGCGGAC-3′
   D. 5′-UAGGAUCCC-3′

34. Translation in a eukaryotic cell is associated with each of the following organelles or locations EXCEPT:

   A. the mitochondrial matrix.
   B. the cytosol.
   C. the nucleus.
   D. the rough endoplasmic reticulum.

35. Which of the following is true concerning the genetic code?

   A. There are more amino acids than codons.
   B. Any change in the nucleotide sequence of a codon must result in a new amino acid.
   C. The genetic code varies from species to species.
   D. There are 64 codons.

36. The large subunit of an 80S ribosome is made from:

   A. rRNA only.
   B. protein only.
   C. rRNA and protein only.
   D. rRNA and protein bound by a phospholipid bilayer.

37. A tRNA molecule attaches to histidine. The anticodon on the tRNA is 5′-AUG-3′. Which of the following nucleotide sequences in an mRNA molecule might contain the codon for histidine?

   A. 3′-GCUAGGCCU-5′
   B. 3′-GGTACCTAC-5′
   C. 5′-CATTCTTAC-3′
   D. 5′-UCAUGGAUC-3′

38. One difference between prokaryotic and eukaryotic translation is:

   A. eukaryotic ribosomes are larger containing more subunits.
   B. prokaryotic translation may occur simultaneously with transcription while eukaryotic translation cannot.
   C. prokaryotes don't contain supra molecular complexes such as ribosomes.
   D. prokaryotic DNA is circular so does not require a termination sequence.

39. During translation the growing polypeptide can be found attached to a tRNA at which site on the ribosome?

   A. the E site
   B. the P site
   C. the A site
   D. the Z site

40. During translation, a signal peptide is synthesized and attaches to an SRP complex in order to:

   A. inactivate the new protein.
   B. activate the new protein.
   C. prevent the ribosome from attaching to the endoplasmic reticulum.
   D. direct the ribosome to attach to the endoplasmic reticulum.

2.12 Chromosomes

If a double strand of all the DNA in a single human cell were stretched out straight, it would measure around 5 ft. Since the nucleus is much smaller than this, the sections of DNA that are not in use are wrapped tightly around globular proteins called **histones**. Eight histones wrapped in DNA form a **nucleosome**. Nucleosomes, in turn, wrap into coils called *solenoids*, which wrap into *supercoils*. The entire DNA/protein complex (including a very small amount of RNA) is called **chromatin** (Greek: chroma: color) (Figure 2.26). By mass, chromatin is about one third DNA, two thirds protein, and a small amount of RNA. Chromatin received its name because it absorbs basic dyes due to the large basic amino acid content in histones. The basicity of histones gives them a net positive charge at the normal pH of the cell.

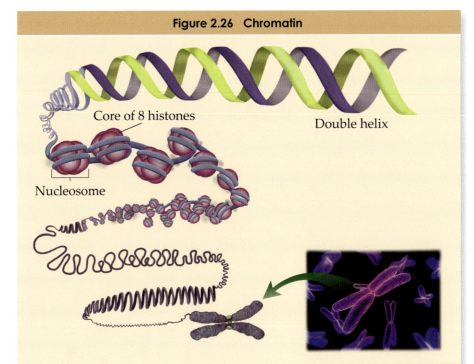

Figure 2.26 Chromatin

Core of 8 histones

Double helix

Nucleosome

Chromatin condensed in the manner described above is called *heterochromatin* (Greek: heteros → other). Some chromatin, called *constitutive heterochromatin*, is permanently coiled. When transcribed, chromatin must be uncoiled. Chromatin that can be uncoiled and transcribed is called euchromatin (Greek: eu → well or properly). Euchromatin is only coiled during nuclear division.

Inside the nucleus of a human somatic cell, there are 46 double stranded DNA molecules. The chromatin associated with each one of these molecules is called a **chromosome** (Greek: chroma → color, soma → body). Each chromosome contains hundreds or thousands of genes.

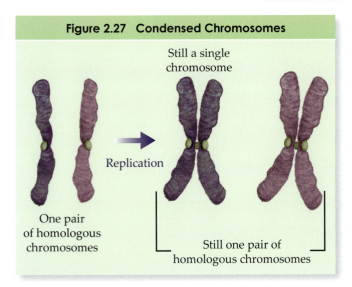

Figure 2.27 Condensed Chromosomes

Still a single chromosome

Replication

One pair of homologous chromosomes

Still one pair of homologous chromosomes

In human cells, each chromosome possesses a partner that codes for the same **traits** as itself. Two such chromosomes are called **homologues** (Greek: homologein → to agree with, homo → same, logia → collection). Humans possess 23 homologous pairs of chromosomes. Although the traits are the same, the actual genes may be different. Different forms of the same gene are called *alleles*. For instance, the trait may be eye color, but the eye color gene on one chromosome may code for blue eyes while the other codes for brown eyes. Any cell that contains homologous pairs is said to be **diploid** (Greek: di- → twice). Any cell that does not contain homologues is said to be **haploid** (Greek: haploos → single or simple).

Cell Life Cycle

Every cell has a life cycle (Figure 2.38) that begins with the birth of the cell and ends with the death or division of the cell. The life cycle of a typical somatic cell of a multicellular organism can be divided into four stages: the first growth phase (G_1); synthesis (S); the second growth phase (G_2); mitosis or meiosis (M); and cytokinesis (C). G_1, S, and G_2 collectively are called **interphase**.

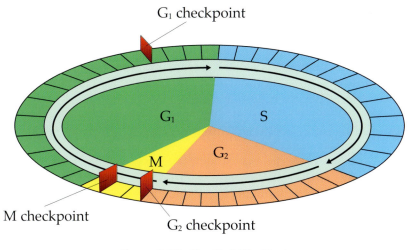

Figure 2.28 The Cell Life Cycle

In **G_1**, the cell has just split, and begins to grow in size producing new organelles and proteins. Regions of heterochromatin have been unwound and decondensed into euchromatin. RNA synthesis and protein synthesis are very active. The cell must reach a certain size, and synthesize sufficient protein in order to continue to the next stage. Cell growth is assessed at the G_1 checkpoint near the end of G_1. If conditions are favorable for division, the cell enters the S phase, otherwise the cell enters the G_0 phase. The main factor in triggering the beginning of S is cell size based upon the ratio of cytoplasm to DNA. G_1 is normally, but not always, the longest stage.

G_0 is a nongrowing state distinct from interphase. The G_0 phase allows for the differences in length of the cell cycle. In humans, enterocytes of the intestine divide more than twice per day, while liver cells spend a great deal of time in G_0 dividing less than once per year. Mature neurons and muscle cells remain in G_0 permanently.

In **S**, the cell devotes most of its energy to replicating DNA. Organelles and proteins are produced more slowly. In this stage, each chromosome is exactly duplicated, but, by convention, the cell is still considered to have the same number of chromosomes, only now, each chromosome is made of two identical sister **chromatids**.

In **G_2**, the cell prepares to divide. Cellular organelles continue to duplicate. RNA and protein (especially tubulin for microtubules) are actively synthesized. G_2 typically occupies 10-20% of the cell life cycle. Near the end of G_2 is the G_2 checkpoint. The G_2 checkpoint checks for mitosis promoting factor (MPF). When the level of MPF is high enough, mitosis is triggered.

There is an M checkpoint during mitosis that triggers the start of G_1.

2.14 Mitosis

Mitosis (Greek: mitos → cell) is nuclear division without genetic change. Mitosis has four stages: prophase, metaphase, anaphase, and telophase (Figure 2.30). These stages in turn are also divided, but this is beyond the MCAT. Mitosis varies among eukaryotes. (For instance, fungi don't have centrioles and never lose their nuclear membranes.) The following stages describe mitosis in a typical animal cell.

Prophase is characterized by the condensation of chromatin into chromosomes. **Centrioles** located in the **centrosomes** move to opposite ends of the cell. First the nucleolus and then the nucleus disappear. The **spindle apparatus** begins to form consisting of **aster** (microtubules radiating from the centrioles), *kinetochore microtubules* growing from the **centromeres** (a group of proteins located toward the center of the chromosome), and **spindle microtubules** connecting the two centrioles. (The **kinetochore** is a structure of protein and DNA located at the centromere of the joined chromtids of each chromosome.)

In **metaphase** (Greek: meta → between) chromosomes align along the equator of the cell.

Anaphase begins when sister chromatids split at their attaching centromeres, and move toward opposite ends of the cell. This split is termed *disjunction*. **Cytokinesis**, the actual separation of the cellular cytoplasm due to constriction of microfilaments about the center of the cell, may commence toward the end of this phase.

In **telophase** (Greek: teleios → complete) the nuclear membrane reforms followed by the reformation of the nucleolus. Chromosomes decondense and become difficult to see under the light microscope. Cytokinesis continues.

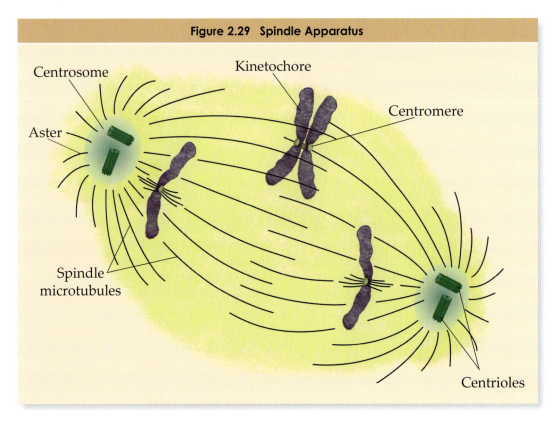

Figure 2.29   Spindle Apparatus

Centrosome

Kinetochore

Centromere

Aster

Spindle microtubules

Centrioles

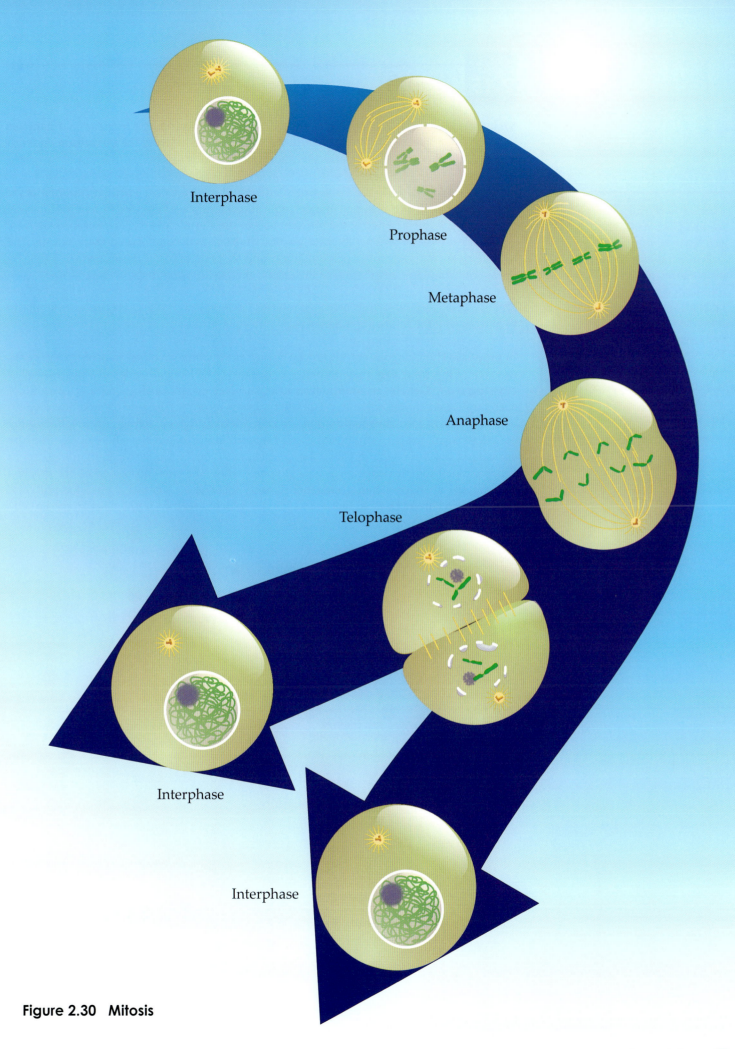

Interphase

Prophase

Metaphase

Anaphase

Telophase

Interphase

Interphase

Figure 2.30 Mitosis

2.15 | *Meiosis*

Meiosis (Figure 2.32) is a double nuclear division which produces four haploid **gametes** (also called **germ cells**). In humans, only the **spermatogonium** and the **oogonium** undergo meiosis. All other cells are somatic cells and undergo mitosis only.

After replication occurs in the S phase of interphase, the cell is called a **primary spermatocyte** or **primary oocyte**. In the human female, replication takes place before birth, and the life cycle of all germ cells are arrested at the primary oocyte stage until puberty. Just before ovulation, a primary oocyte undergoes the first meiotic division to become a secondary oocyte. The secondary oocyte is released upon ovulation, and the penetration of the secondary oocyte by the sperm stimulates anaphase II of the second meiotic division in the oocyte.

Meiosis is two rounds of division called meiosis I and meiosis II. Meiosis I proceeds similarly to mitosis with the following differences.

In **prophase I** homologous chromosomes line up along side each other, matching their genes exactly. At this time, they may exchange sequences of DNA nucleotides in a process called **crossing over**. **Genetic recombination** in eukaryotes occurs during crossing over. Since each duplicated chromosome in prophase I appears as an 'x', the side by side homologues exhibit a total of four **chromatids**, and are called **tetrads** (Greek: tetras → four). If crossing over does occur, the two chromosomes are "zipped" along each other where nucleotides are exchanged, and form what is called the *synaptonemal complex*. Under the light microscope, a synaptonemal complex appears as a single point where the two chromosomes are attached creating an 'x' shape called a **chiasma** (Greek: chiasmata → cross). Genes located close together on a chromosome are more likely to cross over together, and are said to be linked.

> Meiosis is like mitosis except that in meiosis there are two rounds, the daughter cells are haploid, and genetic recombination occurs. You must know the names of the cells at the different stages and whether or not those cells are haploid or diploid. Recognize that, under the light microscope, metaphase in mitosis would appear like metaphase II in meiosis and not like metaphase I.

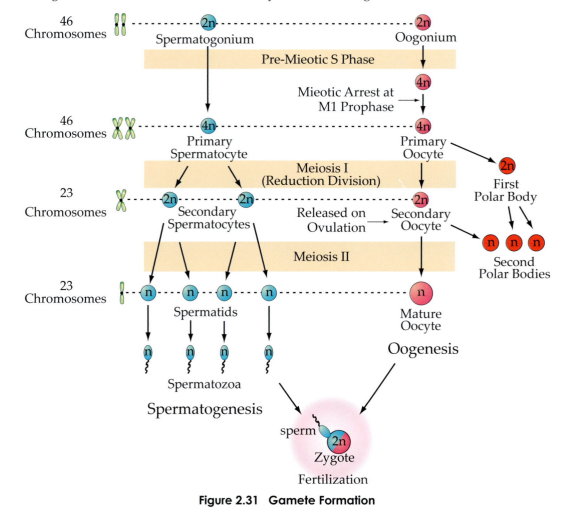

Figure 2.31 Gamete Formation

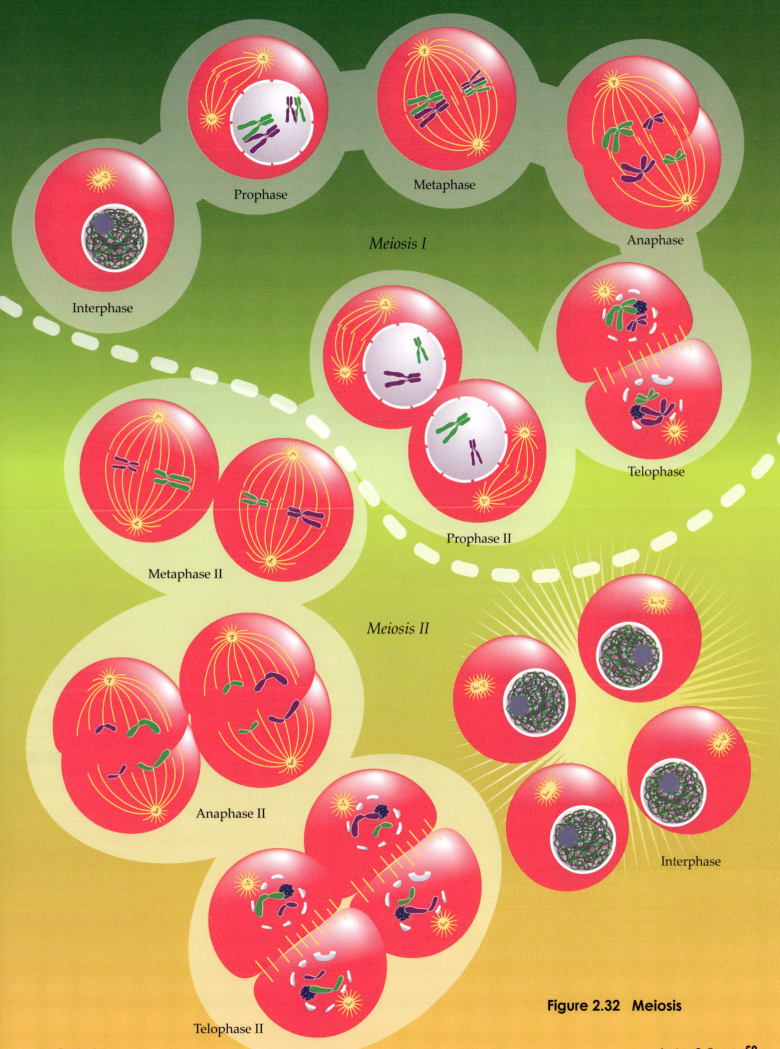

Prophase

Metaphase

Anaphase

Meiosis I

Interphase

Telophase

Prophase II

Metaphase II

Meiosis II

Anaphase II

Interphase

Telophase II

Figure 2.32   Meiosis

In **metaphase I** the homologues remain attached, and move to the metaphase plate. Rather than single chromosomes aligned along the plate as in mitosis, tetrads align in meiosis.

Anaphase I separates the homologues from their partners.

In **telophase I**, a nuclear membrane may or may not reform, and cytokinesis may or may not occur. In humans the nuclear membrane does reform and cytokinesis does occur. If cytokinesis occurs, the new cells are haploid with 23 replicated chromosomes, and are called **secondary spermatocytes** or **secondary oocytes**. In the case of the female, one of the oocytes, called the first **polar body**, is much smaller, and degenerates. This occurs in order to conserve cytoplasm, which is contributed only by the ovum. The first polar body may or may not go through meiosis II producing two polar bodies. These four phases together are called **meiosis I**. Meiosis I is **reduction division**.

Meiosis II proceeds with **prophase II**, **metaphase II**, **anaphase II**, and **telophase II** appearing under the light microscope much like normal mitosis. The final products are haploid gametes each with 23 chromosomes. In the case of the spermatocyte, four sperm cells are formed. In the case of the oocyte, a single ovum is formed. In the female, telophase II produces one gamete and a second polar body.

If during anaphase I or II the centromere of any chromosome does not split, this is called **nondisjunction**. As a result of primary nondisjunction (nondisjunction in anaphase I), one of the cells will have two extra chromatids (a complete extra chromosome) and the other will be missing a chromosome. The extra chromosome will typically line up along the metaphase plate and behave normally in meiosis II. Nondisjunction in anaphase II will result in one cell having one extra chromatid and one cell lacking one chromatid. Nondisjunction can also occur in mitosis but the ramifications are less severe since the genetic information in the new cells is not passed on to every cell in the body. Down syndrome may be caused by non disjunction of chromosome 21. An abnormal gamete with two chromosome 21s may combine with a normal gamete and the resulting zygote will have three copies of chromosome 21. This is sometimes referred to as trisomy 21.

Notes:

Terms	
Adenine (A)	Gametes (Germ Cells)
Anaphase I	Gene
Anaphase II	Genetic Code
Anneal	Genetic Recombination
Anticodon	Genome
Antiparallel	Guanine (G)
Base-Pairing (BP)	Haploid
Base-Pair Substitution Mutation	Histones
Carcinogens	Homologues
Centromeres	Insertion
Chiasma	Initiation
Chromatids	Initiation Complex
Chromatin	Interphase
Chromosome	Introns
Chromosomal Mutation	Inversion
Clone	Kinetochore
Clone Library	Lagging Strand
Codon	Large Subunit
Complementary DNA (cDNA)	Leading Strand
Complementary Strands	Meiosis I
Crossing Over	Meiosis II
Cytokinesis	Metaphase
Cytosine (C)	Missense Mutation
Denatured	Mitosis
Degenerative	Mutagens
Deletion	Mutation
Diploid	Nondisjunction
DNA	Nonsense Mutation
DNA Ligase	Northern Blot
DNA Polymerase	Nucleic Acid Hybridization
Double Helix	Nucleolus
Elongation	Nucleosome
Exons	Okazaki Fragments
Forward Mutation	Oncogenes
Frameshift Mutation	Oogonium

Terms	
Operon	snRNPS
Palindromic	Southern Blotting
Phosphodiester Bond	Spermatogonium
Poly-A Tail	Spindle Apparatus
Plasmid	Spindle Microtubules
Point Mutation	Stop Codons
Polar Body	Telomeres
Polymerase Chain Reaction (PCR)	Telophase
Post-Translational Modifications	Termination
Primary Oocyte	Tetrads
Primary Spermatocyte	Thymine (T)
Primary Transcript	Traits
Primer	Transcription
Probe	Transcription Activators
Prokaryote	Translation
Promoter	Translocation
Prophase	Transposons
Recombinant DNA	Vector
Reduction Division	Western Blot
Restriction Enzymes	Wild Type
RFLP	
Ribosome	
RNA Polymerase	
Secondary Oocyte	
Secondary Spermatocytes	
Semiconservative DNA Replication	
Semidiscontinuous Replication	
Signal Peptide	
Small Subunit	

Answer from page 45:

20 possible amino acids (you should know that) and 100 positions gives 20^{100} possible sequences.

41. How many chromosomes does a human primary spermatocyte contain?

    A. 23
    B. 46
    C. 92
    D. 184

42. In which of the following life cycle phases does translation, transcription, and replication take place?

    A. G1
    B. S
    C. G2
    D. M

43. A scientist monitors the nucleotide sequence of the third chromosome as a cell undergoes normal meiosis. What is the earliest point in meiosis at which the scientist can deduce with certainty the nucleotide sequence of the third chromosome of each gamete?

    A. prophase I
    B. metaphase I
    C. prophase II
    D. telophase II

44. Which of the following is a process undergone by germ cells only?

    A. meiosis
    B. mitosis
    C. interphase
    D. cytokinesis

45. Which of the following represents a germ cell in metaphase I?

    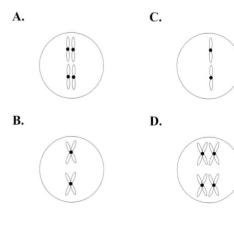

46. Which of the following characterizes mitotic prophase?

    A. chromosomal alignment along the equator of the cell
    B. separation of sister chromatids
    C. centriole migration to the cell poles
    D. cytokinesis

47. All of the following might describe events occurring in prophase I of meiosis EXCEPT:

    A. tetrad formation.
    B. spindle apparatus formation.
    C. chromosomal migration.
    D. genetic recombination.

48. When a human female is born, the development of her oocytes is arrested in:

    A. prophase of mitosis.
    B. prophase I of meiosis.
    C. prophase II of meiosis.
    D. interphase.

_____

STOP.

Notes:

MICROBIOLOGY

3.1 Viruses

Viruses are tiny infectious agents, much smaller than bacteria. They are comparable in size to large proteins (Some viruses are larger and some are much smaller.) In its most basic form, a virus consists of a protein coat, called a **capsid**, and from one to several hundred genes in the form of DNA or RNA inside the capsid. No virus contains both DNA and RNA. Most animal viruses, some plant viruses, and very few bacterial viruses surround themselves with a lipid-rich **envelope** either borrowed from the membrane of their host cell or synthesized in the host cell cytoplasm. The envelope typically contains some virus-specific proteins. A mature virus outside the host cell is called a *virion*. All organisms experience viral infections.

Although there is debate as to the vitality of viruses, viruses are not currently classified as living organisms; they do not belong to any of the taxonomical kingdoms of organisms. Viruses differ from living organisms in the following ways. Although viruses can reproduce through a process involving the transfer of genetic information, they always require the host cell's reproductive machinery in order to do so. Viruses do not metabolize organic nutrients. Instead they use the ATP made available by the host cell. Unlike living organisms, viruses in their active form are not separated from their external environment by some type of barrier such as a cell wall or membrane. All living organisms possess both DNA and RNA; viruses possess either DNA or RNA, but never both. Viruses can be crystallized without losing their ability to infect.

A viral infection begins when a virus adsorbs to a specific chemical receptor site on the **host**. The host is the cell that is being infected. The chemical **receptor** is usually a specific glycoprotein on the host cell membrane. The virus cannot infect the cell if the specific receptor is not available. Next, the nucleic acid of the virus penetrates into the cell. In a **bacteriophage** (Greek: phagein: to eat), a virus that infects bacteria (Figure 3.1), the nucleic acid is normally injected through the **tail** after viral enzymes have digested a hole in the cell wall (Figure 3.2). (Notice that this indicates that some viruses also include enzymes within their capsids.) Most viruses that

Capsid or Head
(*Contains Nucleic Acid*)

Tail

Tail Fiber

Figure 3.1   Bacteriophage

| Landing | Attachment | Tail contraction | Penetration and Injection |

Figure 3.2   Adsorption and Injection

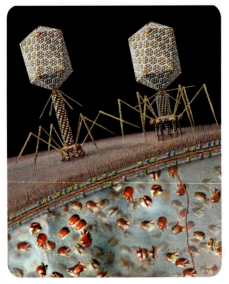

A bacteriophage lands on the surface of a bacterium and injects its DNA into the cell.

infect eukaryotes are engulfed by an **endocytotic** process. Once inside the cell, there are two possible paths: a lysogenic infection, or a lytic infection (Figure 3.3).

In a **lytic** (Greek: lysis → separation) infection, the virus commandeers the cell's reproductive machinery and begins reproducing new viruses. There is a brief period before the first fully formed virion appears. This period is called the *eclipse period*. The cell may fill with new viruses until it lyses or bursts, or it may release the new viruses one at a time in a reverse endocytotic process. The period from infection to lysis is called the latent period. The **latent period** encompasses the eclipse period. A virus following a lytic cycle is called a **virulent virus**.

In a **lysogenic** infection, the viral DNA is incorporated into the host genome, or, if the virus is an RNA virus and it possesses the enzyme **reverse transcriptase**, DNA is actually reverse-transcribed from RNA and then incorporated into the host cell genome. When the host cell replicates its DNA, the viral DNA is replicated as well. A virus in a lysogenic cycle is called a **temperate virus**. A host cell infected with a temperate virus may show no symptoms of infection. While the viral DNA remains incorporated in the host DNA, the virus is said to be **dormant** or **latent**, and is called a **provirus** (a **prophage** [Greek: pro → before, phagein → eat] if the host cell is a bacterium). The dormant virus may become active when the host cell is under some type of stress. Ultraviolet light or carcinogens also may activate the virus. When the virus becomes active, it becomes virulent.

Be sure to know the differences between the two cycles. Pay particularly close attention to how the viral genetic material is converted to proteins and also how it is incorporated into the host cell genome, since the MCAT seems to enjoy testing this. Feel free to look back to genetics for information about transcription and translation.

Most animal viruses do not leave capsids outside the cell, but enter the cell through receptor-mediated endocytosis.

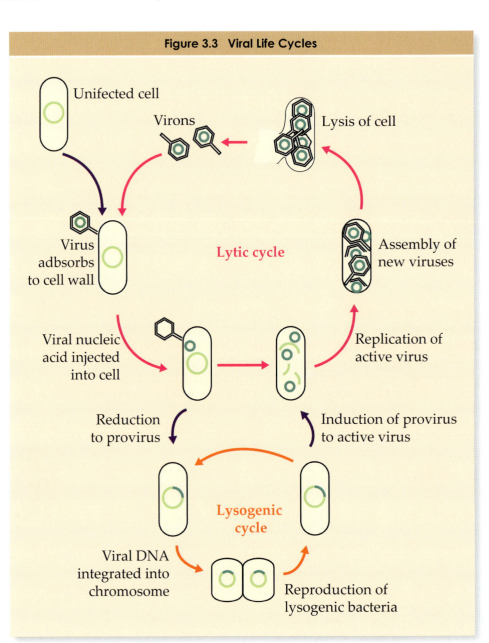

Figure 3.3   Viral Life Cycles

Unifected cell

Virons

Lysis of cell

Virus adbsorbs to cell wall

Lytic cycle

Assembly of new viruses

Viral nucleic acid injected into cell

Replication of active virus

Reduction to provirus

Induction of provirus to active virus

Lysogenic cycle

Viral DNA integrated into chromosome

Reproduction of lysogenic bacteria

There are many of types of viruses. One way to classify them is by the type of nucleic acid that they possess. A virus with *unenveloped* **plus-strand RNA** is responsible for the common cold. (Therefore, not all animal viruses are enveloped.) The "plus-strand" indicates that proteins can be directly translated from the RNA. Enveloped plus-strand RNA viruses include **retroviruses** such as the virus that causes AIDS. A retrovirus carries the enzyme **reverse transcriptase** in order to create DNA from its RNA. The DNA is then incorporated into the genome of the host cell. **Minus-strand RNA** viruses include measles, rabies, and the flu. Minus-strand RNA is the complement to mRNA and must be transcribed to plus-RNA before being translated. There are even **double stranded RNA viruses**, and **single and double stranded DNA viruses**.

Viroids are a related form of infectious agent. Viroids are small rings of naked RNA without capsids. Viroids only infect plants. There also exists naked proteins called *prions* that cause infections in animals. Prions are capable of reproducing themselves, apparently without DNA or RNA.

UAGGCAUCUUUCGCA
$\Downarrow$ (translation)
protein
mRNA

UAGGCAUCUUUCGCA
$\Downarrow$ (translation)
protein
Plus-strand RNA

AUCCGUAGAAAGCGU
$\Downarrow$ (transcription)
UAGGCAUCUUUCGCA
$\Downarrow$ (translation)
protein
Minus-strand RNA

Figure 3.4   A virus may have plus-strand or minus-strand RNA.

Table 3.1 Some Types of Viruses		
Disease	*Pathogen*	*Genome*
A.I.D.S.	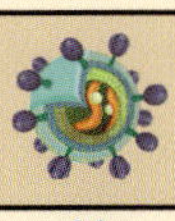 HIV	(+) Single-stranded RNA (two copies)
Chicken Pox, Shingles	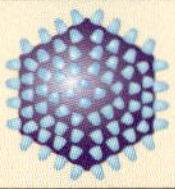 Varicella-zoster virus	Double-stranded DNA
Ebola	 Filoviruses	(-) Single-stranded RNA
Hepatitis B (viral)	 Hepadnavirus	Double-stranded DNA
Herpes	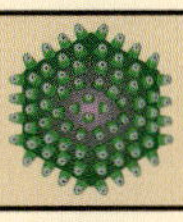 Herpes simplex virus	Double-stranded DNA
Influenza	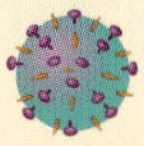 Influenza virus	(-) Single-stranded RNA (two copies)
Measles	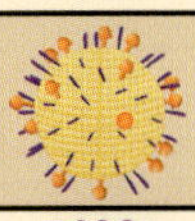 Paramyxoviruses	(-) Single-stranded RNA
Mononucleosis	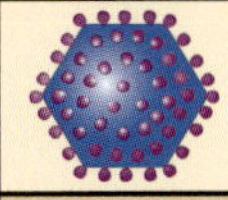 Epstein-Barr virus	Double-stranded DNA
Polio	 Enterovirus	(+) Single-stranded RNA
Rabies	 Rhabdovirus	(-) Single-stranded RNA
SARS	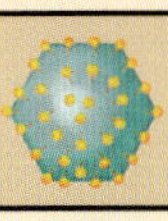 Coronavirus	(-) Single-stranded RNA
Small Pox	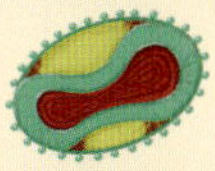 Variola virus	Double-stranded DNA
Yellow Fever	 Flavivirus	(+) Single-stranded RNA

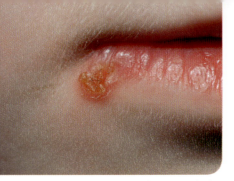

Herpes is a viral infection of the epidermal cells.

Defense Against Viral Infection

The human body fights viral infections with antibodies, which bind to a viral protein, and with cytotoxic T cells, which destroy infected cells. Although the envelope is borrowed from the host cell, *spike proteins* encoded from the viral nucleic acids protrude from the envelope. These proteins bind to receptors on a new host cell causing the virus to be infectious. However, it is also the spike proteins that human antibodies recognize when fighting the infection. Since RNA polymerase does not contain a proofreading mechanism, changes in the spike proteins are common in RNA viruses. When the spike proteins change, the antibodies fail to recognize them, and the virus may avoid detection until new antibodies are formed. A **vaccine** can be either an injection of antibodies or an injection of a nonpathogenic virus with the same capsid or envelope. The later allows the host immune system to create its own antibodies. Vaccines against rapidly mutating viruses are generally not very effective.

Another difficulty of fighting viral infections is that more than one animal may act as a **carrier population**. Even if all viral infections of a certain type were eliminated in humans, the virus may continue to thrive in another animal, thus maintaining the ability to reinfect the human population. For instance, ducks carry the flu virus, apparently without any adverse symptoms. One of the reasons that the fight against small pox was so successful was because the virus can only infect humans.

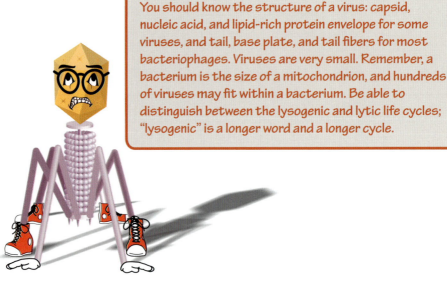

You should know the structure of a virus: capsid, nucleic acid, and lipid-rich protein envelope for some viruses, and tail, base plate, and tail fibers for most bacteriophages. Viruses are very small. Remember, a bacterium is the size of a mitochondrion, and hundreds of viruses may fit within a bacterium. Be able to distinguish between the lysogenic and lytic life cycles; "lysogenic" is a longer word and a longer cycle.

Viral Salty

49. Which of the following events does NOT play a role in the life cycle of a typical retrovirus?

A. Viral DNA is injected into the host cell.
B. Viral DNA is integrated into the host genome.
C. The gene for reverse transcriptase is transcribed and the mRNA is translated inside the host cell.
D. Viral DNA incorporated into the host genome may be replicated along with the host DNA.

50. A mature virus outside the host cell is called a virion. A virion may contain all of the following EXCEPT:

A. a capsid.
B. an envelope made from a phospholipid bilayer.
C. core proteins.
D. both RNA and DNA.

51. Prior to infecting a bacterium, a bacteriophage must:

A. reproduce, making copies of the phage chromosome.
B. integrate its genome into the bacterial chromosome.
C. penetrate the bacterial cell wall completely.
D. attach to a receptor on the bacterial cell membrane.

52. Most viruses that infect animals:

A. enter the host cell via endocytosis.
B. do not require a receptor protein to recognize the host cell.
C. leave their capsid outside the host cell.
D. can reproduce independently of a host cell.

53. Viruses most closely resemble:

A. facultative anaerobes.
B. aerobes.
C. saprophytes.
D. parasites.

54. Which of the following describes a lysogenic cell?

A. a cell that harbors an inactive virus in its genome
B. a cell that has developed immunity from viral infection
C. any cell infected with a virus
D. a cell that is about to lyse as a result of viral infection

55. A bacteriophage is easily recognizable due to:

A. a lysogenic life cycle.
B. a protein capsid.
C. circular nucleic acids.
D. a tail and fibers.

56. Which of the following would never be found in the capsid of a virion?

A. single stranded DNA
B. double stranded RNA
C. ribosomes
D. reverse transcriptase

_____

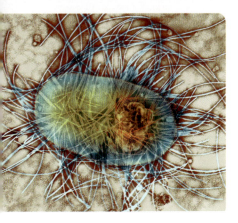

Most Gram-negative rod-shaped bacteria have flagellae. Many are a normal part of the gut flora found in the intestines, but some are pathogenic, such as Salmonella enterica and certain strains of Escherichia coli, and can cause foodborne illnesses.

Don't get caught up in all the minute differences between Archaea and Bacteria. Just know that there is a distinction and that, although both Archaea and Bacteria are prokaryotes, Archaea have similarities to eukaryotes.

MCAT expects you to know two aspects of the classification system:

1) energy source and

2) carbon source.

Autotrophs and heterotrophs differ in their source of carbon: autotrophs use CO_2 and heterotrophs use organic matter. 'Photo' and 'chemo' refer to where the organism derives its energy; 'photo' from light and 'chemo' from chemicals. Only prokaryotes can acquire energy from an inorganic source other than light.

3.3 Prokaryotes

Prokaryotes do not have a membrane bound nucleus. They are split into two domains called Bacteria and Archaea. **Archaea** have as much in common with eukaryotes as they do with bacteria. They are typically found in the extreme environments such as salty lakes and boiling hot springs. Unlike bacteria, the cell walls of archaea are not made from peptidoglycan. Most known prokaryotes are members of the domain **Bacteria** (Greek: bakterion: small rod). The introduction of the two domains makes the kingdom Monera obsolete. The kingdom monera was the kingdom containing all prokaryotes.

In order to grow, all organisms require the ability to acquire carbon, energy, and electrons (usually from hydrogen). Organisms can be classified according to the sources from which they gather these commodities.

A carbon source can be organic or inorganic. Most carbon sources also contribute oxygen and hydrogen. CO_2 is a unique inorganic carbon source because it has no hydrogens. To some degree, all microorganisms are capable of **fixing CO_2** (reducing it and using the carbon to create organic molecules usually through a process called the *Calvin cycle*). However, the reduction of CO_2 is energy expensive and most microorganisms cannot use it exclusively as their carbon source. **Autotrophs** (Greek: autotrophos: supplying one's own food, aut-:self, trephein:to nourish) are organisms that are capable of using CO_2 as their sole source of carbon. **Heterotrophs** (Greek: heteros: different or other) use preformed organic molecules as their source of carbon. Typically these organic molecules come from other organisms both living and dead, but it is believed that at the dawn of life they formed spontaneously in the environment of primitive Earth.

All organisms acquire energy from one of two sources:

1. light; or
2. oxidation of organic or inorganic matter.

Organisms that use light as their energy source are called **phototrophs**; those that use oxidation of organic or inorganic matter are called **chemotrophs**.

Electrons or hydrogens can be acquired from inorganic matter by *lithotrophs*, or organic matter by *organotrophs*.

All organisms can be classified as one of each of the three types. For instance, a flesh eating bacterium is a chemoorganotrophic heterotroph. Most heterotrophs also use organic matter as their energy source, making them some type of chemotrophic heterotrophs. Bacteria are found in all classifications.

Some bacteria are capable of *fixing nitrogen*. Atmospheric nitrogen is abundant, but in a strongly bound form that is useless to plants. Nitrogen fixation is the process by which N_2 is converted to ammonia. Most plants are unable to use ammonia, and must wait for other bacteria to further process the nitrogen in a process called *nitrification*. Nitrification is a two step process that creates nitrates, which are useful to plants, from ammonia. Nitrification requires two genera of chemoautotrophic prokaryotes. The relevant reactions are shown below.

$$NH_4^+ + 1\tfrac{1}{2}O_2 \rightarrow NO_2^- + H_2O + 2H^+$$
$$NO_2^- + \tfrac{1}{2}O_2 \rightarrow NO_3^-$$

Chemoautotrophy is an inefficient mechanism for acquiring energy, so chemoautotrophs require large amounts of substrate. This means that chemoautotrophs have a large environmental impact, which is reflected in processes like nitrification. All known chemoautotrophs are prokaryotes.

3.4 Structure of Prokaryotes

The typical structure of prokaryotes (Figure 3.5) is simpler than that of eukaryotes. The most basic distinction between eukaryotes and prokaryotes is that prokaryotes don't have a nucleus, and eukaryotes always have at least one nucleus. Instead of a nucleus, prokaryotes usually have a single, circular double stranded molecule of DNA. This molecule is twisted into *supercoils* and is associated with histones in Archaea and with proteins that are different from histones in Bacteria. The DNA, RNA and protein complex in prokaryotes forms a structure visible under the light microscope called a **nucleoid** (also called the chromatin body, nuclear region, or nuclear body). The nucleoid is not enclosed by a membrane.

There are two major shapes of bacteria: **cocci** (round) and **bacilli** (rod shaped). There are many other shapes, including helical. Helically shaped bacteria are called **spirilla** if they are rigid. Otherwise they are called **spirochetes**. Certain species of spirochetes may have given rise to eukaryotic flagella through a symbiotic relationship.

Prokaryotes have no complex, membrane-bound organelles. All living organisms contain both DNA and RNA, so prokaryotes have RNA. Since they translate proteins, prokaryotes have **ribosomes**. Prokaryotic ribosomes are smaller than eukaryotic ribosomes. They are made from a 50S and a 30S subunit to form a 70S ribosome.

A prokaryote may or may not contain a *mesosome*. Mesosomes are invaginations of the plasma membrane. They may be in the shape of tubules, lamellae, or vesicles. Under the light microscope mesosomes may appear as bubbles inside the bacterium. They may be involved in cell wall formation during cellular division.

Prokaryotes also have *inclusion bodies*. Inclusion bodies are granules of organic or inorganic matter that may be visible under a light microscope. Inclusion bodies may or may not be bound by a single layer membrane.

MRSA (Methicillin-resistant *Staphylococcus aureus*) is a Gram-positive spherical (coccus) bacteria. Some strains are resistant to most antibiotic drug agents. MRSA is common in hospitals infecting wounds of patients.

E. coli bacteria, Gram-negative bacilli (rod-shaped) bacteria, are normal inhabitants of the human intestine, and are usually harmless. However, under certain conditions their numbers may increase to such an extent that they cause infection. They cause 80% of all urinary tract infections, travellers' diarrhea, particularly in tropical countries & gastroenteritis in children. They are also widely used in genetic research.

> Recognize that the name of the bacteria often reveals the shape, like: *spiroplasma*, *staphylococcus*, or *pneumococcus*.

> Prokaryotes lack a nucleus. In fact, they have no complex, membrane-bound organelles at all. The key words are 'complex' and 'membrane-bound'. They have organelles: ribosomes, nucleoid, mesosomes etc...; just not complex, membrane-bound organelles.

The thick capsule or slime layer (glycocalyx) is a slimy or gummy material secreted by many bacteria onto their surfaces. The capsule can aid in attaching to host cells and may also protect the bacterium from immune cells such as phagocytes.

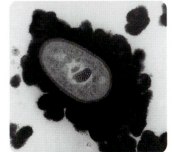

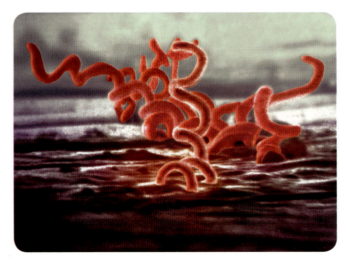

Treponema pallidum, the cause of syphilis, is a spirochete that wiggles vigorously when viewed under a microscope.

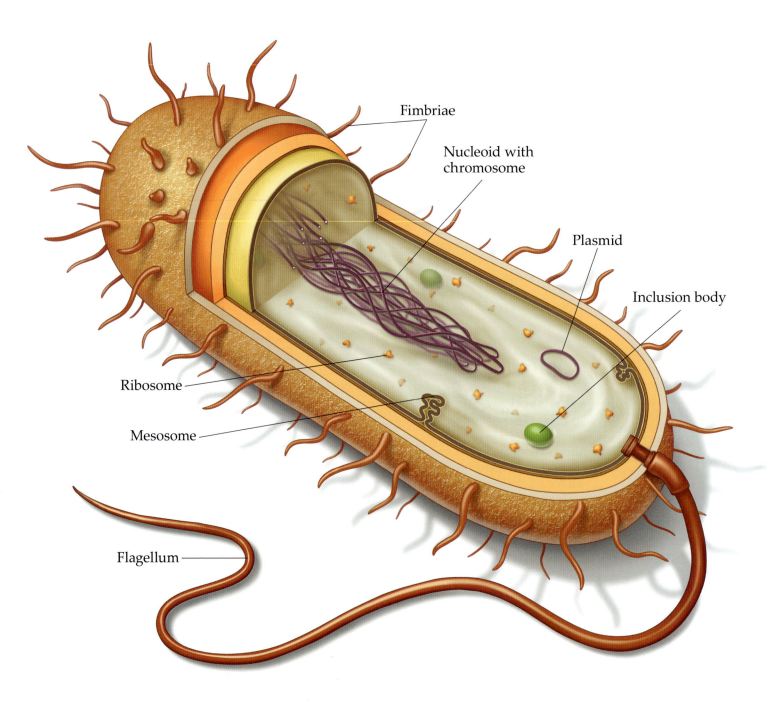

Figure 3.5   Bacterium

The cytosol of nearly all prokaryotes is surrounded by a phospholipid bilayer called the **plasma membrane**. (The membranes of archaea differ in their lipid structure. This is unlikely to be on the MCAT.) The **phospholipid** (Figure 3.8) is composed of a **phosphate group**, **two fatty acid chains**, and a **glycerol** backbone. The phospholipid is often drawn as a balloon with two strings. The balloon portion represents the phosphate group, and the strings represent the fatty acids. The phosphate group is polar, while the fatty acid chains are nonpolar, making the molecule **amphipathic** (having both a polar and a nonpolar portion). When placed in aqueous solution, amphipathic molecules spontaneously aggregate, turning their polar ends toward the solution, and their nonpolar ends toward each other. The resulting spherical structure is called a **micelle** (Figure 3.7). If enough phospholipids exist, and the solution is subjected to ultrasonic vibrations, *liposomes* may form. A liposome is a vesicle surrounded and filled by aqueous solution. It contains a lipid bilayer like that of a plasma membrane. The inner and outer layers of a membrane are referred to as *leaflets*. As well as phospholipids, the plasma membrane contains other types of lipids such as glycolipids. Unlike eukaryotic membranes, prokaryotic plasma membranes usually do not contain steroids such as cholesterol. Instead, some bacterial membranes contain steroid-like molecules called hopanoids (see Figure 3.6.) Different lipid types are arranged asymmetrically between the leaflets. For instance, glycolipids are found on the outer leaflet only.

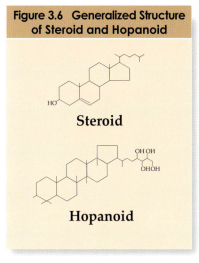

Figure 3.6 Generalized Structure of Steroid and Hopanoid

Steroid

Hopanoid

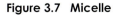

Figure 3.7 Micelle

> Micelles form spontaneously whereas membranes must be actively assembled. So, if you dump some phospholipids into an aqueous solution, expect that a micelle will form because it is the most thermodynamically stable conformation.

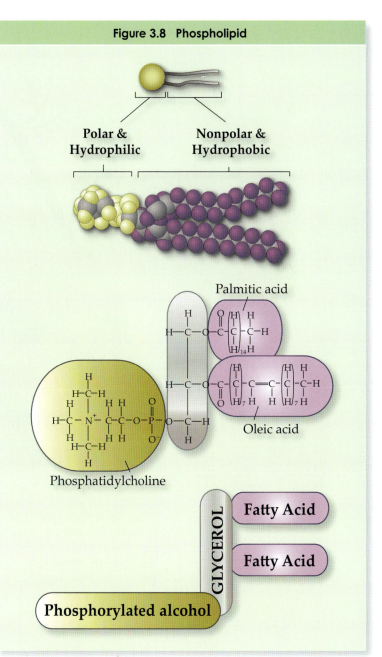

Figure 3.8 Phospholipid

Polar & Hydrophilic

Nonpolar & Hydrophobic

Palmitic acid

Oleic acid

Phosphatidylcholine

GLYCEROL

Fatty Acid

Fatty Acid

Phosphorylated alcohol

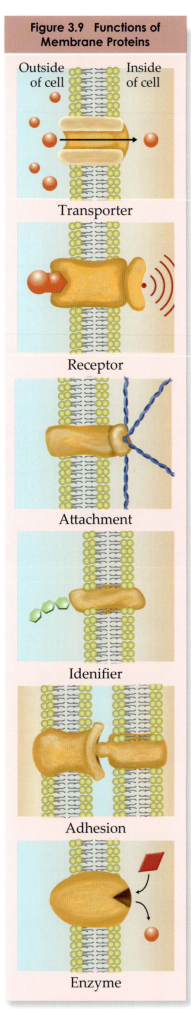

Figure 3.9 Functions of Membrane Proteins

Outside of cell   Inside of cell

Transporter

Receptor

Attachment

Idenifier

Adhesion

Enzyme

Also embedded within the plasma membrane are proteins. Most of the functional aspects of membranes are due to their proteins. Membrane proteins act as transporters, receptors, attachment sites, identifiers, adhesive proteins, and enzymes (Figure 3.9). As transporters, membrane proteins select which solutes enter and leave the cell. Other membrane proteins act as receptors by receiving chemical signals from the cellular environment. Some membrane proteins are attachment sites that anchor to the cytoskeleton. Membrane proteins can act as identifiers which other cells recognize. Adhesion by one cell to another is accomplished by membrane proteins. Many of the chemical reactions that occur within a cell are governed by membrane proteins on the inner surface. Amphipathic proteins that traverse the membrane from the inside of the cell to the outside are called **integral** or **intrinsic proteins**. **Peripheral** or **extrinsic proteins** are situated entirely on the surfaces of the membrane. They are ionically bonded to integral proteins or the polar group of a lipid. Both integral and peripheral proteins may contain carbohydrate chains making them glycoproteins. The carbohydrate portion of membrane *glycoproteins* always protrudes toward the outside of the cell. *Lipoproteins* (sometimes called *lipid anchored proteins*) also exist in some plasma membranes with their lipid portions embedded in the membrane and their protein portions at the surfaces. Membrane proteins are distributed asymmetrically throughout the membrane and between the leaflets. Neither proteins nor lipids flip easily from one leaflet to the other.

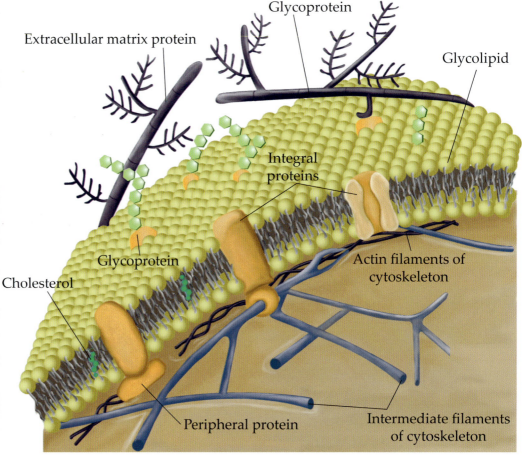

Glycoprotein

Extracellular matrix protein

Glycolipid

Integral proteins

Glycoprotein

Cholesterol

Actin filaments of cytoskeleton

Peripheral protein

Intermediate filaments of cytoskeleton

Figure 3.10 Fluid Mosaic Model

Since the forces holding the entire membrane together are intermolecular, the membrane is fluid; its parts can move laterally but cannot separate. The model of the membrane as just described is known as the **fluid mosaic model**. A mosaic is a picture made by placing many small, colored pieces side by side. The mosaic aspect of the membrane is reflected in the asymmetrical layout of its proteins. In eukaryotic membranes, cholesterol moderates membrane fluidity. In the prokaryotic plasma membrane, hopanoids probably reduce the fluidity of the membrane.

> You should be familiar with the fluid mosaic model of a membrane. Most prokaryotic membranes differ only slightly from eukaryotic membranes.

Membrane Transport

A membrane is not only a barrier between two aqueous solutions of different composition; it actually creates the difference in the compositions of the solutions. At normal temperatures for living organisms, all molecules move rapidly in random directions frequently colliding with one another. This random movement is called *Brownian motion*. Brownian motion creates the tendency of compounds to mix completely with each other over time. If two compounds, X and Y, are placed on opposite sides of the same container, the net movement of X will be toward Y. This movement is called **diffusion**. For molecules without an electric charge, diffusion occurs in the direction of lower concentration. In Figure 3.11, X diffuses in the direction of lower concentration of X, and Y toward lower concentration of Y. A gradual change in concentration of a compound over a distance is called a chemical concentration gradient. The **chemical concentration gradient** is a series of vectors pointing in the direction of lower concentration. For molecules with a charge, there is also an **electrical gradient** pointing in the direction that a positively charged particle will tend to move. The two gradients can be added to form a single **electrochemical gradient** for a specific compound. The electrochemical gradient for compound X points in the direction that particle X will tend to move. (Note: There are other factors affecting the direction of diffusion, including heat and pressure. In this text, we shall assume these factors to be included in the electrochemical gradient. In strict terms, diffusion occurs in the direction of decreasing free energy, or in the strictest terms, in the direction of increasing universal entropy.)

If compounds X and Y are separated by an impermeable membrane, diffusion is stopped. However, if the molecules of X can wiggle their way across the membrane, then diffusion is only slowed. Since the membrane slows the diffusion of X, but does not stop it, the membrane is **semipermeable** to compound X.

Blue food coloring begins to diffuse in a beaker of water. Diffusion is caused by the tiny, random movements of molecules in a solution. Over a period of time these movements cause the dissolved substance to become evenly dispersed throughout the solvent.

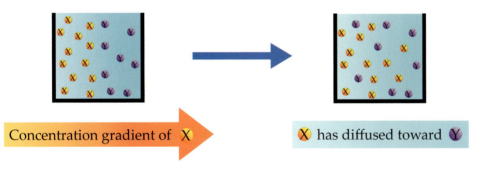

Concentration gradient of X

X has diffused toward Y

Figure 3.11   Diffusion

Figure 3.12 shows examples of membrane transport. Natural membranes are semipermeable to most compounds, but there are degrees of semipermeability. There are two aspects of a compound that affect its semipermeability: **size** and **polarity**. The larger the molecule, the less permeable the membrane to that molecule. A natural membrane is generally impermeable to polar molecules with a molecular weight greater than 100 without some type of assistance. The greater the polarity of a molecule (or if the molecule has a charge), the less permeable the membrane to that molecule. Very large lipid soluble (nonpolar) molecules like steroid hormones can move right through the membrane. When considering permeability, it is important to consider both size and polarity. For example, water is larger than a sodium ion, but water is polar, while the sodium ion possesses a complete charge. Therefore, a natural membrane is more permeable to water than to sodium; in this case, the charge difference outweighs the size difference. A natural membrane is, in fact, highly permeable to water. However, if the membrane were made only of a phospholipid bilayer, and did not contain proteins, the rate of diffusion for water would be very slow. Most of

Figure 3.12 Membrane Transport

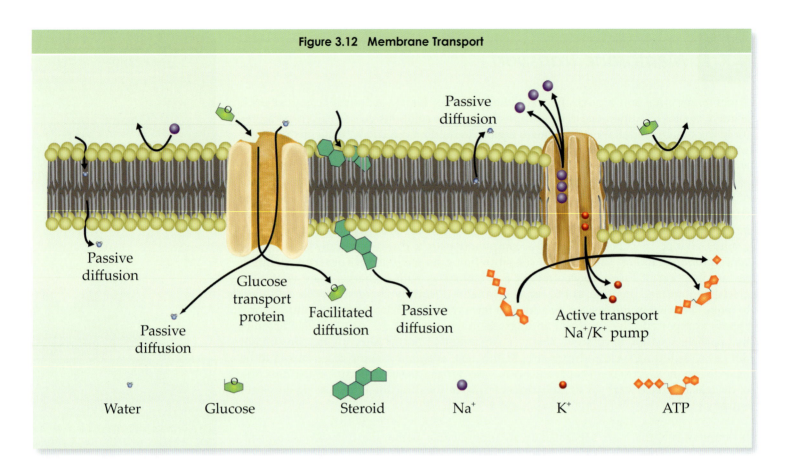

the diffusion of polar or charged molecules across a natural membrane takes place through incidental holes (sometimes called *leakage channels*) created by the irregular shapes of integral proteins. The function of these proteins is not to aid in diffusion. This is merely an incidental contribution.

The diffusion just described, where molecules move through leakage channels across the membrane due to random motion, is called **passive diffusion**. As mentioned previously, some molecules are too large or too charged to passively diffuse, yet they are needed for the survival of the cell. To assist these molecules in moving across the membrane, specific proteins are embedded into the membrane. These proteins, called **transport** or **carrier proteins**, are designed to facilitate the diffusion of specific molecules across the membrane. There are several mechanisms used by transport proteins in facilitated diffusion, but, in order for the passage to be called **facilitated diffusion**, diffusion must occur down the electro-chemical gradient of all species involved. Most, but not all, human cells rely on facilitated diffusion for their glucose supply. Facilitated diffusion is said to make the membrane **selectively permeable** because it is able to select between molecules of similar size and charge.

A living organism must be able to concentrate some nutrients against their electrochemical gradients. Of course, diffusion will not do this. Movement of a compound against its electrochemical gradient requires **active transport**. Active transport requires expenditure of energy. Active transport can be accomplished by the direct expenditure of ATP to acquire or expel a molecule against its electrochemical gradient. It can also be accomplished indirectly by using ATP to create an electrochemical gradient, and then using the energy of the electrochemical gradient to acquire or expel a molecule. The latter method is called *secondary active transport*.

There are some important concepts here that require understanding as well as memorization. The membrane stuff is worth a second read through. Basically, passive diffusion through the membrane depends on lipid solubility (Are you nonpolar enough to slide right through the phospholipid bilayer?) and size (Can you fit through the cracks around the integral proteins?). If you're big and polar, you must rely upon facilitated diffusion, a helper protein to open up a space designed just for you. To move against your electrochemical gradient, it doesn't matter if you are large or small, polar or nonpolar, you need active transport. Only active transport can move something against its concentration gradient.

3.7 Bacterial Envelope

The bacterial plasma membrane and everything inside it is called the *protoplast*. A protoplast is not a complete bacterium. Surrounding the protoplast is the **bacterial envelope** (Figure 3.13). The component of the envelope adjacent to the plasma membrane is the cell wall. (*Archaea* possess cell walls with a different chemical composition than will be described here. This will not be on the MCAT unless it is explained in a passage.) One of the functions of the cell wall is to prevent the protoplast from bursting. Most bacteria are **hypertonic** (Greek: hyper: above) to their environment. This means that the aqueous solution of their cytosol contains more particles than the aqueous solution surrounding them. This is compared to **isotonic** (Greek: iso: same) where the cytosol contains the same amount of particles and **hypotonic** (Greek: hypo: below) where the cytosol contains less particles. When there are more particles on one side of a barrier than the other, the particles want to move down their concentration gradient to the other side of the barrier. If the particles are prevented from crossing the barrier, water will try to cross in the opposite direction. (This is actually water moving down its electrochemical gradient.) The cell wall is strong and able to withstand high pressure. As the cell fills with water and the **hydrostatic pressure** builds, it eventually equals the **osmotic pressure**, and the filling stops (see Physics Lecture 5 for more on hydrostatic and osmotic pressure). Water continues to move in and out of the cell very rapidly, but an equilibrium is reached. If the cell wall is removed, the plasma membrane cannot withstand the pressure, and the bacterium will burst.

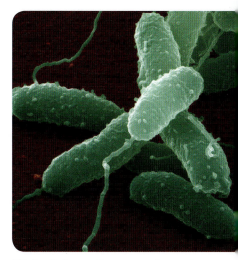

Vibrio cholerae, Gram-negative rod-shaped bacteria, have a single polar flagellum (long, thin), which they use to propel themselves through water. They are the cause of cholera, an infection of the small intestine that is transmitted to humans via contaminated food or water.

The cell wall is made of **peptidoglycan** (also called *murein*). (Archaea do not have peptidoglycan cell walls.) Peptidoglycan is a series of disaccharide polymer chains with amino acids, three of which are not found in proteins. These chains are connected by their amino acids, or crosslinked by an *interbridge* of more amino acids. The chains are continuous, forming a single molecular sac around the bacterium. Peptidoglycan is more elastic than cellulose. (Cellulose is the component of plant cell walls. [see Biology Lecture 1]). It is also porous, so it allows large molecules to pass through. Many antibiotics such as *penicillin* attack the amino acid crosslinks of peptidoglycan. Lysozyme, an enzyme produced naturally by humans, attacks the disaccharide linkage in peptidoglycan. In both cases the cell wall is disrupted and the cell lyses, killing the bacterium.

One method of classification of bacteria is according to the type of cell wall that they possess. **Gram staining** is a staining technique used to prepare bacteria for viewing under the light microscope which stains two major cell wall types differently. The first type is called **gram-positive bacteria** because its thick peptidoglycan cell wall prevents the gram stain from leaking out. These cells show up as purple when stained with this process. Gram-positive bacteria have a cell wall that is approximately four times thicker than the plasma membrane. The space between the plasma membrane and the cell wall is called the *periplasmic space*. The periplasmic space contains many proteins that help the bacteria acquire nutrition, such as hydrolytic enzymes.

Gram-negative bacteria appear pink when gram stained. Their thin peptidoglycan cell wall allows most of the gram stain to be washed off. The peptidoglycan of gram-negative bacteria is slightly different from that of gram-positive. Outside the cell wall, gram-negative bacteria have a phospholipid bilayer. This second membrane is more permeable than the first, even allowing molecules the size of glucose to pass right through. It is similar in structure to the plasma membrane, but also possesses *lipopolysaccharides*. The polysaccharide is a long chain of carbohydrates which protrudes outward from the cell. These polysaccharide chains can form a protective barrier from antibodies and many antibiotics. A lipoprotein in the outer membrane called *Braun's lipoprotein* points inward toward the cell wall and attaches covalently to the peptidoglycan. In gram-negative bacteria the *periplasmic space* is the space between the two membranes. (Different species of Archaea may stain positive or negative.)

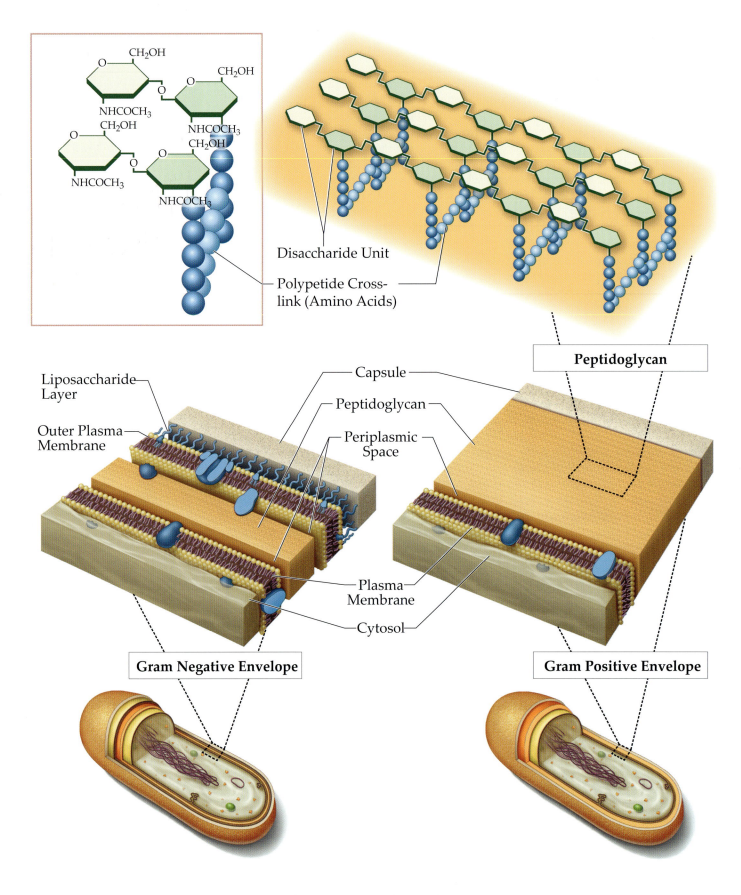

Figure 3.13 Bacterial Envelope

Many bacteria are wrapped in either a *capsule* or a *slime layer*. Both capsules and slime layers are usually made of polysaccharide. Slime layers are easily washed off, while capsules are not. A capsule can protect the bacterium from phagocytosis, desiccation, some viruses, and some components of the immune response of an infected host.

Some gram-negative bacteria possess *fimbriae* or *pili* (not to be confused with the sex pilus discussed below). Fimbriae are short tentacles, usually numbering in the thousands, that can attach a bacterium to a solid surface. They are not involved in cell motility.

Bacterial flagella are long, hollow, rigid, helical cylinders made from a globular protein called flagellin; these should not be confused with eukaryotic flagella, which composed of microtubules. They rotate counterclockwise (from the point of view of looking at the cell from the outside) to propel the bacterium in a single direction. When they are rotated clockwise, the bacterium *tumbles*. The tumbling acts to change the orientation of the bacterium allowing it to move forward in a new direction. The flagellum is propelled using the energy from a proton gradient rather than by ATP. Some bacteria can move via a gliding motion that has not yet been explained. Spirochetes, the flexible, helical shaped bacteria, can move through viscous fluids by flexing and spinning.

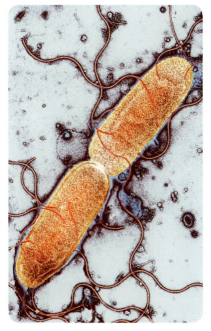

You should have some idea about the differences between gram positive and gram negative bacteria, but most of the details on the structure of bacteria are just good background information.

3.8 Bacterial Reproduction

Sexual reproduction is one method of recombining the genetic information between individuals of the same species to produce a genetically different individual. Sexual reproduction requires meiosis. Bacteria do not undergo meiosis or mitosis, and cannot reproduce sexually. However, they have three alternative forms of genetic recombination: conjugation, transformation, and transduction. They are also capable of undergoing a type of cell division called binary fission (Latin: fissus:split). Binary fission is a type of asexual reproduction (Figure 3.14).

In binary fission, the circular DNA is replicated in a process similar to replication in eukaryotes. (See Biology Lecture 2 for the replication process.) Two DNA polymerases begin at the same point on the circle (origin of replication) and move in opposite directions making complementary single strands that combine with their template strands to form two complete DNA double stranded circles. The cell then divides, leaving one circular chromosome in each daughter cell. The two daughter cells are genetically identical.

A single bacterium divides into identical daughter bacteria. Under optimal conditions, some bacteria can grow and divide extremely rapidly, and bacterial populations can double as quickly as every 10 minutes.

Binary fission results in two genetically identical daughter cells.

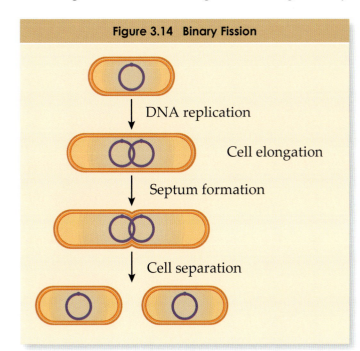

Figure 3.14 Binary Fission

DNA replication

Cell elongation

Septum formation

Cell separation

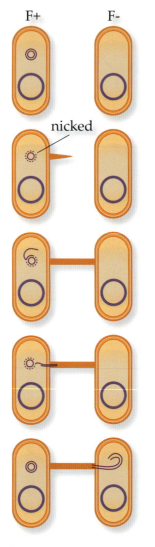

F+   F-

nicked

Figure 3.15 Conjugation

The first method of genetic recombination, **conjugation** (Figure 3.15), requires that one of the bacterium have a **plasmid** with the gene that codes for the **sex pilus**. Plasmids are small circles of DNA that exist and replicate independently of the bacterial chromosome. If the plasmid can integrate into the chromosome it is also called an *episome*. Plasmids are not essential to a bacterium which carries them. Not all bacteria with plasmids can conjugate. In order for a bacterium to initiate conjugation, it must contain a *conjugative plasmid*. Conjugative plasmids possess the gene for the sex pilus. The sex pilus is a hollow, protein tube that connects two bacteria to allow the passage of DNA. The passage of DNA is always from the cell containing the conjugative plasmid to the cell that does not. The plasmid replicates differently than the circular chromosome. One strand is *nicked*, and one end of this strand begins to separate from its complement as its replacement is replicated. The loose strand is then replicated and fed through the pilus.

There are two important plasmids that may be mentioned on the MCAT: the **F plasmid** and the *R plasmid*. The F plasmid is called the **fertility factor** or **F factor**. It was the first plasmid to be described. A bacterium with the F factor is called F+, one without the F factor is called F–. The F plasmid can be in the form of an episome, and if the pilus is made while the F factor is integrated into the chromosome, some or all of the rest of the chromosome may be replicated and transferred. The R plasmid donates resistance to certain antibiotics. It is also a conjugative plasmid. It was once common practice to prescribe multiple antibiotics for patients to take at one time. Such conditions promote conjugation of different R plasmids providing different resistances to antibiotics to produce a super-bacterium that contains many antibiotic resistances on one or more R plasmids. Some R plasmids are readily transferred between species further promoting resistance and causing serious health problems for humans.

Transformation is the process by which bacteria may incorporate DNA from their external environment into their genome. DNA may be added to the external environment in the lab, or it may occur due to lyses of other bacteria. The typical experimental procedure which demonstrates transformation is when heat-killed virulent bacteria are mixed with harmless living bacteria. The living bacteria receive the genes of the heat-killed bacteria through transformation, and become virulent.

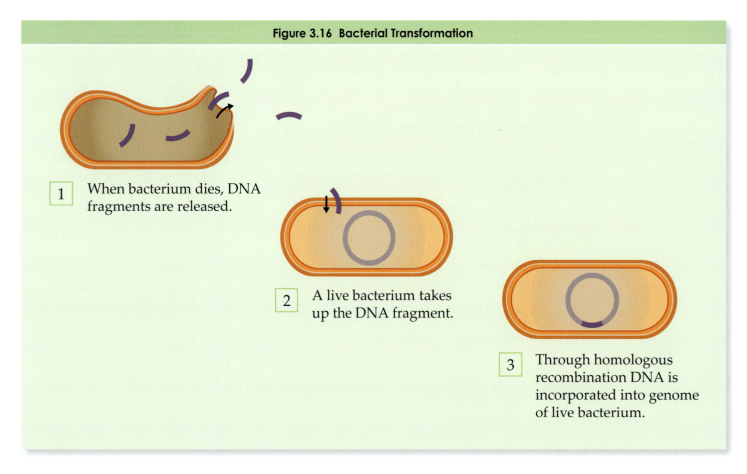

Figure 3.16 Bacterial Transformation

1 When bacterium dies, DNA fragments are released.

2 A live bacterium takes up the DNA fragment.

3 Through homologous recombination DNA is incorporated into genome of live bacterium.

Sometimes, the capsid of a bacteriophage will mistakenly encapsulate a DNA fragment of the host cell. When these virions infect a new bacterium, they inject harmless bacterial DNA fragments instead of virulent viral DNA fragments. This type of genetic recombination is called **transduction**. The virus that mediates transduction is called the **vector**. Transduction can be mediated artificially in the lab.

Figure 3.17 Transduction

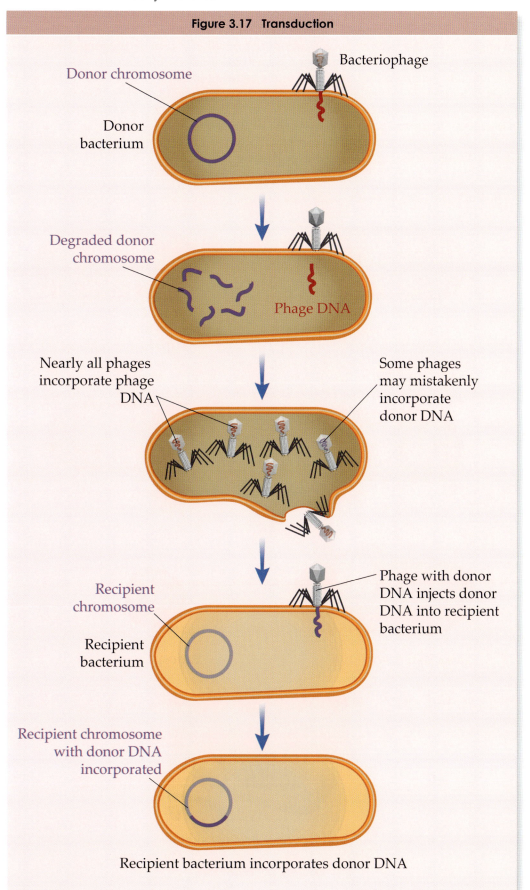

Donor chromosome

Donor bacterium

Bacteriophage

Degraded donor chromosome

Phage DNA

Nearly all phages incorporate phage DNA

Some phages may mistakenly incorporate donor DNA

Recipient chromosome

Recipient bacterium

Phage with donor DNA injects donor DNA into recipient bacterium

Recipient chromosome with donor DNA incorporated

Recipient bacterium incorporates donor DNA

Some gram-positive bacteria can form *endospores* (Greek: endon: within, speirein:to seed or sow) that can lie dormant for hundreds of years. Endospores are resistant to heat, ultraviolet radiation, chemical disinfectants, and desiccation. Endospores can survive in boiling water for over an hour. Endospore formation is usually triggered by a lack of nutrients. In endospore formation, the bacterium divides within its cell wall. One side then engulfs the other. The chemistry of the cell wall of the engulfed bacterium changes slightly to form the cortex of the endospore. Several protein layers lie over the cortex to form the resistant structure called the *spore coat*. A delicate covering, called the *exosporium*, sometimes surrounds the spore coat. The outer cell then lyses, releasing the dormant endospore. The endospore must be *activated* before it can *germinate* and grow. Activation usually involves heating. Germination is triggered by nutrients.

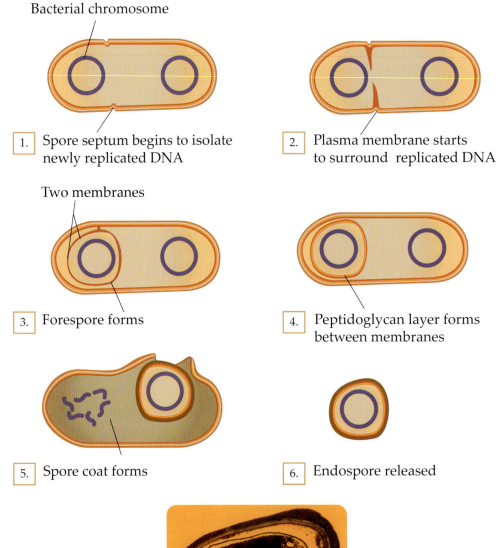

Bacterial chromosome

1. Spore septum begins to isolate newly replicated DNA

2. Plasma membrane starts to surround replicated DNA

Two membranes

3. Forespore forms

4. Peptidoglycan layer forms between membranes

5. Spore coat forms

6. Endospore released

Endospore of Bacillus subtilis

Figure 3.18 Endospore

57. Which of the following structures are found in prokaryotes?

I. A cell wall containing peptidoglycan.
II. A plasma membrane lacking cholesterol.
III. Ribosomes

A. I only
B. II only
C. I and II only
D. I, II, and III

58. DNA from phage resistant bacteria is extracted and placed on agar with phage-sensitive E. coli. After incubation it is determined that these E. coli are now also resistant to phage attack. The most likely mechanism for their acquisition of resistance is:

A. transduction.
B. sexual reproduction.
C. transformation.
D. conjugation.

59. Bacteriophages are parasites that infect bacterial cells in order to carry on their life function. When a phage transfers bacterial DNA from one host to another this process is called:

A. transformation.
B. transduction.
C. conjugation.
D. transmission.

60. The lipopolysaccharide layer outside the peptidoglycan cell wall of a gram negative bacterium:

A. absorbs and holds gram stain.
B. protects the bacterium against certain antibiotics.
C. does not contain phospholipids.
D. allows the bacterium to attach to solid objects.

61. The exponential growth that occurs in *E. coli* at 37°C following inoculation of a sterile, nutrient-rich solution results from:

A. conjugation.
B. transformation.
C. sporulation.
D. binary fission.

62. A staphylococcus infection is most likely caused by an organism that is:

A. rod-shaped.
B. spherical.
C. a rigid helix.
D. a non-rigid helix.

63. Which of the following is a mechanism for reproduction in Bacteria?

A. transduction
B. conjugation
C. binary fission
D. transformation

64. Penicillin interferes with peptidoglycan formation. Penicillin most likely inhibits bacterial growth by disrupting the production of:

A. bacterial plasma membranes.
B. prokaryotic cell walls.
C. the bacterial nucleus.
D. bacterial ribosomes.

3.10 Fungi

Fungi represent a distinct kingdom of organisms with tremendous diversity. Three divisions exist within this kingdom: *Zygomycota, Ascomycota,* and *Basidiomycota* (Greek: mykes: mushroom). (*Oomycota,* which are slime molds and water molds, are not true fungi, and are in the protista kingdom.) Like plants, fungi are separated into divisions not phyla. All fungi are eukaryotic heterotrophs that obtain their food by absorption rather than by ingestion: they secrete their digestive enzymes outside their bodies and then absorb the products of digestion. Although most fungi are considered **saprophytic** (Greek: sapros: rotten or decayed), many fungi do not distinguish between living and dead matter, and thus can be potent pathogens (disease causing). (Saprophytic means to live off dead organic matter.) Most fungi possess cell walls, called **septa** (Latin: saeptum:fence or wall), made of the polysaccharide, **chitin**. Chitin is more resistant to microbial attack than is cellulose. It is the same substance of which the exoskeleton of *arthropods* is made. (Arthropods are insects and crustaceans.) Septa are usually perforated to allow exchange of cytoplasm between cells, called cytoplasmic streaming. *Cytoplasmic streaming* allows for very rapid growth. One division of fungi, zygomycota, possesses no cell walls at all, except in their sexual structures. With the exception of yeasts, fungi are multicellular. (Yeasts are unicellular fungi.) A fungal cell may contain one or more nuclei. The nuclei in a single cell may or may not be identical. Fungi lack centrioles, and mitosis in fungi takes place entirely within the nucleus; the nuclear envelope never breaks down. In their growth state, fungi consist of a tangled mass (called a **mycelium**) of multiply branched thread-like structures, called **hyphae**.

The white areas on this mold fungus are mycellium, a mass of thread-like structures (hyphae) that absorb nutrients.

The fungi are far too diverse and complex to justify memorizing a lot of specifics. The MCAT will not require knowledge of specifics beyond what is in bold and red. Just remember the basics about fungi. Fungi are eukaryotic heterotrophs and spend most of their lives in the haploid state. They can reproduce sexually or asexually; remember when and why.

Fungus can grow on living things.

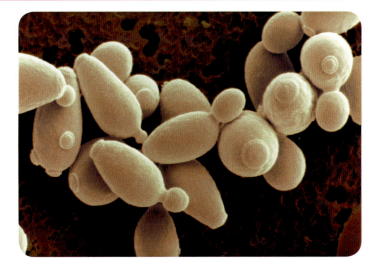

Yeast ferments sugar and produces ethanol and carbon dioxide. Yeast is essential to the making of wine, beer, and bread.

3.11 Fungal Reproduction and Life Cycle

Like most organisms, fungi alternate between haploid and diploid stages in their life cycle; however, the **haploid** stage predominates and is their growth stage. Hyphae are haploid. Some hyphae may form reproductive structures called *sporangiophores* in *Zygomycota*, *condiophores* in *Ascomycota*. These structures release haploid **spores** that give rise to new mycelia in asexual reproduction. *Basidiomycota*, which rarely reproduce asexually, produce *basidiospores* via sexual reproduction. (Note that spore formation is not always via asexual reproduction.) Spores are borne by air currents, water, or animals to locations suitable for new mycelial growth. Yeasts rarely produce sexually by producing spores. More often in yeasts, asexual reproduction occurs by **budding** (also called **cell fission**), in which a smaller cell pinches off from the single parent cell.

When sexual reproduction occurs (Figure 3.20), it is between hyphae from two mycelia of different mating types + and –. These two hyphae grow towards one another, eventually touching and forming a conjugation bridge. The tip of each hypha forms a complete septum in all divisions of fungi, and becomes a gamete-producing cell, called a *gametangium*. In Zygomycota, the gametangia remain attached to the parent hyphae and the nuclei fuse with one another to produce a diploid zygote, called a *zygospore*. The zygospore separates from the parent hyphae and usually enters a dormant phase. When activated by the appropriate environmental conditions, the zygospore undergoes meiosis to produce haploid cells, one of which immediately grows a short sporangiophore to asexually reproduce many spores. Except for the zygospore, all cells of Zygomycota are haploid.

Sexual reproduction in Ascomycota and Basidiomycota is similar, but slightly more complicated. The important thing to understand about fungal reproduction is that asexual reproduction normally occurs when conditions are good; sexual reproduction normally occurs when conditions are tough. This is because if conditions are good for the parent, they will be good for asexually reproduced offspring that are exactly like the parent, but if conditions are bad for the parent, they may not be bad for sexually reproduced offspring that are different from the parent.

Figure 3.19   Septum

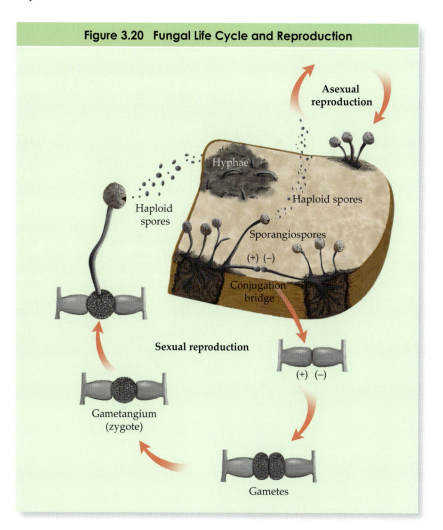

Figure 3.20   Fungal Life Cycle and Reproduction

Terms		
Active Transport	Extrinsic Proteins	Nucleoid
Amphipathic	F Plasmid	Origin of Replication
Archaea	Facilitated Diffusion	Osmotic Pressure
Autotrophs	Fertility Factor	Passive Diffusion
Bacilli	Flagella	Peptidoglycan
Bacteria	Flagellin	Phospholipid
Bacterial Envelope	Fluid Mosaic Model	Phototrophs
Bacteriophage	Fungi	Plasma Membrane
Binary Fission	Glycerol	Plasmid
Budding	Gram-negative	Plus-strand RNA
Capsid	Bacteria	Polarity
Carrier Population	Gram-positive Bacteria	Prokaryotes
Carrier Proteins	Gram Staining	Prophage
Chitin	Haploid	Provirus
Chemical	Heterotrophs	Retroviruses
Concentration	Host	Reverse Transcriptase
Gradient	Hydrostatic Pressure	Saprophytic
Chemotrophs	Hypertonic	Selectively Permeable
Cocci	Hyphae	Septa
Conjugation	Hypotonic	Sex Pilus
Diffusion	Intrinsic Proteins	Spirilla
DNA Viruses	Isotonic	Spirochetes
Dormant Viruses	Lysogenic	Spores
Electrical Gradient	Lytic	Transduction
Electrochemical	Membrane Transport	Transformation
Gradient	Micelle	Vaccine
Endocytotic	Minus-strand RNA	Vector
Envelope	Mycelium	Virus

65. Fungi are classified as a distinct kingdom because:

    A. they don't undergo mitosis.
    B. they reproduce asexually.
    C. sexual reproduction involves the union of different mating types, plus and minus strains.
    D. they have characteristics that are both plant-like and animal-like.

66. Which of the following is not a result of sexual reproduction in fungi?

    A. hyphae of + and – mycelia meet and fuse
    B. fertilization produces a diploid state
    C. diploid cell undergoes meiosis
    D. cell division lengthens mycelia

67. Fungi are considered saprophytic because:

    A. they reproduce by budding.
    B. they reproduce by sporulation.
    C. they absorb chemicals through a mass of tiny threads.
    D. they acquire energy from the break down of the dead remains of living organisms.

68. All of the following statements are true concerning most fungi EXCEPT:

    A. They have cell walls made of chitin.
    B. They are autotrophs.
    C. Their growth stage is composed of filaments containing many nuclei.
    D. Their life cycle alternates between a haploid and diploid stage.

69. What selective advantage is offered by the haploid state of fungi?

    A. The haploid state can reproduce more quickly than the diploid state under favorable conditions.
    B. The haploid state is more genetically diverse.
    C. The haploid state produces a large number of cells.
    D. The haploid state requires less energy to sustain itself.

70. Which of the following is the best explanation for why fungicides tend to cause more side effects in humans than do bacterial antibiotics?

    A. Chitin is more difficult to break down than peptidoglycan.
    B. Fungus doesn't respond to penicillin.
    C. Fungal cells are more similar to human cells than are bacterial cells.
    D. Fungus is a topical infection.

71. The Kingdom of Fungi is divided into:

    A. phyla.
    B. divisions.
    C. orders.
    D. species.

72. Which of the following is true of Fungi?

    A. Fungi prey upon only dead organic matter.
    B. Fungi digest their food outside their bodies.
    C. Fungi undergo meiosis during asexual reproduction.
    D. All fungi are obligate aerobes.

Notes:

THE EUKARYOTIC CELL; THE NERVOUS SYSTEM

4.1 The Nucleus

The major feature distinguishing eukaryotic (Greek: eu → well, karyos → kernel) cells from prokaryotic cells is the **nucleus** of the eukaryote (Figure 4.1). The nucleus contains all of the DNA in an animal cell (except for a small amount in the mitochondria). The aqueous 'soup' inside the nucleus is called the nucleoplasm. The nucleus is wrapped in a double phospholipid bilayer called the **nuclear envelope** or **membrane**. The nuclear envelope is perforated with large holes called nuclear pores. RNA can exit the nucleus through the **nuclear pores**, but DNA cannot. Within the nucleus is an area called the **nucleolus** where rRNA is transcribed and the subunits of the ribosomes are assembled. The nucleolus is not separated from the nucleus by a membrane.

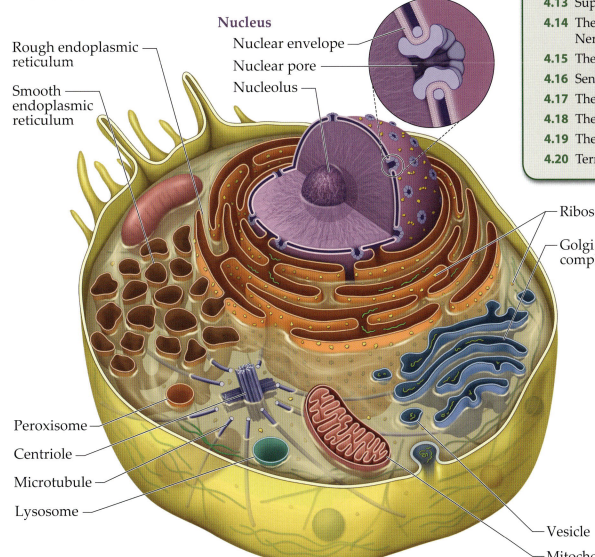

Nucleus
Nuclear envelope
Nuclear pore
Nucleolus

Rough endoplasmic reticulum

Smooth endoplasmic reticulum

Ribosomes

Golgi complex

Peroxisome

Centriole

Microtubule

Lysosome

Vesicle

Mitochondria

Figure 4.1   Eukaryotic Cell

Only eukaryotes have nuclei. If you remember that DNA cannot leave the nucleus, then you will remember that transcription must take place in the nucleus. RNA leaves the nucleus through nuclear pores.

4.2 *The Membrane*

Besides transport across the membrane (discussed in Biology Lecture 3), cells can acquire substances from the extracellular environment through **endocytosis** (Figure 4.2). There are several types of endocytosis: **phagocytosis** (Greek: phagein: to eat), pinocytosis (Greek: pinein: to drink), and *receptor mediated endocytosis*. In phagocytosis, the cell membrane protrudes outward to envelope and engulf particulate matter. Only a few specialized cells are capable of phagocytosis. The impetus for phagocytosis is the binding of proteins on the particulate matter to protein receptors on the phagocytotic cell. In humans, *antibodies* or *complement proteins* (discussed in Biology Lecture 8) bind to particles and stimulate receptor proteins on *macrophages* and *neutrophils* to initiate phagocytosis. Once the particulate matter is engulfed, the membrane bound body is called a *phagosome*.

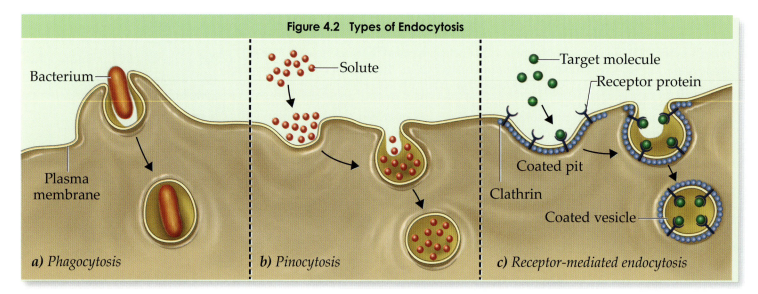

Figure 4.2 Types of Endocytosis

a) *Phagocytosis* — Bacterium, Plasma membrane

b) *Pinocytosis* — Solute

c) *Receptor-mediated endocytosis* — Target molecule, Receptor protein, Coated pit, Clathrin, Coated vesicle

> On the MCAT, you will prob-ably not have to distinguish between the different types of endocytosis, but you should understand the basic concept, and be aware that multiple methods for par-ticles to gain access to the interior of a cell.

In pinocytosis, extracellular fluid is engulfed by small invaginations of the cell membrane. This process is performed by most cells, and in a random fashion; it is nonselective.

Receptor mediated endocytosis refers to specific uptake of macromolecules such as hormones and nutrients. In this process, the ligand binds to a receptor protein on the cell membrane, and is then moved to a *clathrin coated pit*. Clathrin is a protein that forms a polymer adding structure to the underside of the coated pit. The coated pit invaginates to form a *coated vesicle*. One way that this process differs from phagocytosis is that its purpose is to absorb the ligands, whereas the ligands in phagocytosis exist only to act as signals to initiate phagocytosis of other particles.

Exocytosis is simply the reverse of endocytosis.

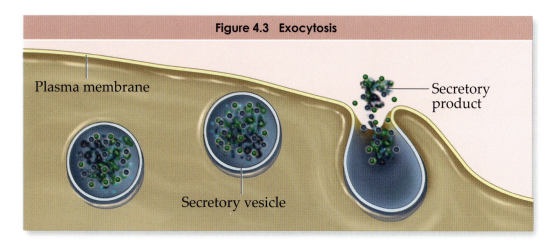

Figure 4.3 Exocytosis

Plasma membrane, Secretory product, Secretory vesicle

The structure of the phospolipid bilayer of the eukaryotic membrane is similar to the prokaryotic plasma membrane discussed in Biology Lecture 3. However, in the eukaryotic cell, the membrane invaginates and separates to form individual, membrane bound compartments and organelles. The eukaryotic cell contains a thick maze of membranous walls called the **endoplasmic reticulum (ER)** separating the **cytosol** (the aqueous solution inside the cell) from the **ER lumen** or **cisternal space** (the "extracellular fluid" side of the ER). In many places the ER is contiguous with the cell membrane and the nuclear membrane. The ER lumen is contiguous in places with the space between the double bilayer of the nuclear envelope.

ER near the nucleus has many ribosomes attached to it on the cytosolic side, giving it a granular appearance, hence the name **granular** or **rough ER**. Translation on the rough ER propels proteins into the ER lumen as they are created. These proteins are tagged with a *signal sequence* of amino acids and sometimes *glycosylated* (carbohydrate chains are added). The newly synthesized proteins are moved through the lumen toward the **Golgi apparatus** or **Golgi complex**. The Golgi apparatus is a series of flattened, membrane bound sacs. Small *transport vesicles* bud off from the ER and carry the proteins across the cytosol to the Golgi. The Golgi organizes and concentrates the proteins as they are shuttled by transport vesicles progressively outward from one compartment or *cisterna* of the Golgi to the next. Proteins are distinguished based upon their signal sequence and carbohydrate chains. Those proteins not possessing a signal sequence are packaged into secretory vesicles and expelled from the cell in a process called *bulk flow*. The Golgi may change proteins chemically by *glycosylation* or by removing amino acids. Some polysaccharide formation also takes place within the Golgi. The end-product of the Golgi is a vesicle full of proteins. These protein filled vesicles may either be expelled from the cell as **secretory vesicles**, released from the Golgi to mature into **lysosomes**, or transported to other parts of the cell such as the mitochondria or even back to the ER (Figure 4.4 and 4.5).

Secretory vesicles (sometimes called *zymogen granules*) may contain enzymes, growth factors, or extracellular matrix components. Secretory vesicles release their contents through exocytosis. Since exocytosis incorporates vesicle membranes into the cell membrane, secretory vesicles also act as the vehicle with which to supply the cell membrane with its integral proteins and lipids, and as the mechanism for membrane expansion. In the reverse process, endocytotic vesicles made at the cell membrane are shuttled back to the Golgi for recycling of the cell membrane. Secretory vesicles are continuously released by most cells in a process called *constitutive secretion*. Some specialized cells can release secretory vesicles in response to a certain chemical or electrical stimulus. This is called *regulated secretion*. Some proteins are activated within the secretory vesicles. For instance, proinsulin is cleaved to insulin only after the secretory vesicle buds off the Golgi.

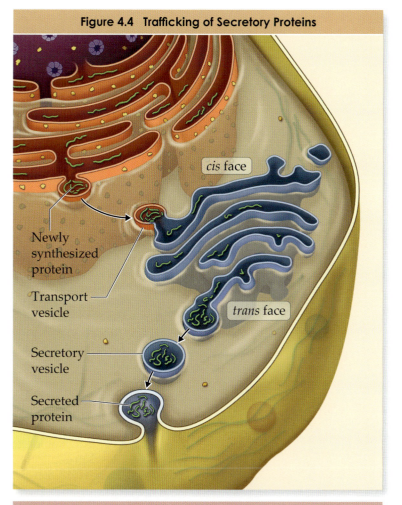

Figure 4.4   Trafficking of Secretory Proteins

cis face

Newly synthesized protein

Transport vesicle

trans face

Secretory vesicle

Secreted protein

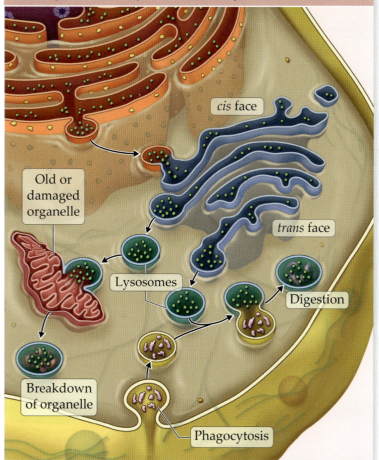

Figure 4.5   Transport and Fate of Lysosomal Proteins

cis face

Old or damaged organelle

trans face

Lysosomes

Digestion

Breakdown of organelle

Phagocytosis

Lysosomes contain *acid hydrolases* (hydrolytic enzymes that function best in an acid environment) such as *proteases*, *lipases*, *nucleases* and *glycosidases*. Together, these enzymes are capable of breaking down every major type of macromolecule within the cell. Lysosomes generally have an interior pH of 5. They fuse with endocytotic vesicles (the vesicles formed by phagocytosis and pinocytosis), and digest their contents. Any material not degraded by the lysosome is ejected from the cell through exocytosis. Lysosomes also take up and degrade cytosolic proteins in an endocytotic process. Under certain conditions lysosomes will rupture and release their contents into the cytosol killing the cell in a process called *autolysis*. Autolysis is useful in the formation of certain organs and tissues, like in the destruction of the tissue between the digits of a human fetus in order to form fingers.

Endoplasmic reticulum, which lacks ribosomes, is called **agranular** or **smooth endoplasmic reticulum**. Rough ER tends to resemble flattened sacs, whereas smooth ER tends to be tubular. Smooth ER plays several important roles in the cell. Smooth ER contains *glucose 6-phosphatase*, the enzyme used in the liver, the intestinal epithelial cells, and renal tubule epithelial cells, to hydrolyze glucose 6-phosphate to glucose, an important step in the production of glucose from glycogen. Triglycerides are produced in the smooth ER and stored in fat droplets. **Adipocytes** are cells containing predominately fat droplets. Such cells are important in energy storage and body temperature regulation. The smooth ER and the cytosol share in the role of cholesterol formation and its subsequent conversion to various steroids. Most of the phospholipids in the cell membrane are originally synthesized in the smooth ER. The phospholipids are all synthesized on the cytosol side of the membrane and then some are flipped to the other side by proteins called phospholipid translocators located exclusively in the smooth ER. Finally, smooth ER oxidizes foreign substances, detoxifying drugs, pesticides, toxins, and pollutants.

Peroxisomes are vesicles in the cytosol. They grow by incorporating lipids and proteins from the cytosol. Rather than budding off membranes like lysosomes from the golgi, peroxisomes self-replicate. They are involved in the production and breakdown of hydrogen peroxide. Peroxisomes inactivate toxic substances such as alcohol, regulate oxygen concentration, play a role in the synthesis and the breakdown of lipids, and in the metabolism of nitrogenous bases and carbohydrates.

Delivery for Mr. Membrane.

Golgi Salty

There is a lot of background information here that is not required by the MCAT. For the MCAT you should know the following:

1. There are many internal compartments in a cell (organelles) separated from the cytosol by membranes. In order to enter into a cell, a substance must be transported via passive transport, active transport, or facilitated diffusion or it can be bulk transported via endocytosis.

2. Rough ER has ribosomes attached to its cytosol side, and it synthesizes virtually all proteins not used in the cytosol. Proteins synthesized on the rough ER are pushed into the ER lumen and sent to the Golgi.

3. The Golgi modifies and packages proteins for use in other parts of the cell and outside the cell.

4. Lysosomes contain hydrolytic enzymes that digest substances taken in by endocytosis. Lysosomes come from the Golgi.

5. Smooth ER is the site of lipid synthesis including steroids. The smooth ER also helps to detoxify some drugs.

The structure and motility of a cell is determined by a network of filaments known as the **cytoskeleton**. The cytoskeleton anchors some membrane proteins and other cellular components, moves components within the cell, and moves the cell itself. Two major types of filaments in the cytoskeleton are **microtubules** (Figure 4.6) and microfilaments. Microtubules are larger than **microfilaments**. They are rigid hollow tubes made from a protein called **tubulin**. Although tubulin is a globular protein, under certain cellular conditions it polymerizes into long straight filaments. Thirteen of these filaments lie alongside each other to form the tube. The spiral appearance is due to the two types of tubulin, α and β, used in the synthesis. The **mitotic spindle** (see Biology Lecture 2) is made from microtubules.

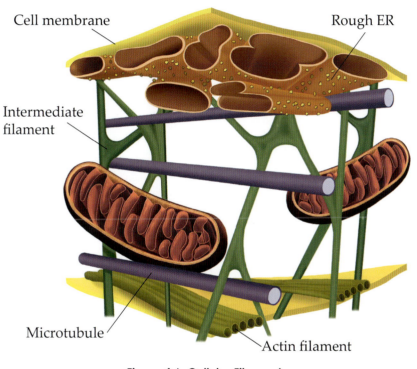

Cell membrane

Rough ER

Intermediate filament

Microtubule

Actin filament

Figure 4.6 Cellular Filaments

You must know the difference between microtubules and microfilaments. Microtubules are larger and are involved in flagella and cilia construction, and the spindle apparatus. In humans, cilia are found only in the fallopian tubes and the respiratory tract. Microfilaments squeeze the membrane together in phagocytosis and cytokinesis. They are also the contractile force in microvilli and muscle.

Flagella and **cilia** are specialized structures also made from microtubules. The major portion of each flagellum and cilium, called the **axoneme**, contains nine pairs of microtubules forming a circle around two lone microtubules in an arrangement known as **9+2**. Cross bridges made from a protein called **dynein** connect each of the outer pairs of microtubules to their neighbor. The cross bridges cause the microtubule pairs to slide along their neighbors creating a whip action in cilia causing fluid to move laterally, or a wiggle action in flagella causing fluid to move directly away from the cell.

Microtubules have a + and − end. The − end attaches to a *microtubule-organizing center (MTOC)* in the cell. A microtubule grows away from an MTOC at its + end. The major MTOC in animal cells is the **centrosome**. The **centrioles** function in the production of flagella and cilia, but are not necessary for microtubule production.

Microfilaments are smaller than microtubules. The polymerized protein **actin** forms a major component of microfilaments. Microfilaments produce the contracting force in muscle (discussed in Biology Lecture 8) as well as being active in **cytoplasmic streaming** (responsible for amoeba-like movement), phagocytosis, and microvilli movement.

There is also a third class of cellular filaments that you should be aware of: intermediate filaments. When fully formed, they have a diameter that is smaller than microtubules but larger than microfilaments. Intermediate filaments are not nearly as dynamic as microtubules or microfilaments and primarily serve to impart structural rigidity to the cell. Keratin, a type of intermediate filament found in epithelial cells, is associated with hair and skin.

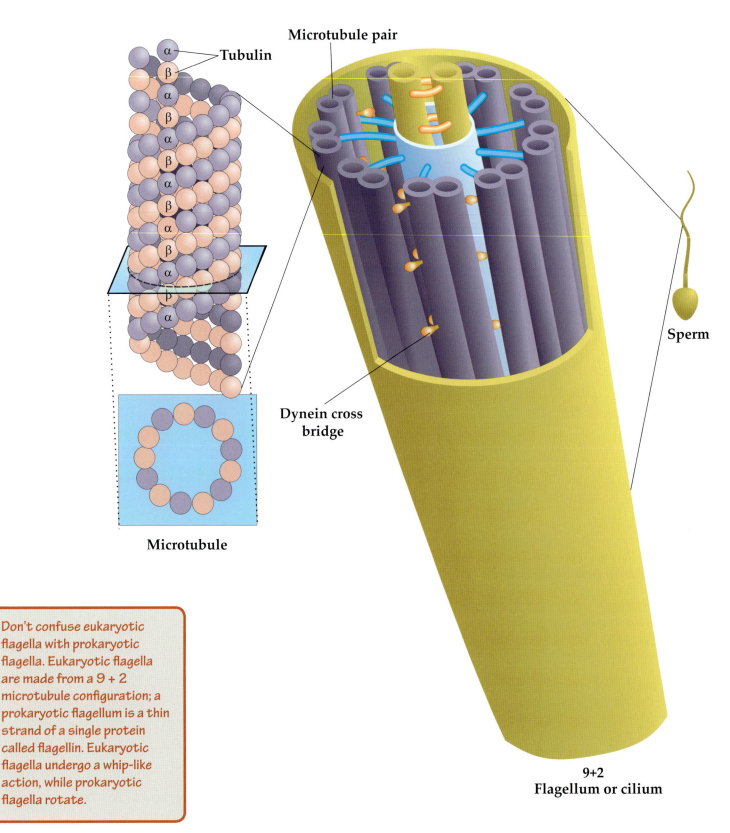

Don't confuse eukaryotic flagella with prokaryotic flagella. Eukaryotic flagella are made from a 9 + 2 microtubule configuration; a prokaryotic flagellum is a thin strand of a single protein called flagellin. Eukaryotic flagella undergo a whip-like action, while prokaryotic flagella rotate.

Figure 4.7 Structure of Flagella and Cilia

4.4 Cellular Junctions

There are three types of junctions or attachments that connect animal cells: tight junctions, desmosomes, and gap junctions (Figure 4.8). Each junction performs a different function. **Tight junctions** form a watertight seal from cell to cell that can block water, ions, and other molecules from moving around and past cells. Tissue held together by tight junctions may act as a complete fluid barrier. Epithelial tissue in organs like the bladder, the intestines, and the kidney are held together by tight junctions in order to prevent waste materials from seeping around the cells and into the body. Since proteins have some freedom to move laterally about the cell membrane, tight junctions also act as a barrier to protein movement between the *apical* and the *basolateral* surface of a cell. (The part of a cell facing the lumen of a cavity is called the apical surface. The opposite side of a cell is called the basolateral surface.)

Desmosomes join two cells at a single point. They attach directly to the cytoskeleton of each cell. Desmosomes do not prevent fluid from circulating around all sides of a cell. Desmosomes are found in tissues that normally experience a lot of stress, like skin or intestinal epithelium. Desmosomes often accompany tight junctions.

Gap junctions are small tunnels connecting cells. They allow small molecules and ions to move between cells. Gap junctions in cardiac muscle provide for the spread of the action potential from cell to cell.

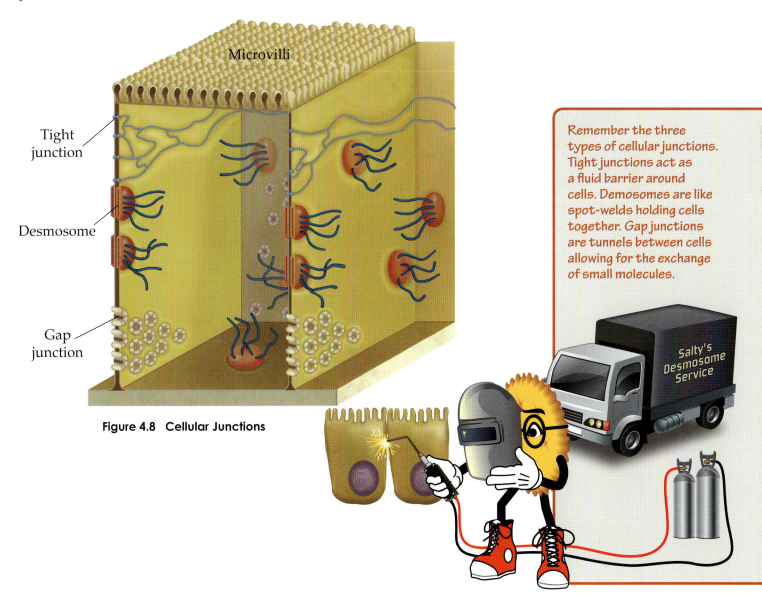

Figure 4.8 Cellular Junctions

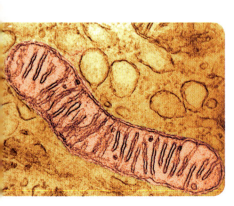

The inner membrane on mitochondria folds inward to form cristae.

4.5 Mitochondria

Mitochondria (Figure 4.9) are the powerhouses of the eukaryotic cell. We have already seen that the Krebs cycle takes place inside the mitochondria. According to the **endosymbiont theory**, mitochondria may have evolved from a symbiotic relationship between ancient prokaryotes and eukaryotes. Like prokaryotes, mitochondria have their own circular DNA that replicates independently from the eukaryotic cell. This DNA contains no histones or nucleosomes. Most animals have a few dozen to several hundred molecules of circular DNA in each mitochondrion. The genes in the mitochondrial DNA code for mitochondrial RNA that is distinct from the RNA in the rest of the cell. Thus mitochondria have their own ribosomes with a sediment coefficient of 55-60S in humans. However, most proteins used by mitochondria are coded for by nuclear DNA, not mitochondrial DNA. Antibiotics that block translation by prokaryotic ribosomes but not eukaryotic ribosomes, also block translation by mitochondrial ribosomes. Interestingly, some of the codons in mitochondria differ from the codons in the rest of the cell, presenting an exception to the universal genetic code! Mitochondrial DNA is passed maternally (from the mother) even in organisms whose male gamete contributes to the cytoplasm.

Mitochondria are surrounded by two phospholipid bilayers. The **inner membrane** invaginates to form **cristae**. It is the inner membrane that holds the electron transport chain. Between the inner and **outer membrane** is the **intermembrane space**.

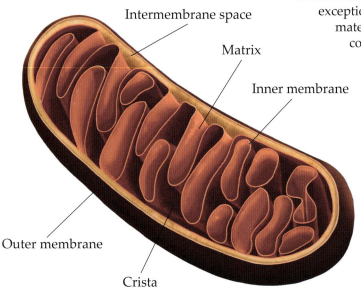

Intermembrane space

Matrix

Inner membrane

Outer membrane

Crista

Figure 4.9 Mitochondrion

> Memorize the parts of the mitochondrion and its purpose. Relate the parts to respiration discussed in Biology Lecture 1.

4.6 The Extracellular Matrix

Most cells in multicellular organisms form groups of similar cells, or cells that work together for a common purpose, called *tissue*. In some tissues, cells called fibroblasts secrete fibrous proteins such as elastin and collagen that form a molecular network that holds tissue cells in place called an *extracellular matrix*. Different tissues form dramatically different matrices. The matrix can constitute most of the tissue as in bone, where a few cells are interspersed in a large matrix, or the matrix may be only a small part of the tissue. The consistency of the matrix may be liquid as in blood, or solid as in bone.

An extracellular matrix may provide structural support, help to determine the cell shape and motility, and affect cell growth.

Three classes of molecules make up animal cell matrices:

1. *glycosaminoglycans* and *proteoglycans*;
2. structural proteins;
3. adhesive proteins.

Glycosaminoglycans are polysaccharides that typically have proteoglycans attached. They make up over 90% of the matrix by mass. This first class of molecules provides pliability to the matrix. Structural proteins provide the matrix with strength. The most common extracellular matrix structural protein in the body is collagen. Collagen is the structural protein that gives cartilage and bone their tensile strength. Adhesive proteins help individual cells within a tissue to stick together.

You may see basal lamina (which along with the reticular lamina forms the basement membrane) in an MCAT passage. The basal lamina is a thin sheet of matrix material that separates epithelial cells from *support tissue*. (Epithelial cells separate the outside environment from the inside of the body. Support tissue is composed of the cells adjacent to the epithelial cells on the inside of the body.) Basal lamina is also found around nerves, and muscle and fat cells. Basal lamina typically acts as a sieve type barrier, selectively allowing the passage of some molecules but not others.

Many animal cells contain a carbohydrate region analogous to the plant cell wall or bacterial cell wall, called the *glycocalyx*. The glycocalyx separates the cell membrane from the extracellular matrix; however, a part of the glycocalyx is made from the same material as the matrix. Thus, the glycocalyx is often difficult to identify. The glycocalyx may be involved in cell-cell recognition, adhesion, cell surface protection, and permeability.

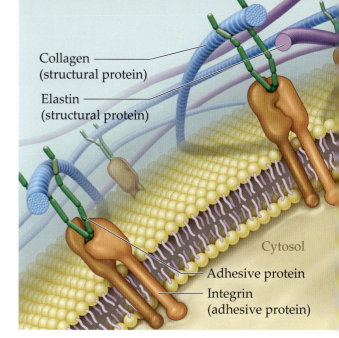

Figure 4.10 The Extracellular Matrix

Don't worry about memorizing anything about the extracellular matrix. Just know that it's the stuff that surrounds the cell and that it is formed by the cell itself.

4.7 *Organization in Multicellular Eukaryotes*

In multicellular eukaryotes, groups of cells work together, each type of cell performing a unique function that contributes to the specialized function of the group. These groups of cells are called *tissues*. Cells in the same tissue usually have similar embryology; they arise from the same embryonic germ layer. There are four basic types of tissue in animals: *epithelial tissue, muscle tissue, connective tissue,* and *nervous tissue*. Epithelial tissue separates free body surfaces from their surroundings. *Simple epithelium* is one layer thick, while *stratified epithelium* is two or more layers thick. Simple epithelium includes *endothelium* lining the various vessels of the body including the heart. Connective tissue is characterized by an extensive matrix. Examples include: blood, lymph, bone, cartilage, and connective tissue proper making up tendons and ligaments. Muscle and nervous tissue will be discussed later.

Different tissue types work together to form *organs*. For example, the stomach is an organ with an outer layer made from epithelial tissue and connective tissue, a second layer of muscle tissue, and an innermost layer of epithelial tissue.

Organs that work together to perform a common function are called *systems*. The remainder of this manual (except for Lecture 9) is devoted to the study of biological systems in the human body. Many of the details are not required knowledge for the MCAT. However, if you keep in mind the holistic concept (the body is not simply many disconnected parts but an entire organism with systems that work in conjunction with each other), you will attain a stronger recall of the details and a greater understanding of material.

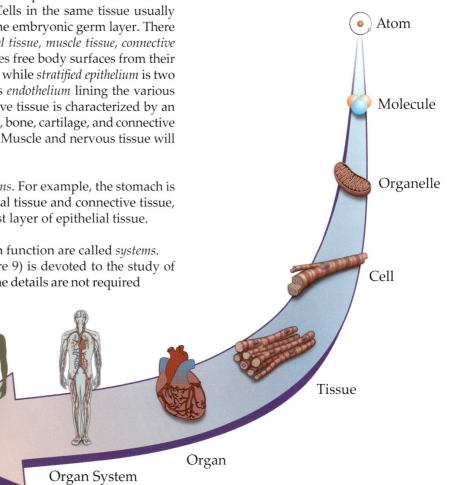

73. All of the following are composed of microtubules EXCEPT:

    A. the tail of a sperm cell.
    B. the spindle apparatus.
    C. the cilia of the fallopian tubes.
    D. the flagella of bacteria.

74. Which of the following is true concerning the nucleolus?

    A. It is bound by a phospholipid membrane.
    B. It disappears during prophase.
    C. It is the site of translation of ribosomal RNA.
    D. It is found in most bacteria.

75. In some specialized cells, glucose is transported against its concentration gradient via an integral protein using the energy of the sodium ion electrochemical gradient. If no ATP is used for this transport, it is most likely:

    A. active transport.
    B. facilitated transport.
    C. passive transport.
    D. osmosis.

76. Which of the following cells would be expected to contain the most smooth endoplasmic reticulum?

    A. a liver cell
    B. an islet cell from the pancreas
    C. a mature sperm
    D. a zygote

77. Which of the following statements is true concerning tight junctions?

    I. They connect adjacent cells.
    II. They may form a barrier to extracellular fluids.
    III. They have the greatest strength of all cellular adhesions.

    A. I only
    B. I and II only
    C. II and III only
    D. I, II, and III

78. Which of the following is not a membrane bound organelle?

    A. the golgi body
    B. the nucleus
    C. the smooth endoplasmic reticulum
    D. the ribosome

79. One function of the liver is to detoxify alcohol taken into the body. The organelle within the liver cell that most directly affects this process is:

    A. the smooth endoplasmic reticulum
    B. the nucleus
    C. the Golgi apparatus
    D. the rough endoplasmic reticulum

80. When a primary lysosome fuses with a food vesicle to become a secondary lysosome:

    A. its pH drops via active pumping of protons into its interior.
    B. its pH drops via active pumping of protons out of its interior.
    C. its pH rises via active pumping of protons into its interior.
    D. its pH rises via active pumping of protons out of its interior.

4.8 Intercellular Communication

In multicellular organisms, cells must be able to communicate with each other so that the organism can function as a single unit. Communication is accomplished chemically via three types of molecules: 1) neurotransmitters; 2) local mediators; 3) hormones. These methods of communication are governed by the nervous system, the paracrine system, and the endocrine system respectively.

There are several distinctions between the methods of communication, the major distinction being the distance traveled by the mediator. Neurotransmitters travel over very short intercellular gaps; local mediators function in the immediate area around the cell from which they were released; hormones travel throughout the organism via the blood stream.

For the MCAT, you should focus on the distinctions between neurotransmitter and hormonal mediated communication. Neurotransmitters are released by neurons. **Neuronal communication** tends to be **rapid**, direct, and **specific**. **Hormonal communication**, on the other hand, tends to be **slower, spread throughout the body, and affect many cells and tissues in many different ways.**

4.9 Paracrine System

Local mediators are released by a variety of cells into the **interstitial fluid** (fluid between the cells) and act on neighboring cells a few millimeters away. Local mediators may be proteins, other amino acid derivatives, or even fatty acids. *Prostaglandins* are fatty acid derivative local mediators. Prostaglandins affect smooth muscle contraction, platelet aggregation, inflammation and other reactions. Aspirin inhibits prostaglandin synthesis and thus is an anti-inflammatory. Growth factors and lymphokines, discussed later, are other examples of local mediators.

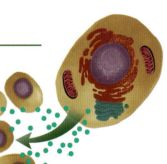

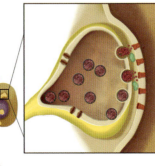

Specific knowledge of the paracrine system will not be tested by the MCAT. However, you should be aware of the existence of this intermediate communication system.

The remainder of this lecture and the next are devoted mainly to the nervous system and the endocrine system. Besides memorizing the details of each system, keep in mind that they represent two different methods of cellular communication. After reading about the two systems, compare them carefully, and think about why some types of communication are better served by one method over the other.

Dendrites

Mitochondrion

Nucleus

Axon hillock

Node of Ranvier

Myelin sheath

Axon terminals

Layers of myelin sheath

Figure 4.11 Myelinated Neuron

You must know the basic anatomy of a neuron. Remember that a signal travels from the dendrites to the axon hillock, where an action potential is generated and moves down the axon to the synapse. Neurons do not depend upon insulin to obtain glucose.

4.10 Nervous System

The nervous system allows for rapid and direct communication between specific parts of the body resulting in changes in muscular contractions or glandular secretions. Included within the nervous system are the brain, spinal cord, nerves and neural support cells, and certain sense organs such as the eye, and the ear.

The functional unit of the nervous system is the **neuron** (Figure 4.12). A neuron is a highly specialized cell capable of transmitting an electrical signal from one cell to another via electrical or chemical means. The neuron is so highly specialized that it has lost the capacity to divide. In addition, it depends almost entirely upon glucose for its chemical energy. Although the neuron uses facilitated transport to move glucose from the blood into its cytosol, unlike most other cells, the neuron is not dependent upon insulin for this transport. The neuron depends heavily on the efficiency of aerobic respiration. However, it has low stores of glycogen and oxygen, and must rely on blood to supply sufficient levels of these nutrients. Neurons in different parts of the body have a different appearance, but all neurons have a basic anatomy consisting of many dendrites, a single cell body, and usually one axon with many small branches.

The **dendrites** receive a signal to be transmitted. Typically, the cytosol of the cell body is highly conductive and any electrical stimulus creates a disturbance in the electric field that is transferred immediately to the axon hillock. If the stimulus is great enough, the **axon hillock** generates an action potential in all directions, including down the **axon**. The membrane of the cell body usually does not contain enough ion channels to sustain an action potential. The axon, however, carries the action potential to a synapse, which passes the signal to another cell.

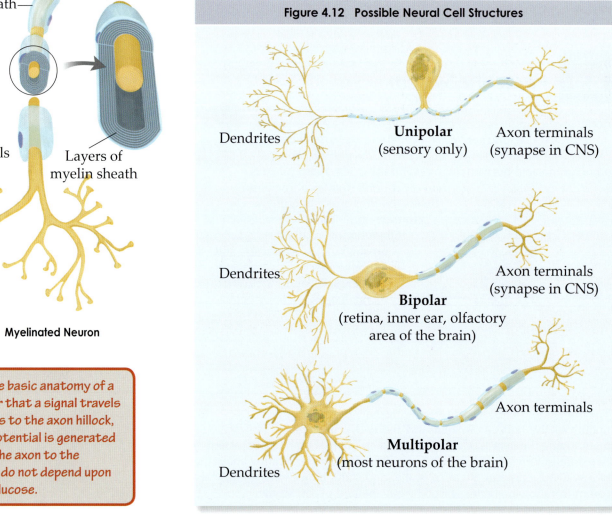

Figure 4.12 Possible Neural Cell Structures

Dendrites

Unipolar
(sensory only)

Axon terminals
(synapse in CNS)

Dendrites

Bipolar
(retina, inner ear, olfactory area of the brain)

Axon terminals
(synapse in CNS)

Axon terminals

Multipolar
(most neurons of the brain)

Dendrites

The Action Potential

The **action potential** is a disturbance in the electric field across the membrane of a neuron. To understand the action potential, we first must understand the **resting potential**. The resting potential is established mainly by an equilibrium between passive diffusion of ions across the membrane and the Na⁺/K⁺ pump. The Na⁺/K⁺ pump moves three positively charged sodium ions out of the cell while bringing two positively charged potassium ions into the cell. This action increases the positive charge along the membrane just outside the cell relative to the charge along the membrane on the inside of the cell. As the electrochemical gradient of Na⁺ increases, the force pushing the Na⁺ back into the cell also increases. The rate at which Na⁺ passively diffuses back into the cell increases until it equals the rate at which it is being pumped out of the cell. The same thing happens for potassium. When all rates reach equilibrium, the inside of the membrane has a negative potential difference (voltage) compared to the outside. This potential difference is called the resting potential. Although other ions are involved, Na⁺ and K⁺ are the major players in establishing the resting potential.

The membrane of a neuron also contains integral membrane proteins called **voltage gated sodium channels**. These proteins change configuration when the voltage across the membrane is disturbed. Specifically, they allow Na⁺ to flow through the membrane for a fraction of a second as they change configuration. As Na⁺ flows into the cell, the voltage changes still further, causing more sodium channels to change configuration, allowing still more sodium to flow into the cell in a positive feedback mechanism. Since the Na⁺ concentration moves toward equilibrium, and the K⁺ concentration remains higher inside the cell, the membrane potential actually reverses polarity so that it is positive on the inside and negative on the outside. This process is called **depolarization**. The neuronal membrane also contains **voltage gated potassium channels**. The potassium channels are less sensitive to voltage change so they take longer to open. By the time they begin to open, most of the sodium channels are closing. Now K⁺ flows out of the cell, making the inside more negative in a process called **repolarization**. The potassium channels are so slow to close that, for a fraction of a second, the inside membrane becomes even more negative than the resting potential. This portion of the process is called **hyperpolarization**. Passive diffusion returns the membrane to its resting potential. The entire process just described is called the action potential (Figure 4.14). Throughout the action potential, the Na⁺/K⁺ pump keeps working.

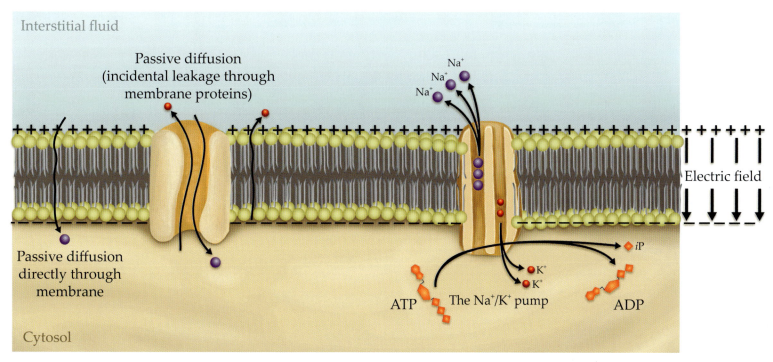

Figure 4.13   Resting Potential

The action potential occurs at a point on a membrane and propagates along that membrane by depolarizing the section of membrane immediately adjacent to it. In Figure 4.14, the protein channels marked 1 are about to recieve the action potential while the protein channels marked 5 have already received the action potential. Therefore, the action potential is traveling from right to left along the membrane (Figure 4.14 and 4.15); the synapse would be to the left of the portion of the membrane shown. The voltage as a function of time at any given point on the membrane is given by the wave shown. The entire action potential as measured at one point on the membrane of a neuron takes place in a fraction of a millisecond.

Figure 4.14   Steps of the Action Potential

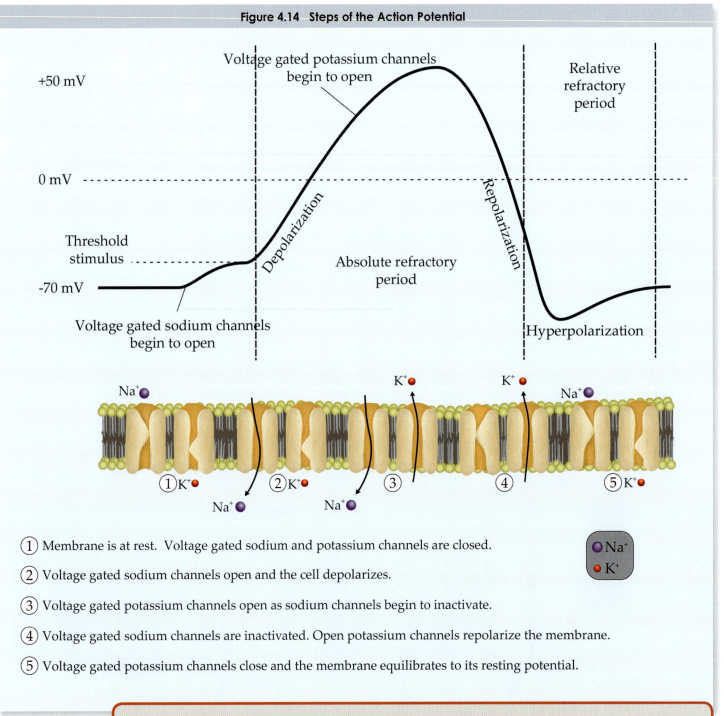

① Membrane is at rest. Voltage gated sodium and potassium channels are closed.

② Voltage gated sodium channels open and the cell depolarizes.

③ Voltage gated potassium channels open as sodium channels begin to inactivate.

④ Voltage gated sodium channels are inactivated. Open potassium channels repolarize the membrane.

⑤ Voltage gated potassium channels close and the membrane equilibrates to its resting potential.

You must understand the dynamics of the membrane potential. Use the Na$^+$/K$^+$ pump to help you remember that the inside of the membrane is negative with respect to the outside. Different cells have different action potentials. If you understand the principle, you should be able to understand any action potential. Remember that an action potential originates at the axon hillock. This section is important for the MCAT. If you don't thoroughly understand it, then you should reread it.

An action potential is **all-or-nothing**; the membrane completely depolarizes or no action potential is generated. In order to create an action potential, the stimulus to the membrane must be greater than the **threshold stimulus**. Any stimulus greater than the threshold stimulus creates the same size action potential. If the threshold stimulus is reached, but is reached very slowly, an action potential still may not occur. This is called *accommodation*. Once an action potential has begun, there is a short period of time called the *absolute refractory period* in which no stimulus will create another action potential. The *relative refractory period* gives the time during which only an abnormally large stimulus will create an action potential.

Other cells, such as skeletal and cardiac muscle cells also conduct action potentials. Although these action potentials are slightly different in duration, shape, and even the types of ions, they work on the same principles.

The action potential in Figure 4.15 moves along the axon from right to left.

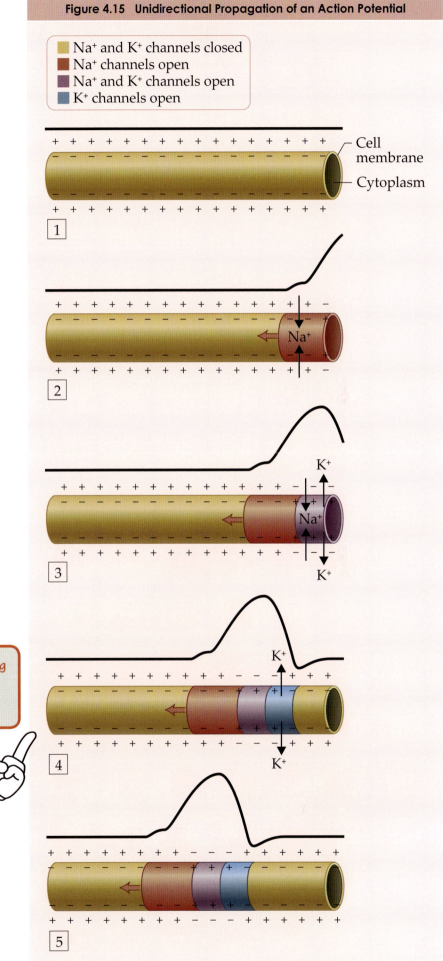

Figure 4.15 Unidirectional Propagation of an Action Potential

Na⁺ and K⁺ channels closed
Na⁺ channels open
Na⁺ and K⁺ channels open
K⁺ channels open

4.12 The Synapse

Neural impulses are transmitted from one cell to another chemically or electrically via a **synapse**. The transmission of the signal from one cell to another is the slowest part of the process of nervous system cellular communication, yet it occurs in a fraction of a second.

Electrical synapses are uncommon. They are composed of gap junctions between cells. Cardiac muscle, visceral smooth muscle, and a very few neurons in the central nervous system contain electrical synapses. Since they don't involve diffusion of chemicals, they transmit signals much faster than chemical synapses and in both directions.

A more common synapse, a **chemical synapse** (Figure 4.16) (called a motor end plate when connecting a neuron to a muscle), is **unidirectional**. In a chemical synapse, small vesicles filled with neurotransmitter rest just inside the presynaptic membrane. The membrane near the synapse contains an unusually large number of Ca^{2+} voltage gated channels. When an action potential arrives at a synapse, these channels are activated allowing Ca^{2+} to flow into the cell. In a mechanism not completely understood, the sudden influx of calcium ions causes some of the neurotransmitter vesicles to be released through an exocytotic process into the **synaptic cleft**. The neurotransmitter diffuses across the synaptic cleft via **Brownian motion** (the random motion of the molecules). The postsynaptic membrane contains neurotransmitter receptor proteins. When the neurotransmitter attaches to the receptor proteins, the postsynaptic membrane becomes more permeable to ions. Ions move across the postsynaptic membrane through proteins called *ionophores*, completing the transfer of the neural impulse. In this way, the impulse is not attenuated by electrical resistance as it moves from one cell to the next. If a cell is fired too often (hundreds of times per second for several minutes) it will not be able to replenish its supply of neurotransmitter vessicles, and the result is *fatigue* (the impulse will not pass to the postsynaptic neuron).

The **neurotransmitter** attaches to its receptor for only a fraction of a second, and is released back into the synaptic cleft. If the neurotransmitter remains in the synaptic cleft, the postsynaptic cell may be stimulated over and over. There are several mechanisms by which the cell deals with this problem. The neurotransmitter may be destroyed by an enzyme in the matrix of the synaptic cleft and its parts recycled by the presynaptic cell. It may be directly absorbed by the presynaptic cell via active transport. The neurotransmitter may also diffuse out of the synaptic cleft.

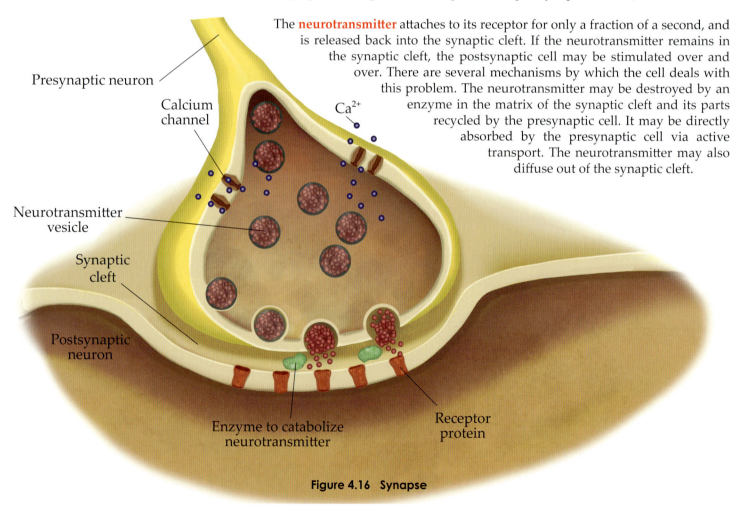

Figure 4.16 Synapse

Figure 4.17 G-protein

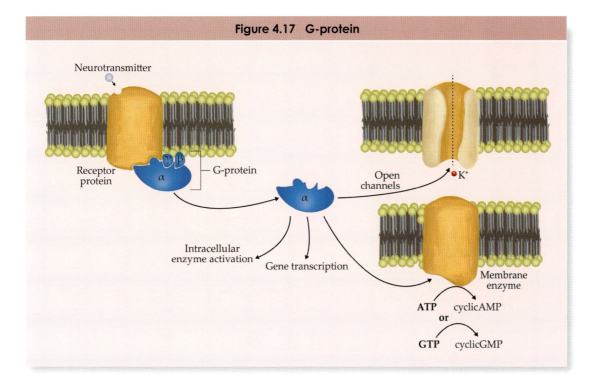

Over 50 types of neurotransmitters have been identified. Different neurotransmitters are characteristic of different parts of the nervous system (i.e. certain neurotransmitters are found in certain areas of the nervous system). A single synapse usually releases only one type of neurotransmitter and is designed either to inhibit or to excite, but not both. A single synapse cannot change from inhibitory to excitatory, or vice versa. On the other hand, some neurotransmitters are capable of inhibition or excitation depending upon the type of receptor in the postsynaptic membrane. Acetylcholine, a common neurotransmitter, has an inhibitory effect on the heart, but an excitatory effect on the visceral smooth muscle of the intestines.

Receptors may be ion channels themselves, which are opened when their respective neurotransmitter attaches, or they may act via a **second messenger system** activating another molecule inside the cell to make changes. For prolonged change, such as that involved in memory, the second messenger system is preferred. *G-proteins* (Figure 4.17) commonly initiate second messenger systems. A G-protein is attached to the receptor protein along the inside of the postsynaptic membrane. When the receptor is stimulated by a neurotransmitter, part of the G-protein, called the *α-subunit*, breaks free. The α-subunit may:

1. activate separate specific ion channels;
2. activate a second messenger (i.e. cyclic AMP or cyclic GMP);
3. activate intracellular enzymes;
4. activate gene transcription.

A single neuron may make a few to as many as 200,000 synapses (Figure 4.18). Most synapses contact dendrites, but some may directly contact other cell bodies, other axons, or even other synapses. The firing of one or more of these synapses creates a change in the neuron cell potential. This change in the cell potential is called either the *excitatory postsynaptic potential (EPSP)* or the *inhibitory postsynaptic potential (IPSP)*. Normally, 40-80 synapses must fire simultaneously on the same neuron in order for an EPSP to create an action potential within that neuron.

The chemical synapse is the important synapse for the MCAT. Understand that it is the slowest step in the transfer of a nervous signal, and that it can only transfer a signal in one direction. Also recognize what a second messenger system is.

Figure 4.18 Motor Neuron with Synaptic Terminals

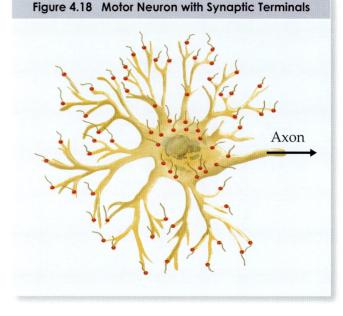

Axon

4.13 Support Cells

Besides neurons, nervous tissue contains many support cells called *glial cells* or *neuroglia*. In fact, in the human brain, glial cells typically outnumber neurons 10 to 1. Neuroglia are capable of cellular division, and, in the case of traumatic injury to the brain, it is the neuroglia that multiply to fill any space created in the central nervous system.

There are six types of glial cells: *microglia; ependymal cells; satellite cells; astrocytes; oligodendrocytes;* and *neurolemmocytes* or Schwann cells. Microglia arise from white blood cells called monocytes. They phagocytize microbes and cellular debris in the central nervous system. Ependymal cells are epithelial cells that line the space containing the cerebrospinal fluid. Ependymal cells use cilia to circulate the cerebrospinal fluid. Satellite cells support *ganglia* (groups of cell bodies in the peripheral nervous system). Astrocytes are star-shaped neuroglia in the central nervous system that give physical support to neurons, and help maintain the mineral and nutrient balance in the interstitial space. Oligodendrocytes wrap many times around axons in the central nervous system creating electrically insulating sheaths called **myelin**. In the peripheral nervous system, myelin is produced by **Schwann cells**. Myelin increases the rate at which an axon can transmit signals. To the naked eye, myelinated axons appear white while the neuronal cell bodies appear gray. Hence the name **white matter** and **gray matter**. Tiny gaps between myelin are called **nodes of Ranvier**. When an action potential is generated down a myelinated axon, the action potential jumps from one node of Ranvier to the next as quickly as the disturbance moves through the electric field between them. This is called **saltatory conduction** (Latin *saltus*: a jump).

> For support cells, you should understand how myelin increases the speed with which the action potential moves down the axon. Memorizing the names of support cells is not necessary, but know the general functions of support cells. Only vertebrates have myelinated axons.

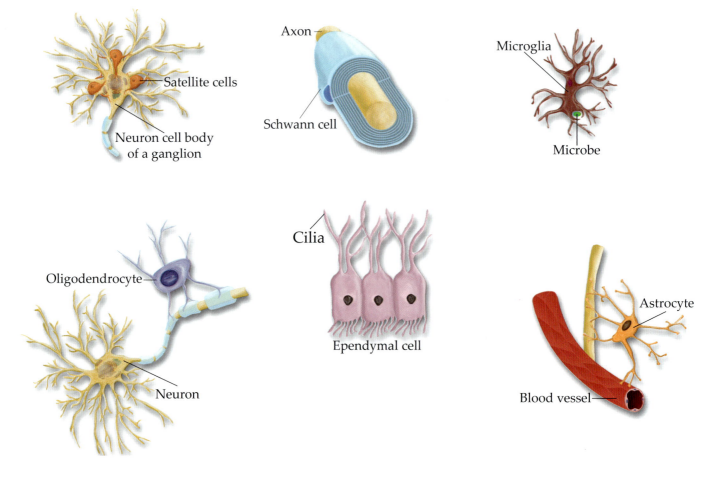

Figure 4.19 Neuroglia

81. Which of the following gives the normal direction of signal transmission in a neuron?

    A. from the axon to the cell body to the dendrites
    B. from the dendrites to the cell body to the axon
    C. from the cell body to the axon and dendrites
    D. from the dendrites to the axon to the cell body

82. Novocaine is a local anesthetic used by many dentists. Novocaine most likely inhibits the action potential of a neuron by:

    A. stimulating calcium voltage gated channels at the synapse.
    B. increasing chloride ion efflux during an action potential.
    C. uncoiling Schwann cells wrapped around an axon.
    D. blocking sodium voltage gated channels.

83. A cell membrane is normally slightly passively permeable to potassum ions. If a neuronal membrane were to become suddenly impermeable to potassium ions but retain an active Na^+/K^+-ATPase, the neurons resting potential would:

    A. become more positive because potassium ion concentration would increase inside the neuron.
    B. become more positive because potassium ion concentration would increase outside the neuron.
    C. become more negative because potassium ion concentration would increase inside the neuron.
    D. become more negative because potassium ion concentration would increase outside the neuron.

84. If an acetylcholinesterase inhibitor were administered into a cholinergic synapse, what would happen to the activity of the postsynaptic neuron?

    A. It would decrease, because acetylcholine would be degraded more rapidly than normal.
    B. It would decrease, because acetylcholinesterase would bind to postsynaptic membrane receptors less strongly.
    C. It would increase, because acetylcholine would be produced more rapidly than normal.
    D. It would increase, because acetylcholine would be degraded more slowly than normal.

85. White matter in the brain and spinal cord appears white because:

    A. it contains large amounts of myelinated axons.
    B. it does not contain any myelinated axons.
    C. it is composed primarily of cell bodies.
    D. it contains a high concentration of white blood cells to protect the central nervous system from infection.

86. The jumping of an action potential from one node of Ranvier to the next is known as:

    A. Brownian motion.
    B. saltatory conduction.
    C. a threshold stimulus.
    D. an all-or-nothing response.

87. What is the ratio of sodium ions to potassium ions transferred by the Na+/K+ pump out of and into the cell?

    A. 2 sodium ions in; 3 potassium ions out
    B. 3 sodium ions in; 2 potassium ions out
    C. 3 sodium ions out; 2 potassium ions in
    D. 2 sodium ions in; 3 potassium ions in

88. Which of the following is found in vertebrates but NOT in invertebrates?

    A. a dorsal, hollow nerve chord
    B. mylenation to increase the speed of nervous impulse transmission along the axon
    C. axons through which the nervous impulse is conducted
    D. Na^+/K^+-pump

4.14 The Structure of the Nervous System

Neurons may perform one of three functions.

1. **Sensory (afferent) neurons** receive signals from a receptor cell that interacts with its environment. The sensory neuron then transfers this signal to other neurons. 99% of sensory input is discarded by the brain.

2. **Interneurons** transfer signals from neuron to neuron. 90% of neurons in the human body are interneurons.

3. **Motor (efferent) neurons** carry signals to a muscle or gland called the **effector**. Sensory neurons are located dorsally (toward the back) from the spinal cord, while motor neurons are located ventrally (toward the front or abdomen).

Figure 4-20 shows a simple reflex arc using all three types of neurons. Some reflex arcs do not require an interneuron. Neuron processes (axons and dendrites) are typically bundled together to form **nerves** (called tracts in the CNS as discussed below).

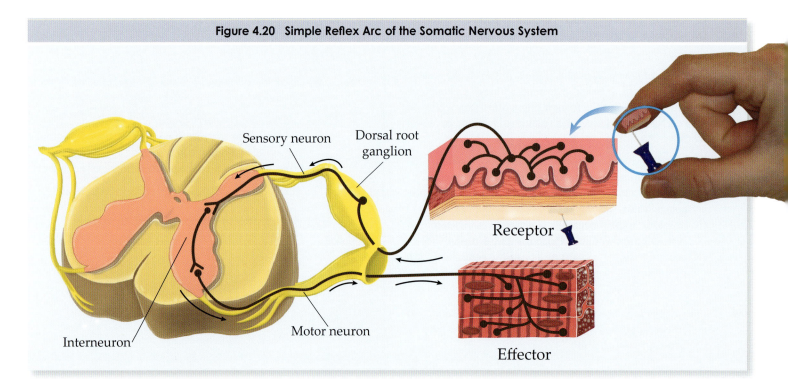

Figure 4.20 Simple Reflex Arc of the Somatic Nervous System

> For the MCAT, think of the CNS as the brain and spinal cord, and the PNS as everything else.

The nervous system has two major divisions: the **central nervous system (CNS)** and the **peripheral nervous system (PNS)**. The CNS consists of the interneurons and support tissue within the brain and the spinal cord. The function of the CNS is to integrate nervous signals between sensory and motor neurons.

The CNS is connected to the peripheral parts of the body by the PNS. Parts of the PNS, such as the *cranial nerves* and the *spinal nerves*, project into the brain and spinal cord. The PNS handles the sensory and motor functions of the nervous system. The PNS can be further divided into the **somatic nervous system** and **autonomic nervous system (ANS)**. The somatic nervous system is designed primarily to respond to the external environment. It contains sensory and motor functions. Its motor neurons innervate only skeletal muscle. The cell bodies of somatic motor neurons are located in the ventral horns of the spinal cord. These neurons synapse directly on their effectors and use acetylcholine for their neurotransmitter. The motor functions of the somatic nervous system can be consciously controlled and are considered voluntary. The sensory neuron cell bodies are located in the *dorsal root ganglion*.

The sensory portion of the ANS receives signals primarily from the viscera (the organs inside the ventral body cavity). The motor portion of the ANS then conducts these signals to smooth muscle, cardiac muscle, and glands. The function of the ANS is generally involuntary. The motor portion of the ANS is divided into two systems: **sympathetic** and **parasympathetic** (Figure 4.21). Most internal organs ar innervated by both with the two systems working antagonistically. The sympathetic ANS deals with "fight or flight" responses. For instance, its action on the heart would be to increase beat rate and stroke volume; it works to constrict blood vessels around the digestive and excretory systems in order to increase blood flow around skeletal muscles. Parasympathetic action, on the other hand, generally works toward the opposite goal, to "rest and digest". Parasympathetic activity slows the heart rate and increases digestive and excretory activity.

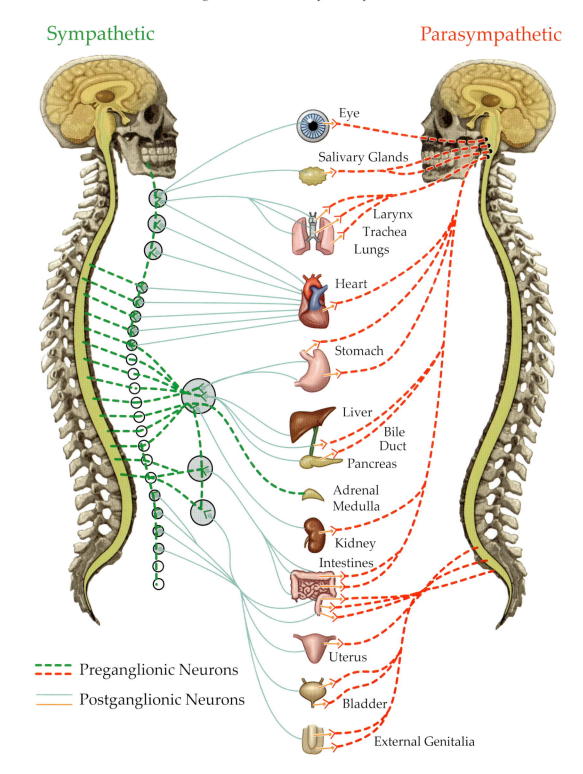

Figure 4.21   The Autonomic Nervous System

Sympathetic signals originate in neurons whose cell bodies are found in the spinal cord, while parasympathetic signals originate in neurons whose cell bodies can be found in both the brain and spinal cord. (A group of cell bodies located in the CNS is called a *nucleus*; if located outside the CNS, it is called a *ganglion*.) These neurons extend out from the spinal cord to synapse with neurons whose cell bodies are located outside the CNS. The former neurons are called preganglionic neurons; the later are called postganglionic neurons. The cell bodies of sympathetic postganglionic neurons lie far from their effectors generally within the paravertebral ganglion, which runs parallel to the spinal cord, or within the prevertebral ganglia in the abdomen. The cell bodies of the parasympathetic postganglionic neurons lie in ganglia inside or near their effectors.

With few exceptions, the neurotransmitter used by all preganglionic neurons in the ANS and by postganglionic neurons in the parasympathetic system is **acetylcholine**; the postganglionic neurons of the sympathetic nervous system use either **epinephrine** or **norepinephrine** (also called **adrenaline** and **noradrenaline**).

Receptors for acetylcholine are called *cholinergic receptors*. There are two types of cholinergic receptors: *nicotinic* and *muscarinic*. Generally, nicotinic receptors are found on the postsynaptic cells of the synapse between ANS preganglionic and postganglionic neurons and on skeletal muscle membranes at the neuromuscular junction. Muscarinic receptors are found on the effectors of the parasympathetic nervous system. The receptors for epinephrine and norepinephrine are called *adrenergic*.

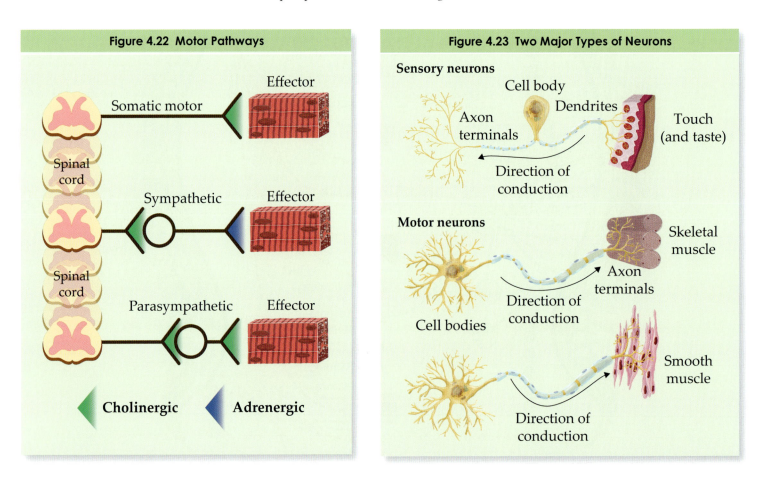

Figure 4.22 Motor Pathways

Figure 4.23 Two Major Types of Neurons

The Central Nervous System

The CNS consists of some of the spinal cord, the lower brain, and all of the higher brain. Although the spinal cord acts mainly as a conduit for nerves to reach the brain, it does possess limited integrating functions such as walking reflexes, leg stiffening, and limb withdrawal from pain. (See Figure 4.20 for the reflex arc.)

The lower brain consists of the **medulla**, *pons*, *mesencephalon*, **hypothalamus**, **thalamus**, **cerebellum**, and *basal ganglia*. It integrates subconscious activities such as the respiratory system, arterial pressure, salivation, emotions, and reaction to pain and pleasure.

The higher brain or cortical brain consists of the **cerebrum** or **cerebral cortex**. The cerebral cortex is incapable of functioning without the lower brain. It acts to store memories and process thoughts.

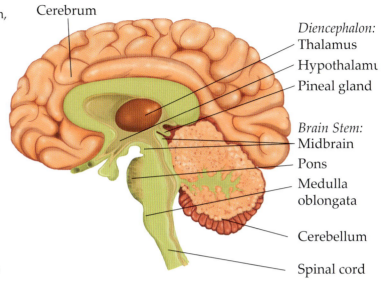

Cerebrum

Diencephalon:
Thalamus
Hypothalamu
Pineal gland

Brain Stem:
Midbrain
Pons
Medulla oblongata

Cerebellum

Spinal cord

Don't worry to much about memorizing the specific parts of the brain. Just realize that the brain is involved in the processing of sensory information, regulation of the body's internal enviroment, and responding to stimuli. You should also know that the forebrain (cerebrum) is resposible for higher level thoughts and consciousness.

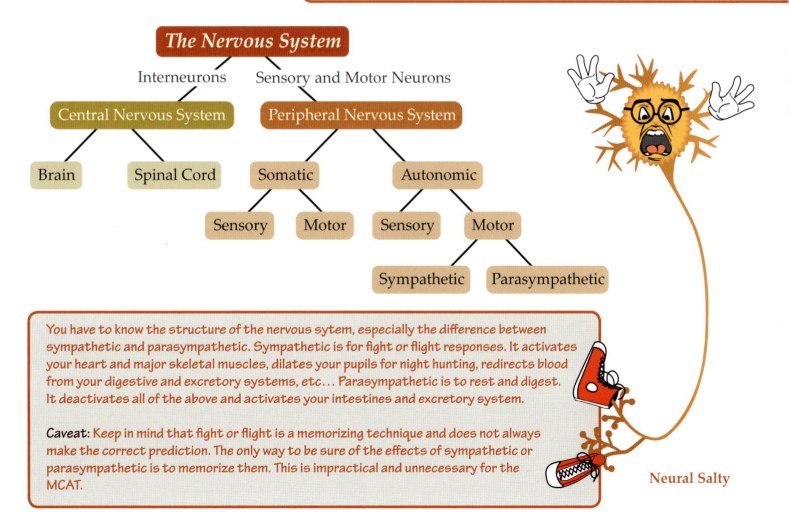

The Nervous System

Interneurons   Sensory and Motor Neurons

Central Nervous System   Peripheral Nervous System

Brain   Spinal Cord   Somatic   Autonomic

Sensory   Motor   Sensory   Motor

Sympathetic   Parasympathetic

You have to know the structure of the nervous sytem, especially the difference between sympathetic and parasympathetic. Sympathetic is for fight or flight responses. It activates your heart and major skeletal muscles, dilates your pupils for night hunting, redirects blood from your digestive and excretory systems, etc... Parasympathetic is to rest and digest. It deactivates all of the above and activates your intestines and excretory system.

Caveat: Keep in mind that fight or flight is a memorizing technique and does not always make the correct prediction. The only way to be sure of the effects of sympathetic or parasympathetic is to memorize them. This is impractical and unnecessary for the MCAT.

Neural Salty

4.16 Sensory Receptors

Although the somatic sensory neurons transfer signals from the external environment to the brain, they are incapable of distinguishing between different types of stimuli, and are not designed to be the initial receptors of such signals. Instead, the body contains *5 types of sensory receptors*:

1. *mechanoreceptors* for touch;
2. *thermoreceptors* for temperature change;
3. *nociceptors* for pain;
4. *electromagnetic receptors* for light; and
5. *chemoreceptors* for taste, smell and blood chemistry.

Each receptor responds strongly to its own type of stimulus and weakly or not at all to other types of stimuli. Each type of receptor has its own neural pathway and termination point in the central nervous system which results in the various sensations.

4.17 The Eye

For the MCAT®, you should know the basic anatomy of the eye (Figure 4.24), and understand the function of a few of its parts. A good way to remember this is to follow the path of light as it enters the eye.

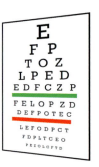

Light reflects off an object in the external environment and first strikes the eye on the **cornea**. (The light first strikes a very thin, protective layer known as the corneal epithelium.) The cornea is nonvascular and made largely from collagen. It is clear with a refractive index of about 1.4, which means that the most bending of light actually occurs at the interface of the air and the cornea and not at the lens.

From the cornea, the light enters the *anterior cavity*, which is filled with *aqueous humor*. Aqueous humor is formed by the *ciliary processes* and leaks out the *canal of Schlemm*. Blockage of the canal of Schlemm increases intraocular pressure resulting in one form of *glaucoma* and possibly blindness.

From the anterior cavity, light enters the **lens**. The lens would have a spherical shape, but stiff suspensory ligaments tug on it and tend to flatten it. These ligaments are connected to the **ciliary muscle**. The ciliary muscle circles the lens. When the ciliary muscle contracts, the opening of the circle decreases allowing the lens to become more like a sphere and bringing its focal point closer to the lens; when the muscle relaxes, the lens flattens increasing the focal distance. The elasticity of the lens declines with age making it difficult to focus on nearby objects as one gets older.

The eye system just described focuses light through the gel-like *vitreous humor* and onto the retina. Since the eye acts as a converging lens, and the object is outside the focal distance, the image on the retina is real and inverted. (See Physics Lecture 8 for more on lenses.)

The **retina** covers the inside of the back (distal portion) of the eye. It contains light sensitive cells called **rods** and **cones**. These cells are named for their characteristic shapes. The tips of these cells contain light sensitive photochemicals called *pigments* that go through a chemical change when one of their electrons is struck by a single photon. The pigment in rod cells is called *rhodopsin*. Rhodopsin is made of a protein bound to a prosthetic group called *retinal* which is derived from vitamin A. The photon isomerizes retinal causing the membrane of the rod cell to become less permeable to sodium ions and hyperpolarize. The hyperpolarization

Very little knowledge concerning the sensory receptors is required for the MCAT. However, you should know some very basic anatomy of the eye and ear. Also realize that sensory receptors transduce physical stimulus to neural signals.

is transduced into a neural action potential and the signal is sent to the brain.

Rods sense all photons with wavelengths in the visible spectrum (390 nm to 700 nm). Thus rods cannot distinguish colors. There are three types of cones, each with a different pigment that is stimulated by a slightly different spectrum of wavelengths. Thus cones distinguish colors. Vitamin A is a precursor to all the pigments in rods and cones.

The *fovea* is a small point on the retina containing mostly cones. The fovea marks the point on the retina where vision is most acute.

One other feature of the eye with which you should be familiar is the **iris** (Greek: irid:colored circle). The iris is the colored portion of the eye that creates the opening called the **pupil**. The iris is made from circular and radial muscles. In a dark environment, the sympathetic nervous system contracts the iris dilating the pupil and allowing more light to enter the eye. In a bright environment, the parasympathetic nervous system contracts the circular muscles of the iris constricting the pupil and screening out light.

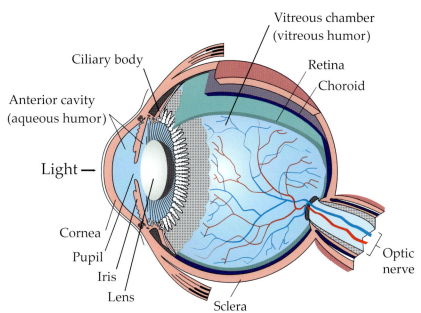

Figure 4.24 The Eye

Remember that cones distinguish colors and rods don't.

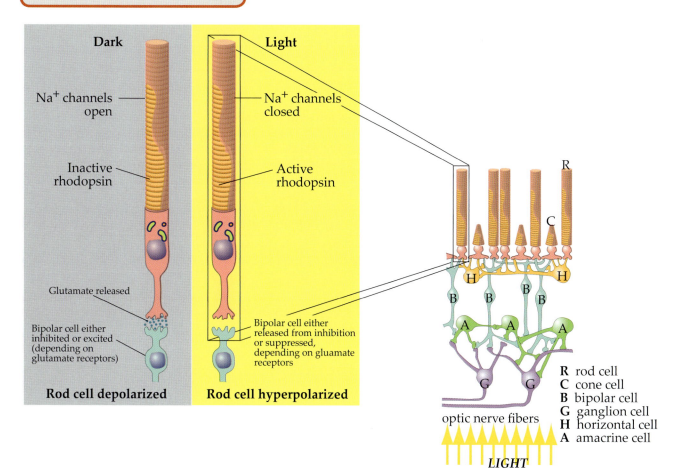

4.18 The Ear

Like the eye, you should know the basic parts of the ear (Figure 4.25) and the functions of these parts. The ear is divided into three parts:

1. the **outer ear**;
2. the **middle ear**; and
3. the **inner ear**.

Following the path of a sound wave through the ear can be helpful in remembering the parts.

The *auricle* or *pinna* is the skin and cartilage flap that is commonly called the ear. The auricle functions to direct the sound wave into the *external auditory canal*. The external auditory canal carries the wave to the **tympanic membrane** or eardrum. The tympanic membrane begins the middle ear.

The middle ear contains the three small bones: the **malleus**, the **incus**, and the **stapes**. These three small bones act as a lever system translating the wave to the *oval window*. Like any lever system, these bones change the combination of force and displacement from the inforce to the outforce. The displacement is actually lessened, which creates an increase in force. In addition, the oval window is smaller than the tympanic membrane, acting to increase the pressure. (See Physics Lecture 4 for more on machines and mechanical advantage.) This increase in force is necessary because the wave is being transferred from the air in the outer ear to a more resistant fluid (the *perilymph*) within the inner ear.

The wave in the inner ear moves through the *scala vestibuli* of the **cochlea** to the center of the spiral, and then spirals back out along the *scala tympani* to the *round window*. As the wave moves through the cochlea, the alternating increase and decrease in pressure moves the *vestibular membrane* in and out. This movement is detected by the **hair cells** of the **organ of Corti** and transduced into neural signals, which are sent to the brain. The hair cells do not actually contain hair, but contain instead a specialized microvilli called *stereocilia*, which detect movement.

Also in the inner ear are the **semicircular canals**. The semicircular canals are responsible for balance. Each canal contains fluid and hair cells. When the body moves or the head position changes with respect to gravity, the momentum of the fluid is changed impacting on the hair cells, and the body senses motion. The canals are oriented at right angles to each other, in order to detect movement in all directions.

4.19 The Nose and Mouth

The senses of smell and taste are called *olfactory* and *gustatory*, respectively. These senses involve chemoreceptors. Different chemoreceptors sense different chemicals. There are only four primary taste sensations:

1. bitter
2. sour
3. salty and
4. sweet.

All taste sensations are combinations of these four.

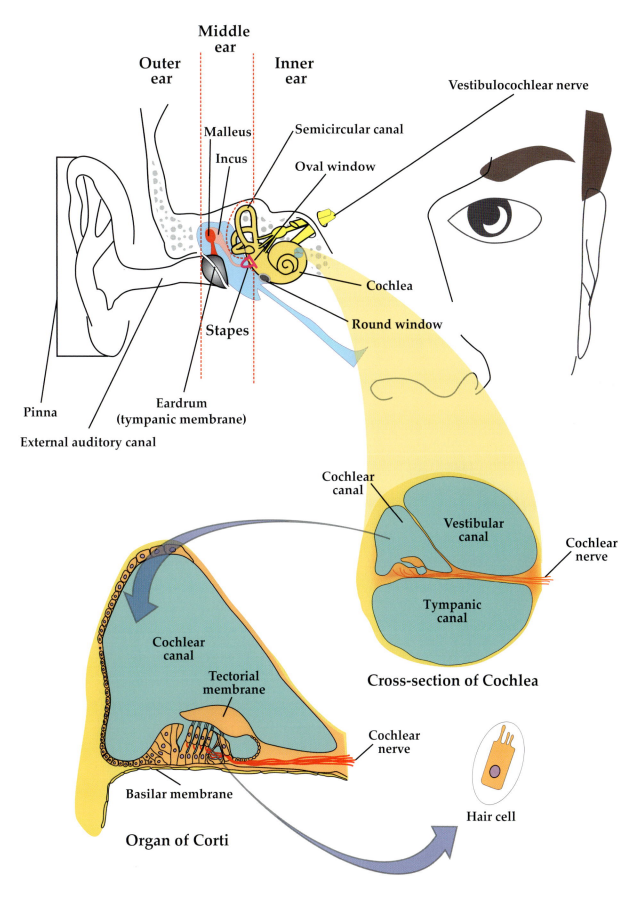

Figure 4.25 The Ear

Terms	
Acetylcholine	Hyperpolarization
Actin	Hypothalamus
Action potential	Incus
Adipocytes	Inner Ear
Autonomic Nervous System (ANS)	Interneurons
Axon	Interstitial Fluid
Axon Hillock	Iris
Central Nervous System (CNS)	Lens
Centriole	Lysosomes
Cerebellum	Malleus
Cerebral Cortex	Medulla
Cerebrum	Microfilaments
Chemical Synapse	Microtubules
Cilia	Middle Ear
Cochlea	Mitochondria
Cones	Mitotic Spindle
Cornea	Motor (Efferent) Neurons
Cytoskeleton	Myelin
Cytosol	Neuron
Dendrites	Neuronal Communication
Depolarization	Neurotransmitter
Desmosomes	Nerves
Dynein	Nodes of Ranvier
Effector	Norepinephrine
Electrical Synapses	Nuclear Envelope
Endocytosis	Nuclear Pores
Endoplasmic Reticulum (ER)	Nucleus
Endosymbiont Theory	Nucleolus
Epinephrine	Organ of Corti
Exocytosis	Outer Ear
Flagella	Parasympathetic
Gap Junctions	Peripheral Nervous
Golgi Apparatus	System (PNS)
Grey Matter	Peroxisomes
Hair Cells	Phagocytosis
Hormonal Communication	Pupil

Terms You Need To Know, Continued...

Terms	
Repolarization	Sympathetic
Resting Potential	Synapse
Retina	Synaptic Cleft
Rods	Thalamus
Saltatory Conduction	Threshold Stimulus
Schwann Cell	Tight Junctions
Second Messenger System	Tubulin
Secretory Vesicles	Tympanic Membrane
Semicircular Canal	Voltage Gated Potassium Channels
Sensory (afferent) Neurons	Voltage Gated Sodium Channels
Somatic Nervous System	White Matter
Stapes	

Notes:

89. Which of the following activities is controlled by the cerebellum?

   A. Involuntary breathing movements
   B. Fine muscular movements during a dance routine
   C. Contraction of the thigh muscles during the knee-jerk reflex
   D. Absorption of nutrients across the microvilli of the small intestine

90. If an acetylcholine antagonist were administered generally into a person, all of the following would be affected EXCEPT:

   A. the neuroeffector synapse in the sympathetic nervous system.
   B. the neuroeffector synapse in the parasympathetic nervous system
   C. the neuromuscular junction in the somatic nervous system.
   D. the ganglionic synapse in the sympathetic nervous system.

91. Which of the following occurs as a result of parasympathetic stimulation?

   A. Vasodilation of the arteries leading to the kidneys
   B. Increased rate of heart contraction.
   C. Piloerection of the hair cells of the skin.
   D. Contraction of the abdominal muscles during exercise.

92. Pressure waves in the air are converted to neural signals at the:

   A. retina
   B. tympanic membrane
   C. cochlea
   D. semicircular canals

93. Reflex arcs:

   A. involve motor neurons exiting the spinal cord dorsally.
   B. require fine control by the cerebral cortex.
   C. always occur independently of the central nervous system.
   D. often involve inhibition as well as excitation of muscle groups.

94. Which of the following structures is NOT part of the central nervous system?

   A. a parasympathetic effector
   B. the medulla
   C. the hypothalamus
   D. the cerebral cortex

95. Which part of the brain controls higher-level thought processes?

   A. the thalamus
   B. the cerebellum
   C. the cerebrum
   D. the medulla

96. A cook touches a hot stove and involuntarily withdraws his hand before he feels pain. Which of the following would not be involved in the stimulus-response pathway described?

   A. a neuron in the cerebellum
   B. a neuron in the spinal cord
   C. a motor neuron
   D. a sensory neuron

THE ENDOCRINE SYSTEM

5.1 Hormone Chemistry

The neurotransmitters and local mediators discussed in Biology Lecture 4 are often referred to as local hormones. General hormones are the hormones released by the endocrine system. They are referred to as 'general' because they are released into the body fluids, often the blood, and may affect many cell types in a tissue, and multiple tissues in the body. Although the following section concentrates on general hormones, the chemistry described is accurate for local hormones as well.

The endocrine glands differ from **exocrine glands** in the following manner. Exocrine glands release enzymes to the external environment through ducts. Exocrine glands include *sudoriferous* (sweat), *sebaceous* (oil), *mucous*, and digestive glands. **Endocrine glands** release hormones directly into body fluids. For instance, the pancreas acts as both an exocrine gland, releasing digestive enzymes through the pancreatic duct, and an endocrine gland releasing insulin and glucagon directly into the blood. (See Biology Lecture 6 for more on the pancreas.)

The effects of the endocrine system tend to be slower, less direct, and longer lasting than those of the nervous system. Endocrine hormones may take anywhere from seconds to days to produce their effects. They do not move directly to their target tissue, but released into the general circulation. All hormones act by binding to proteins called **receptors**. Each receptor is highly specific for its hormone. One method of hormone regulation occurs by the reduction or increase of these receptors in the presence of high or low concentrations of the hormone. Some hormones have receptors on virtually all cells, while other hormones have receptors only on specific tissues. Very low concentrations of hormones in the blood have significant effects on the body.

In general, the effects of the endocrine system are to alter metabolic activities, regulate growth and development, and guide reproduction. The endocrine system works in conjunction with the nervous system. Many endocrine glands are stimulated by neurons to secrete their hormones.

Hormones exist in three basic chemistry types:

1. **peptide hormones**;
2. **steroid hormones**; and
3. **tyrosine derivatives**.

Always remember that hormones need a receptor, either on the membrane or inside the cell. Also, when comparing the endocrine system with the nervous system remember that the endocrine system is slow, indirect, and long lasting.

Figure 5.1 Endocrine and exocrine glands

Exocrine gland
Releases enzymes to external environments through ducts

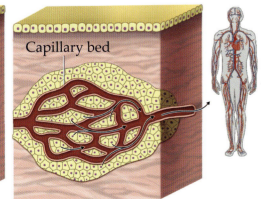

Capillary bed

Endocrine gland
Releases hormones into fluids that circulate throughout the body

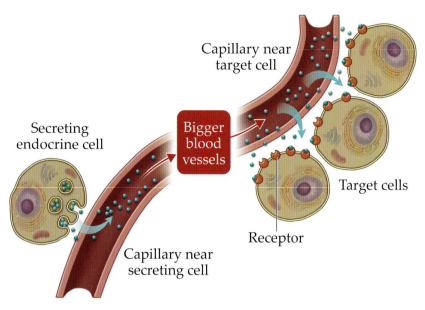

Capillary near target cell

Secreting endocrine cell

Bigger blood vessels

Target cells

Receptor

Capillary near secreting cell

Figure 5.2 Transport of Hormones in the Blood

The endocrine system delivers hormones to target cells throughout the body via the circulatory system.

Peptide hormones are derived from peptides. They may be large or small, and often include carbohydrate portions. All peptide hormones are manufactured in the rough ER, typically as a *preprohormone* that is larger than the active hormone. The preprohormone is cleaved in the ER lumen to become a *prohormone*, and transported to the Golgi apparatus. In the Golgi, the prohormone is cleaved and sometimes modified with carbohydrates to its final form. The Golgi packages the hormone into secretory vesicles, and, upon stimulation by another hormone or a nervous signal, the cell releases the vesicles via exocytosis.

Since they are peptide derivatives, peptide hormones are water soluble, and thus move freely through the blood, but have difficulty diffusing through the cell membrane of the **effector**. (The effector is the **target cell** of the hormone, the cell that the hormone is meant to affect.) Instead of diffusing through the membrane, peptide hormones attach to a membrane-bound receptor. Once bound by a hormone, the receptor may act in several ways. The receptor may itself act as an ion channel increasing membrane permeability to a specific ion, or the receptor may activate or deactivate other intrinsic membrane proteins also acting as ion channels. Another effect of the hormone binding to the receptor may be to activate an **intracellular second messenger** such as cAMP, cGMP, or calmodulin. These chemicals are called second messengers because the hormone is the original, or first messenger, to the cell. The second messenger activates or deactivates enzymes and/or ion channels and often creates a 'cascade' of chemical reactions that amplifies the effect of the hormone. A cascade is one way that a small concentration of hormone can have a significant effect.

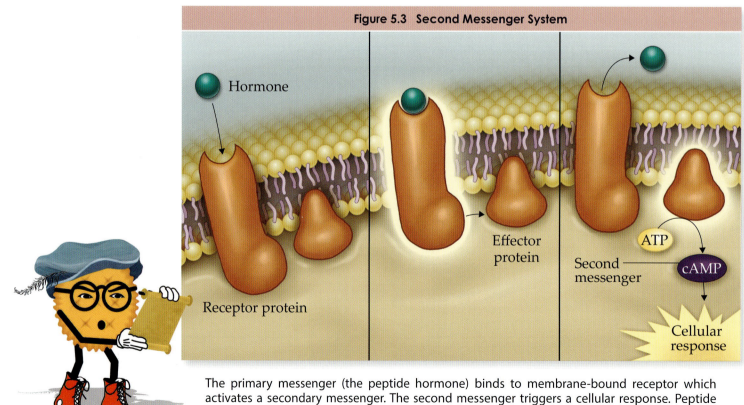

Figure 5.3 Second Messenger System

Hormone

Effector protein

ATP

Second messenger

cAMP

Cellular response

Receptor protein

The primary messenger (the peptide hormone) binds to membrane-bound receptor which activates a secondary messenger. The second messenger triggers a cellular response. Peptide hormones typically utilize a secondary messenger system.

2nd messenger Salty

The peptide hormones that you must know for the MCAT are:

1. **the anterior pituitary hormones: FSH, LH, ACTH, hGH, TSH, Prolactin**;
2. **the posterior pituitary hormones: ADH and oxytocin**;
3. **the parathyroid hormone PTH**;
4. **the pancreatic hormones: glucagon and insulin**.

The specifics of these hormones will be discussed later in this lecture.

Steroid hormones are derived from and are often chemically similar to cholesterol. They are formed in a series of steps taking place mainly in the smooth endoplasmic reticulum and the mitochondria. Since they are lipids, steroids typically require a protein transport molecule (carrier protein) in order to dissolve into the blood stream. (Usually, a fraction of the steroid concentration is bound to a transport molecule and a fraction is freeform in the blood.) Being lipid soluble, steroids diffuse through the cell membrane of their effector. Once inside the cell, they combine with a receptor in the cytosol. The receptor transports the steroid into the nucleus, and the steroid acts at the transcription level. Thus, the typical effect of a steroid hormone is to increase certain membrane or cellular proteins within the effector.

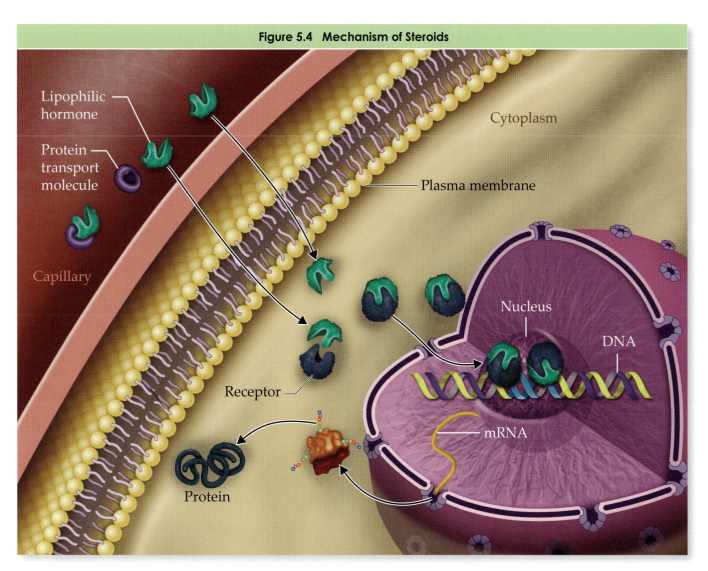

Figure 5.4   Mechanism of Steroids

Non-polar hormones such as steroids and some tyrosine-derivatives are transported in the blood by carrier proteins. Once they reach their target cell, they freely diffuse across the membrane and attach to receptors. The receptor-hormone complex then moves to the nucleus where it regulates transcription of certain genes or gene families.

The important steroid hormones for the MCAT are:

1. **the glucocorticoids and mineral corticoids of the adrenal cortex: cortisol and aldosterone**;

2. **the gonadal hormones: estrogen, progesterone, testosterone**. (Estrogen and progesterone are also produced by the placenta.) The specifics of these steroids will be discussed later in this lecture.

The **tyrosine derivatives** are: **the thyroid hormones T$_3$ (triiodothyronine contains 3 iodine atoms) and T$_4$ (thyroxine contains 4 iodine atoms), and the catecholamines formed in the adrenal medulla: epinephrine and norepinephrine**. All tyrosine derivative hormones are formed by enzymes in the cytosol or on the rough ER.

Thyroid hormones are lipid soluble and must be carried in the blood by plasma protein carriers. They are slowly released to their target tissues and bind to receptors inside the nucleus. Their high affinity to their binding proteins in the plasma and in the nucleus create a latent period in their response and increase the duration of the effect of thyroid hormones. Thyroid hormones increase the transcription of large numbers of genes in nearly all cells of the body.

Epinephrine and norepinephrine are water soluble and dissolve in the blood. They bind to receptors on the target tissue and act mainly through the second messenger cAMP.

The specifics of the tyrosine derivative hormones will be discussed later in this lecture.

You should know which hormones are steroids, which are tyrosines, and which are peptides. Then you should know where and how each of these types of hormones reacts. This isn't really so tough. First, steroid hormones come only from the adrenal cortex, the gonads, or the placenta. Second, tyrosines are the thyroid hormones and the catecholamines (the adrenal medulla hormones). The rest of the hormones discussed in this book are peptide hormones. Since steroids are lipids, they diffuse through the membrane and act in the nucleus. Since peptides are proteins, they can't diffuse through the membrane, so they bind to receptors on the membrane and act through a second messenger. The tyrosines are split: thyroid hormones diffuse into the nucleus and catecholamines act on receptors at the membrane.

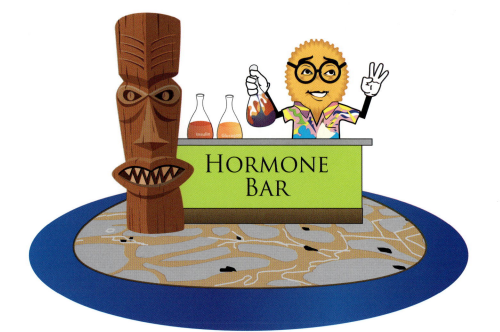

5.2 Negative Feedback

Endocrine glands tend to over-secrete their hormones. Typically, some aspect of their effect on the target tissue will inhibit this secretion. This is an example of negative feedback (discussed in Biology Lecture 1). An important aspect to understand about negative feedback in endocrine glands is that the control point of the feedback is the conduct of the effector, not the concentration of hormone. In other words, the gland lags behind the effector. For instance, high insulin levels do not typically create low blood glucose. Instead, high insulin levels are caused by high blood glucose, and low blood glucose would cause high blood glucagon levels. So if an MCAT question indicates that a patient has high blood glucose and asks whether high levels or insulin or high levels of glucagon would be expected, the correct answer is the hormone that is responding to the condition, not creating it; in this case insulin.

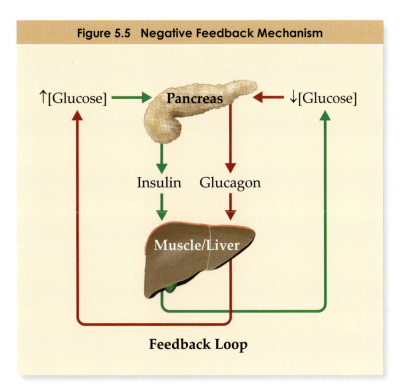

Figure 5.5   Negative Feedback Mechanism

$\uparrow$[Glucose] → **Pancreas** ← $\downarrow$[Glucose]

Insulin   Glucagon

Muscle/Liver

Feedback Loop

There will be a negative feedback question on the MCAT. See if you get the idea. If ADH holds water in the body decreasing urine output and increasing blood pressure, does a person with high blood pressure (holding water) have a high ADH blood level or a low ADH blood level?

If you said high ADH, then you reasoned that the ADH created the high blood pressure. WRONG! If you said low ADH, then you reasoned that the ADH output responded to the body. CORRECT!

How about another? A secondary effect of aldosterone is to increase blood pressure. Would expected aldosterone levels be high or low in a person with low blood pressure?

The answer: since aldosterone increases blood pressure, and the body tries to bring blood pressure back to normal, the adrenal cortex should release more aldosterone into the blood. Expected aldosterone levels would be higher than normal.

97. Aldosterone exerts its effects on target cells by:

    A. binding to a receptor at the cell surface, setting off a second-messenger cascade.
    B. diffusing into adrenal cortical cells, where it influences transcription of certain DNA sequences.
    C. flowing across the synapse, where it binds and initiates an action potential.
    D. entering into target cells, where it increases the rate of production of sodium-potassium pump proteins.

98. A patient develops an abdominal tumor resulting in the secretion of large quantities of aldosterone into the bloodstream. Which of the following will most likely occur?

    A. Levels of renin secreted by the kidney will increase.
    B. Levels of oxytocin secreted by the pituitary will increase.
    C. Levels of aldosterone secreted by the adrenal cortex will decrease.
    D. Levels of aldosterone secreted by the tumor will decrease.

99. Which of the following is true for all endocrine hormones?

    A. They act through a second messenger system.
    B. They bind to a protein receptor.
    C. They dissolve in the blood.
    D. They are derived from a protein precursor.

100. All of the following act as second messengers for hormones EXCEPT:
    A. cyclic AMP.
    B. calmodulin.
    C. acetylcholine.
    D. cyclic GMP.

101. Which of the following is true of all steroids?

    A. The target cells of any steroid include every cell in the body.
    B. Steroids bind to receptor proteins on the membrane of their target cells.
    C. Steroids are synthesized on the rough endoplasmic reticulum.
    D. Steroids are lipid soluble.

102. The pancreas is a unique organ because it has both exocrine and endocrine function. The exocrine function of the pancreas releases:

    A. digestive enzymes straight into the blood.
    B. digestive enzymes through a duct.
    C. hormones straight into the blood.
    D. hormones through a duct.

103. Most steroid hormones regulate enzymatic activity at the level of:

    A. replication.
    B. transcription.
    C. translation.
    D. the reaction.

104. Which of the following side-effects might be experienced by a patient who is administered a dose of thyroxine?

    A. an increase in endogenous TSH production
    B. a decrease in endogenous TSH production
    C. an increase in endogenous thyroxine production
    D. a decrease in endogenous parathyroid hormone production

5.3 Specific Hormones and Their Function

Memorization of several major hormones, their glands, and their target tissues is required for the MCAT. As a memory aid, you should group hormones according to the gland that secretes them. A given gland produces steroids, peptides, or tyrosine derivatives, but not two categories of hormones. (The adrenal glands are really two glands. The cortex produces steroids; the medulla produces catecholamines. The thyroid is a true exception. The thyroid secretes T_3 and T_4, which are tyrosine derivatives, and calcitonin, which is a peptide.) We will start by discussing the hormones of the anterior pituitary.

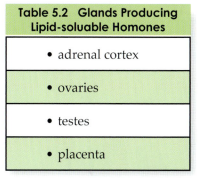

Table 5.1 Glands Producing Water-soluable Homones
• anterior pituitary
• posterior pituitary
• parathyroid
• pancreas
• adrenal medulla

5.4 Anterior Pituitary

The **anterior pituitary** (Figure 5.6) (also called the *adenohypophysis*) is located in the brain beneath the **hypothalamus**. The hypothalamus controls the release of the anterior pituitary hormones with *releasing and inhibitory hormones* of its own. These releasing and inhibitory hormones are carried to the capillary bed of the anterior pituitary by small blood vessels. The release of the releasing and inhibitory hormones is, in turn, controlled by nervous signals throughout the nervous system.

The anterior pituitary releases six major hormones and several minor hormones. All of these are peptide hormones. For the MCAT you should be familiar with the six major hormones, their target tissues, and their functions. The hormones are:

Table 5.2 Glands Producing Lipid-soluable Homones
• adrenal cortex
• ovaries
• testes
• placenta

1. human growth hormone (hGH);
2. adrenocorticotropin (ACTH);
3. thyroid-stimulating hormone (TSH);
4. follicle-stimulating hormone (FSH);
5. luteinizing hormone (LH) and;
6. prolactin.

Figure 5.6 Anterior pituitary

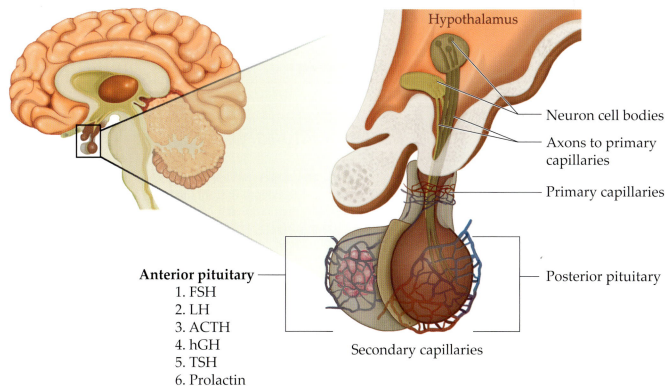

Hypothalamus

Neuron cell bodies

Axons to primary capillaries

Primary capillaries

Posterior pituitary

Secondary capillaries

Anterior pituitary
1. FSH
2. LH
3. ACTH
4. hGH
5. TSH
6. Prolactin

hGH

Human growth hormone (hGH) (also called *somatotropin*), a peptide, stimulates growth in almost all cells of the body. All other hormones of the anterior pituitary have specific target tissues. hGH stimulates growth by increasing episodes of mitosis, increasing cell size, increasing the rate of protein synthesis, mobilizing fat stores, increasing the use of fatty acids for energy, and decreasing the use of glucose. The effect on proteins by hGH is accomplished by increasing amino acid transport across the cell membrane, increasing translation and transcription, and decreasing the breakdown of protein and amino acids.

ACTH

Adrenocorticotropic hormone (ACTH), a peptide, stimulates the adrenal cortex to release glucocorticoids via the second messenger system using cAMP. Release of ACTH is stimulated by many types of stress. Glucocorticoids are stress hormones. (See below for the effects of the adrenal cortical hormones.)

TSH

Thyroid-stimulating hormone (TSH) (also called thyrotropin), a peptide, stimulates the thyroid to release T_3 and T_4 via the second messenger system using cAMP. Among other effects on the thyroid, TSH increases thyroid cell size, number, and the rate of secretion of T_3 and T_4. It is important to note that T_3 and T_4 concentrations have a negative feedback effect on TSH release, both at the anterior pituitary and the hypothalamus. (See below for effects of T_3 and T_4.)

FSH and LH

(These peptides are discussed in this lecture under reproduction.)

Prolactin

Prolactin, a peptide, promotes lactation (milk production) by the breasts. The reason that milk is not normally produced before birth is due to the inhibitory effects of milk production by progesterone and estrogen. Although the hypothalamus has a stimulatory effect on the release of all other anterior pituitary hormones, it mainly inhibits the release of prolactin. The act of suckling, which stimulates the hypothalamus to stimulate the anterior pituitary to release prolactin, inhibits the menstrual cycle. It is not known whether or not this is directly due to prolactin. The milk production effect of prolactin should be distinguished from the milk ejection effect of oxytocin.

5.5 Posterior Pituitary

The **posterior pituitary** is also called the *neurohypophysis* because it is composed mainly of support tissue for nerve endings extending from the hypothalamus. The hormones oxytocin and ADH are synthesized in the neural cell bodies of the hypothalamus, and transported down axons to the posterior pituitary where they are released into the blood. Both oxytocin and ADH are small polypeptides.

Oxytocin

Oxytocin is a small peptide hormone that increases uterine contractions during pregnancy and causes milk to be ejected from the breasts.

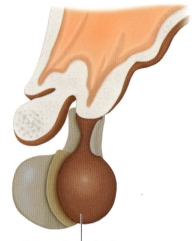

Posterior pituitary
1. Oxytocin
2. ADH

Figure 5.7   Posterior pituitary

ADH

Antidiuretic hormone (ADH) (also called **vasopressin**) is a small peptide hormone which causes the collecting ducts of the kidney to become permeable to water reducing the amount of urine and concentrating the urine. Since fluid is reabsorbed, ADH also increases blood pressure. Coffee and beer are ADH blockers that increase urine volume.

5.6 | *Adrenal Cortex*

The **adrenal glands** (Figure 5.8) are located on top of the kidneys. They are generally separated into the adrenal cortex and the adrenal medulla. The **adrenal cortex** is the outside portion of the gland. The cortex secretes only steroid hormones. There are two types of steroids secreted by the cortex: **mineral corticoids** and **glucocorticoids**. (The cortex also secretes a small amount of sex hormones, significant in the female but not the male.) Mineral corticoids affect the electrolyte balance in the blood stream; glucocorticoids increase blood glucose concentration and have an even greater effect on fat and protein metabolism. About 30 corticoids have been isolated from the cortex, but the major mineral corticoid is aldosterone, and the major glucocorticoid is cortisol.

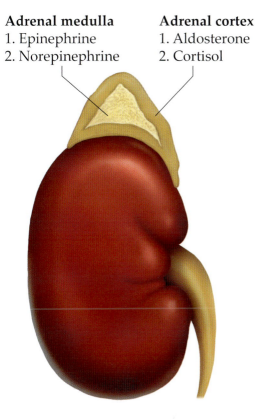

Adrenal medulla
1. Epinephrine
2. Norepinephrine

Adrenal cortex
1. Aldosterone
2. Cortisol

Aldosterone

Aldosterone, a steroid, is a mineral corticoid that acts in the distal convoluted tubule and the collecting duct to increase Na^+ and Cl^- reabsorption and K^+ and H^+ secretion. It creates a net gain in particles in the plasma, which results in an eventual increase in blood pressure. Aldosterone has the same effect, but to a lesser extent, on the sweat glands, salivary glands, and intestines.

For the MCAT, the main effect of aldosterone is the Na^+ reabsorption and K^+ secretion in the collecting tubule of the kidney. The increase in blood pressure is a secondary effect.

Figure 5.8 Adrenal Gland

Cortisol

Cortisol, a steroid, is a glucocorticoid that increases blood glucose levels by stimulating **gluconeogenesis** in the liver. (Gluconeogenesis is the creation of glucose and glycogen, mainly in the liver, from amino acids, glycerol, and/or lactic acid.) Cortisol also degrades adipose tissue to fatty acids to be used for cellular energy. In addition, cortisol causes a moderate decrease in the use of glucose by the cells. Cortisol causes the degradation of nonhepatic proteins, a decrease of nonhepatic amino acids and a corresponding increase in liver and plasma proteins and amino acids.

Cortisol is a stress hormone. The benefit of excess cortisol under stressful situations is not fully understood. One explanation may include anti-inflammatory properties possessed by cortisol. Cortisol also diminishes the capacity of the immune system to fight infection.

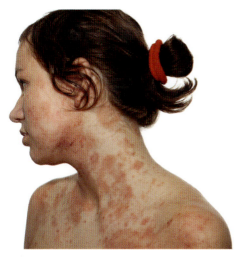

The anti-inflammatory properties of cortisal make it useful for treating eczema.

Catecholamines

The *catecholamines* are the tyrosine derivatives synthesized in the adrenal medulla: **epinephrine** and **norepinephrine** (also called **adrenaline** and **noradrenaline**). The effects of epinephrine and norepinephrine on the target tissues are similar to their effects in the sympathetic nervous system but they last much longer. Epinephrine and norepinephrine are vasoconstrictors (they constrict blood vessels) of most internal organs and skin, but are vasodilators of skeletal muscle (they increase blood flow); this is consistent with the 'fight-or-flight' response of these hormones. Because of their 'fight or flight' response, the catecholamines are also considered stress hormones.

5.7 Thyroid

The thyroid hormones are *triiodothyronine* (T_3), thyroxine (T_4), and calcitonin. The thyroid (Figure 5.9) is located along the trachea just in front of the larynx.

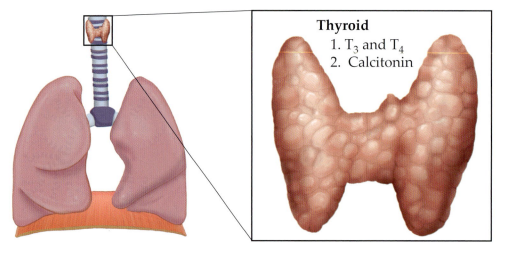

Figure 5.9 Thyroid Hormones

Thyroid
1. T_3 and T_4
2. Calcitonin

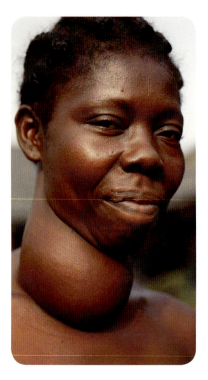

Goiter, which is caused by a dietary deficiency of iodine, results in decreased thyroxine production. Sensing low thyroxine levels, the anterior pituitary responds by releasing excess TSH, which stimulates abnormal growth of the thyroid gland.

Calcitonin works in opposition to PTH, just like insulin works in opposition to glucagon.

T_3 and T_4

T_3 and T_4 are very similar in effect, and no distinction will be made on an MCAT question unless it is thoroughly explained in a passage. T_3 contains three iodine atoms, and T_4 contains four. Both hormones are lipid soluble tyrosine derivatives that diffuse through the lipid bilayer and act in the nucleus of the cells of their effector. Their general effect is to increase the **basal metabolic rate** (the resting metabolic rate). Thyroid hormone secretion is regulated by TSH.

Calcitonin

Calcitonin is a large peptide hormone released by the thyroid gland. Calcitonin slightly decreases blood calcium by decreasing osteoclast activity and number. Calcium levels can be effectively controlled in humans in the absence of calcitonin.

5.8 Pancreas (Islets of Langerhans)

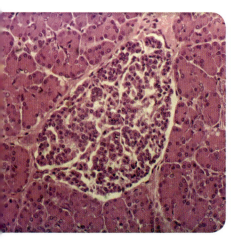

Pancreatic islets of Langerhans release hormones into the blood. They are composed of numerous beta cells, which secrete insulin, and the less numerous alpha cells, which secrete glucagon.

The pancreas (Figure 5.10) acts as both an endocrine and an exocrine gland. For the MCAT, the two important endocrine hormones released into the blood by the pancreas are the peptide hormones insulin and glucagon. *Somatostatin*, not likely to be seen on the MCAT, is released by the δ-cells of the pancreas. Somatostatin inhibits both insulin and glucagon. The role of somatostatin may be to extend the period of time over which nutrients are absorbed.

Insulin

Insulin, a peptide hormone, is released by the *β-cells* of the pancreas. It is associated with energy abundance in the form of high energy nutrients in the blood. Insulin is released when blood levels of carbohydrates or proteins are high. It affects carbohydrate, fat, and protein metabolism. In the presence of insulin, carbohydrates are stored as glycogen in the liver and muscles, fat is stored in adipose tissue, and amino acids are taken up by the cells of the body and made into proteins. The net effect of insulin is to lower blood glucose levels.

Insulin binds to a membrane receptor beginning a cascade of reactions inside the cell. Except for neurons in the brain and a few other cells which are not affected by insulin, the cells of the body become highly permeable to glucose upon the binding of insulin. The insulin receptor itself is not a carrier for glucose. The permeability of the membrane to amino acids is also increased. In addition, intracellular metabolic enzymes are activated and, much more slowly, even translation and transcription rates are affected.

Glucagon

Glucagon, a peptide hormone, is released by the *α-cells* of the pancreas. The effects of glucagon are nearly opposite to those of insulin. Glucagon stimulates glycogenolysis (the breakdown of glycogen), and gluconeogenesis in the liver. It acts via the second messenger system of cAMP. In higher concentrations, glucagon breaks down adipose tissue increasing the fatty acid level in the blood. The net effect of glucagon is to raise blood glucose levels.

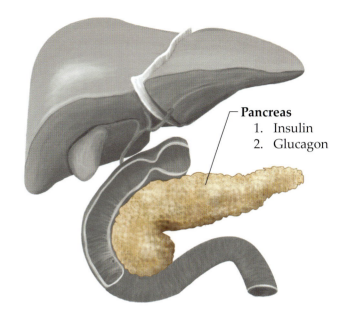

Pancreas
1. Insulin
2. Glucagon

Figure 5.10 Pancreatic Hormones

> Remember, only the pancreas acts as both an endocrine and an exocrine gland.

> In Type I diabetes the islets of Langerhans do not produce insulin.

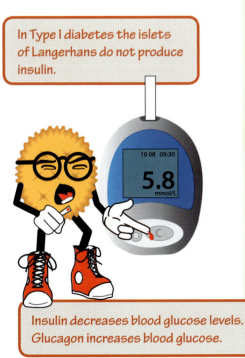

> **Insulin** decreases blood glucose levels.
> **Glucagon** increases blood glucose.

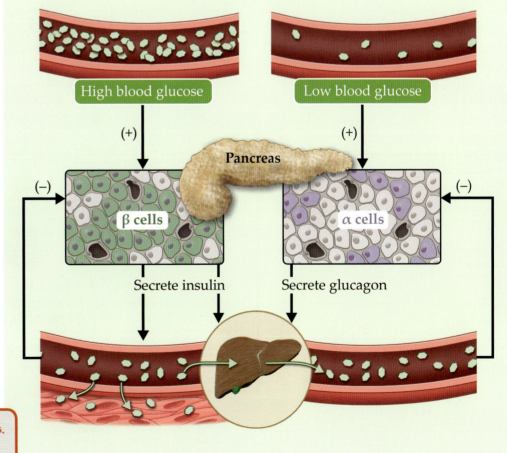

Figure 5.11 The Antagonistic Effects of Insulin and Glucagon

High blood glucose

Low blood glucose

(+)

(+)

Pancreas

(−)

(−)

β cells

α cells

Secrete insulin

Secrete glucagon

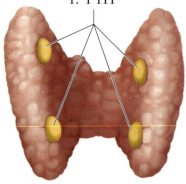

Parathyroid
(behind Thyroid)
1. PTH

Figure 5.12   Parathyroid

Parathyroid

There are four small **parathyroid glands** (Figure 5.12) attached to the back of the thyroid. The parathyroid glands release **parathyroid hormone**.

PTH

Parathyroid hormone (PTH), a peptide, increases blood calcium. It increases osteocyte absorption of calcium and phosphate from the bone and stimulates proliferation of osteoclasts. PTH increases renal calcium reabsorption and renal phosphate excretion. It increases calcium and phosphate uptake from the gut by increasing renal production of the steroid, *1,25 dihydroxycholecalciferol (DOHCC)*, derived from vitamin D. PTH secretion is regulated by the calcium ion plasma concentration, and the parathyroid glands shrink or grow accordingly.

> As with other hormones, PTH has multiple effectors and its effect may be different for each effector.

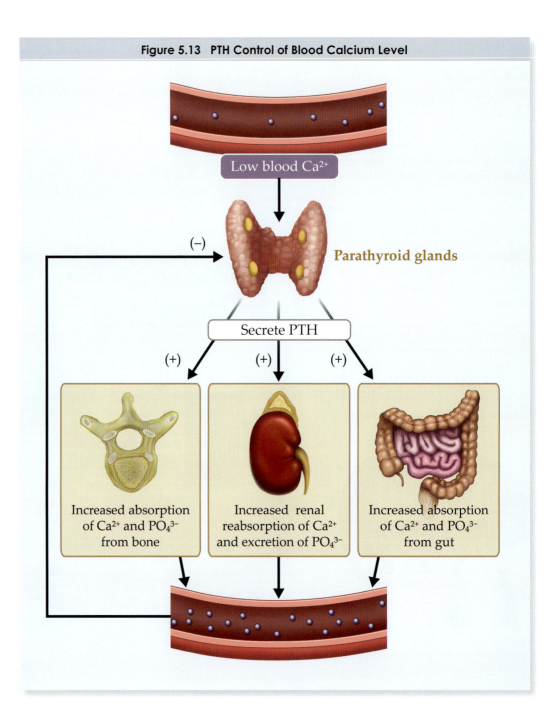

Figure 5.13   PTH Control of Blood Calcium Level

Low blood Ca^{2+}

(−)

Parathyroid glands

Secrete PTH

(+)   (+)   (+)

Increased absorption of Ca^{2+} and PO_4^{3-} from bone

Increased renal reabsorption of Ca^{2+} and excretion of PO_4^{3-}

Increased absorption of Ca^{2+} and PO_4^{3-} from gut

105. Sympathetic stimulation results in responses most similar to release of which of the following hormones?
    A. insulin
    B. acetylcholine
    C. epinephrine
    D. aldosterone

106. When compared with the actions of the nervous system, those of the endocrine system are:

    A. quicker in responding to changes, and longer-lasting.
    B. quicker in responding to changes, and shorter-lasting.
    C. slower in responding to changes, and longer-lasting.
    D. slower in responding to changes, and shorter-lasting.

107. Insulin shock occurs when a patient with diabetes self-administers too much insulin. Typical symptoms are extreme nervousness, trembling, sweating, and ultimately loss of consciousness. The physiological effects of insulin shock most likely include:

    A. a pronounced increase in gluconeogenesis by the liver.
    B. a rise in blood fatty acid levels leading to atherosclerosis.
    C. a dramatic rise in blood pressure.
    D. dangerously low blood glucose levels.

108. Vasopressin, a hormone involved in water balance, is produced in the:

    A. hypothalamus.
    B. posterior pituitary.
    C. anterior pituitary.
    D. kidney.

109. Osteoporosis is an absolute decrease in bone tissue mass, especially trabecular bone. All of the following might be contributory factors to the disease EXCEPT:

    A. increased sensitivity to endogenous parathyroid hormone.
    B. defective intestinal calcium absorption.
    C. menopause.
    D. abnormally high blood levels of calcitonin.

110. All of the following hormones are produced by the anterior pituitary EXCEPT:

    A. thyroxine.
    B. growth hormone.
    C. prolactin.
    D. leutinizing hormone.

111. Which of the following hormonal and physiologic effects of stress would NOT be expected in a marathoner in the last mile of a marathon?

    A. Increased glucagon secretion
    B. Increased heart rate
    C. Decreased ACTH secretion
    D. Decreased blood flow to the small intestine

112. Parathyroid hormone is an important hormone in the control of blood calcium ion levels. Parathyroid hormone directly impacts:

    I. bone density
    II. renal calcium reabsorption
    III. blood calcium concentration

    A. I only
    B. I and II only
    C. I and III only
    D. I, II and III

5.10 Reproduction

Except for **FSH**, **LH**, **HCG**, and *inhibin*, which are peptides, the reproductive hormones discussed below are steroids released from the testes, ovaries and placenta.

5.11 The Male Reproductive System

You should know the basic anatomy of the male and female reproductive systems (Figure 5.14 and 5.19). The male **gonads** are called the **testes**. Production of sperm (Figure 5.15 and 5.16) occurs in the **seminiferous tubules** of the testes. **Spermatogonia** located in the seminiferous tubules arise from epithelial tissue to become spermatocytes, spermatids, and then spermatozoa. *Sertoli cells* stimulated by FSH surround and nurture the spermatocyte and spermatids. *Leydig cells*, located in the interstitium between the tubules, release **testosterone** when stimulated by LH. Testosterone is the primary **androgen** (male sex hormone), and stimulates the germ cells to become sperm. Testosterone is also responsible for the development of secondary sex characteristics such as pubic hair, enlargement of the larynx, and growth of the penis and seminal vesicles. While testosterone helps to initiate the growth spurt at puberty, it also stimulates closure of the epiphyses of the long bones, ending growth in stature. Sertoli cells secrete *inhibin*, a peptide hormone (actually a glycoprotein) which acts on the pituitary gland to inhibit FSH secretion.

> Be aware of basic male anatomy but don't stress too much about it.

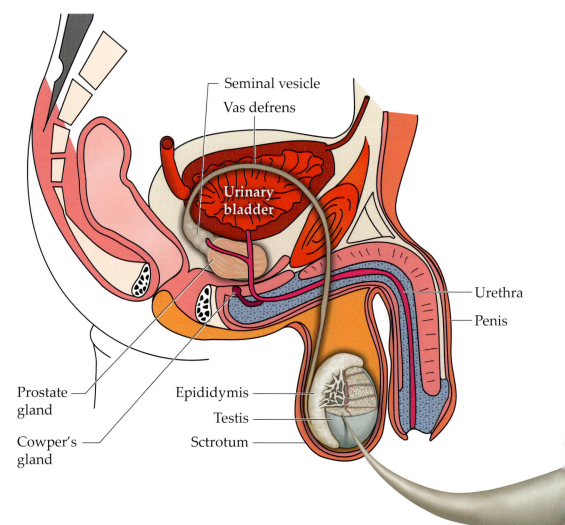

Figure 5.14 Male Reproductive Anatomy

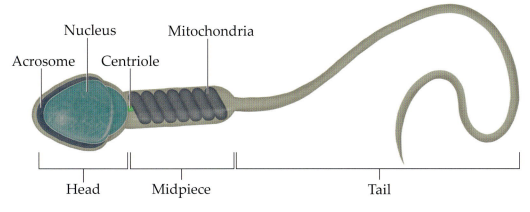

Figure 5.15   Spermatozoan

The spermatid has the characteristics of a typical cell. However, as it becomes a spermatozoon it loses its cytoplasm and forms the *head, midpiece,* and *tail* shown in Figure 5.15. The head is composed of the nuclear material and an *acrosome*. The acrosome contains lysosome-like enzymes for penetrating the egg during fertilization. The midpiece contains many mitochondria to provide energy for movement of the tail. Only the nuclear portion of the sperm enters the egg.

Once freed into the tubule lumen, the spermatozoon is carried to the **epididymis** to mature. Upon ejaculation, spermatozoa are propelled through the **vas deferens** into the **urethra** and out of the penis. **Semen** is the complete mixture of spermatozoa and fluid that leaves the penis upon ejaculation. Semen is composed of fluid from the **seminal vesicles**, the **prostate**, and the **bulbourethral glands** (also called **Cowper's glands**). Spermatozoa become activated for fertilization in a process called *capacitation*, which takes place in the vagina.

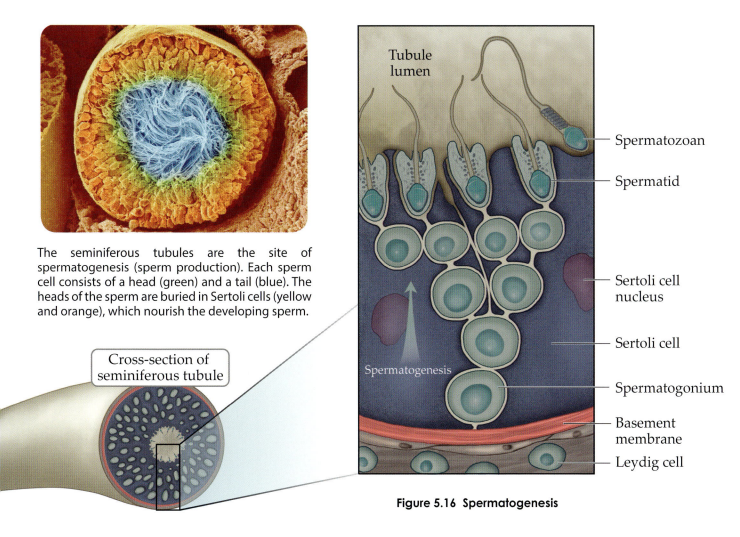

The seminiferous tubules are the site of spermatogenesis (sperm production). Each sperm cell consists of a head (green) and a tail (blue). The heads of the sperm are buried in Sertoli cells (yellow and orange), which nourish the developing sperm.

Figure 5.16   Spermatogenesis

5.12 The Female Reproductive System

Oogenesis begins in the ovaries of the fetus. All the eggs of the female are arrested as primary oocytes at birth. At puberty, FSH stimulates the growth of *granulosa cells* around the primary oocyte (Figure 5.17). The granulosa cells secrete a viscous substance around the egg called the **zona pellucida**. The structure at this stage is called a *primary follicle*. Next, theca cells differentiate from the interstitial tissue and grow around the follicle to form a *secondary follicle*. Upon stimulation by LH, theca cells secrete androgen, which is converted to **estradiol** (a type of **estrogen**) by the granulosa cells in the presence of FSH and secreted into the blood. The Estradiol is a steroid hormone that prepares the uterine wall for pregnancy. The follicle grows and bulges from the ovary. Typically, estradiol inhibits LH secretion by the anterior pituitary. However, just before **ovulation** (the bursting of the follicle), the estradiol level rises rapidly, actually causing a dramatic increase in LH secretion. This increase is called the luteal surge. The **luteal surge** results from a positive feedback loop of rising estrogen levels which increase LH levels, which increase estrogen. The luteal surge causes the follicle to burst, releasing the egg (now a secondary oocyte) into the body cavity. The egg is swept into the **Fallopian (uterine) tube** or **oviduct** by the *fimbriae*. The remaining portion of the follicle is left behind to become the **corpus luteum**. The corpus luteum secretes estradiol and progesterone throughout pregnancy, or, in the case of no pregnancy, for about 2 weeks until the corpus luteum degrades into the **corpus albicans**.

The cycle just described repeats itself approximately every 28 days after puberty unless pregnancy occurs. This cycle is called the **menstrual cycle** (Figure 5.18). With each menstrual cycle, several primordial oocytes may begin the process, but, normally, only one completes the development to ovulation. The cycle is divided into three phases:

1. the *follicular phase*, which begins with the development of the follicle and ends at ovulation;

2. the *luteal phase*, which begins with ovulation and ends with the degeneration of the corpus luteum into the corpus albicans;

3. *flow*, which is the shedding of the uterine lining lasting approximately 5 days.

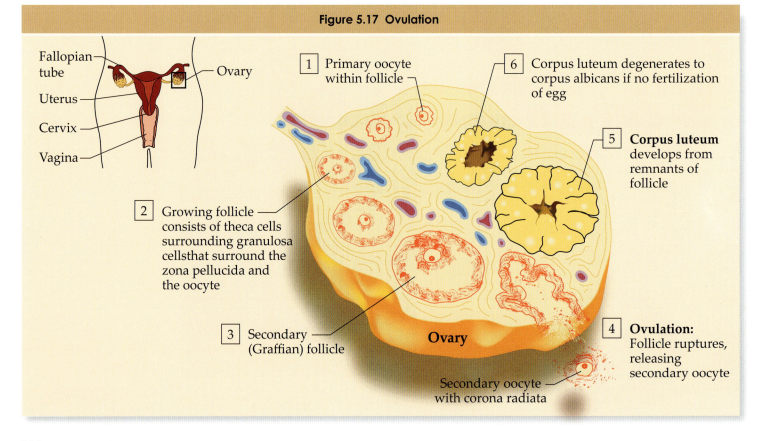

Figure 5.17 Ovulation

Fallopian tube
Ovary
Uterus
Cervix
Vagina

1 Primary oocyte within follicle

2 Growing follicle consists of theca cells surrounding granulosa cells that surround the zona pellucida and the oocyte

3 Secondary (Graffian) follicle

Ovary

Secondary oocyte with corona radiata

6 Corpus luteum degenerates to corpus albicans if no fertilization of egg

5 **Corpus luteum** develops from remnants of follicle

4 **Ovulation:** Follicle ruptures, releasing secondary oocyte

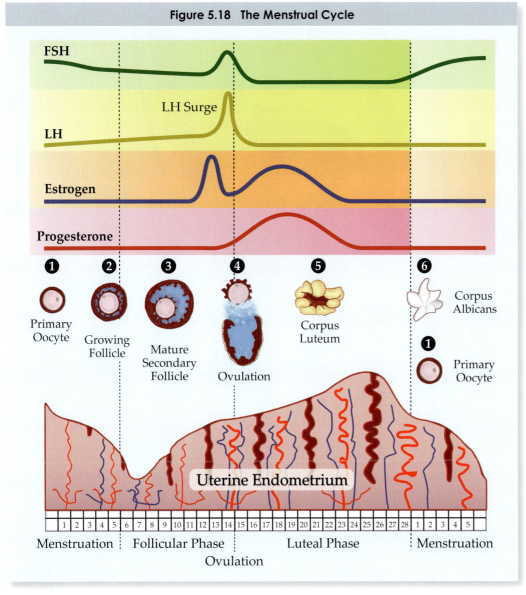

Figure 5.18 The Menstrual Cycle

FSH

LH Surge

LH

Estrogen

Progesterone

① Primary Oocyte
② Growing Follicle
③ Mature Secondary Follicle
④ Ovulation
⑤ Corpus Luteum
⑥ Corpus Albicans
① Primary Oocyte

Uterine Endometrium

| 1 | 2 | 3 | 4 | 5 | 6 | 7 | 8 | 9 | 10 | 11 | 12 | 13 | 14 | 15 | 16 | 17 | 18 | 19 | 20 | 21 | 22 | 23 | 24 | 25 | 26 | 27 | 28 | 1 | 2 | 3 | 4 | 5 |

Menstruation   Follicular Phase   Luteal Phase   Menstruation
                     Ovulation

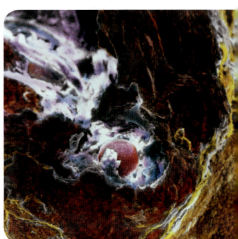

This is a mature ovum (Graafian follicle) at ovulation. The ovum (red) is surrounded by remnants of corona cells & liquid from the ruptured ovarian follicle.

5.13 *Fertilization and Embryology*

Once in the Fallopian tube, the egg is swept toward the uterus by cilia. Fertilization normally takes place in the Fallopian tubes. The enzymes of the acrosome in the sperm are released upon contact with the egg, and digest a path for the sperm through the granulosa cells and the zona pellucida. The cell membranes of the sperm head and the oocyte fuse upon contact, and the sperm nucleus enters the cytoplasm of the oocyte. The entry of the sperm causes the *cortical reaction*, which prevents other sperms from fertilizing the same egg. Now the oocyte goes through the second meiotic division to become an **ovum** and releases a second polar body. Fertilization occurs when the nuclei of the ovum and sperm fuse to form the **zygote**.

Sperm cell fertilizing an egg cell. The sperm cell (brown) is trying to penetrate the surface of the egg cell (blue). Once a sperm has fertilized the egg, rapid chemical changes make the outer layer (zona pellucida) of the egg cell thicken, preventing other sperm cells from entering. The head of the successful sperm cell releases genetic material that mixes with the genetic material in the egg cell.

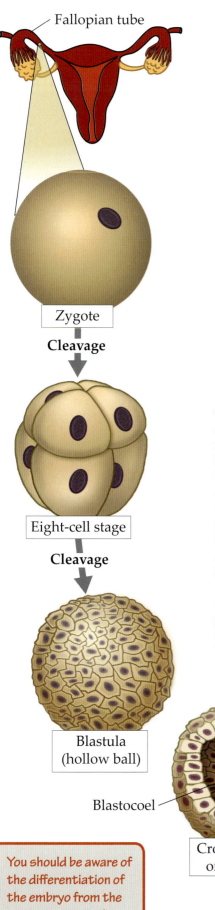

Fallopian tube

Zygote

Cleavage

Eight-cell stage

Cleavage

Blastula
(hollow ball)

Blastocoel

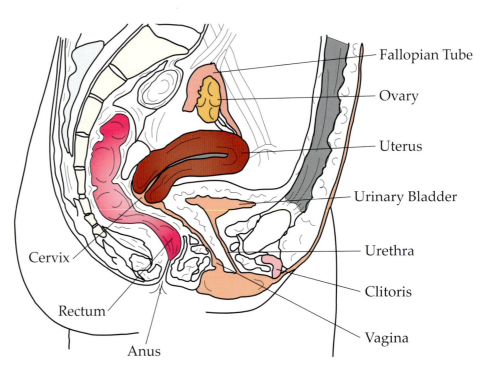

Fallopian Tube

Ovary

Uterus

Urinary Bladder

Urethra

Clitoris

Vagina

Cervix

Rectum

Anus

Figure 5.19 Female Reproductive Anatomy

Cleavage begins while the zygote is still in the Fallopian tube. The zygote goes through many cycles of mitosis; when the zygote is comprised of eight or more cells, it is called a **morula**. The embryo at this stage does not grow during cleavage. The first eight cells formed by cleavage are equivalent in size and shape and are said to be *totipotent*, meaning that they have the potential to express any of their genes. Any one of these eight cells at this stage could produce a complete individual. The cells of the morula continue to divide for four days forming a hollow ball filled with fluid. This fluid filled ball is called a **blastocyst**. It is the blastocyst that lodges in the uterus in a process called **implantation** on about the 5th to 7th day after ovulation. The blastocyst is made up of *embryonic stem cells* that each have the ability to develop into most of the types of cells in the human body. Upon implantation, the female is said to be pregnant.

Upon implantation, the egg begins secreting a peptide hormone called **human chorionic gonadotropin (HCG)**. HCG prevents the degeneration of the corpus luteum, and maintains its secretion of estrogen and progesterone. HCG in the blood and urine of the mother is the first outward sign of pregnancy. A **placenta** is formed from the tissue of the egg and the

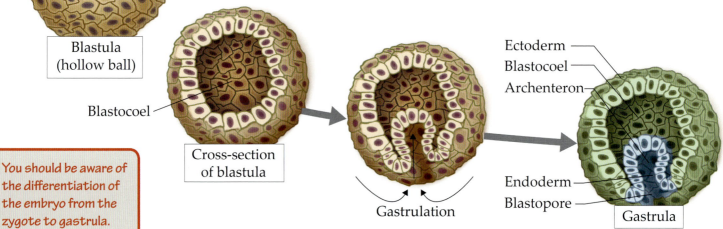

Ectoderm

Blastocoel

Archenteron

Cross-section
of blastula

Gastrulation

Endoderm

Blastopore

Gastrula

Figure 5.20 Early Cleavages in Animal Development

mother, and takes over the job of hormone secretion. The placenta reaches full development by the end of the first trimester, and begins secreting its own estrogen and progesterone while lowering its secretion of HCG.

As the embryo develops past the eight cell stage, the cells become different from each other due to cell-cell interactions. This process where a cell becomes committed to a specialized developmental path is called **determination**. Cells become determined to give rise to a particular tissue early on. The specializaton that occurs at the end of the development forming a specialized tissue cell is called **differentiation**. The fate of a cell is typically determined early on, but that same cell usually doesn't differentiate into a specialized tissue cell until much later at the end of the developmental process. Recent research has shown that the fate of even a fully differentiated cell can be altered given the proper conditions.

The formation of the **gastrula** occurs in the second week after fertilization in a process called **gastrulation**. Cells begin to slowly move about the embryo for the first time. In mammals, a *primitive streak* is formed, which is analogous to the blastopore in aquatic vertebrates. Cells destined to become mesoderm migrate into the primitive streak. During gastrulation, the three primary germ layers are formed:

1. the **ectoderm**;
2. the **mesoderm**;
3. the **endoderm**.

Although there is no absolute rule for memorizing which tissues arise from which germ layer, for the MCAT certain guidelines can be followed. The ectodermal cells develop into the outer coverings of the body, such as the outer layers of skin, nails, and tooth enamel, and into the cells of the nervous system and sense organs. The endodermal cells develop into the lining of the digestive tract, and into much of the liver and pancreas. The mesoderm is the stuff that lies between the inner and outer covering of the body, the muscle, bone, and the rest. (**WARNING:** These are just guidelines, not absolute rules.)

In the third week, the gastrula develops into a **neurula** in a process called **neurulation**. In neurulation, the **notochord** (made from mesoderm) **induces** the overlying ectoderm to thicken and form the *neural plate*. The notochord eventually degenerates, while a *neural tube* forms from the neural plate to become the spinal cord, brain, and most of the nervous system. For the MCAT you must know that induction occurs when one cell type affects the direction of differentiation of another cell type.

Part of normal cell development is programmed cell death or **apoptosis**. Apoptosis is essential for development of the nervous system, operation of the immune system, and destruction of tissue between fingers and toes to create normal hands and feet in humans. Damaged cells may undergo apoptosis as well. Failure to do so may result in cancer. Apoptosis is a complicated process in humans, but it is basically regulated by protein activity as opposed to regulation at the transcription or translation level. The proteins involved in apoptosis are present but inactive in a normal healthy cell. In mammals, mitochondria play an important role in apoptosis.

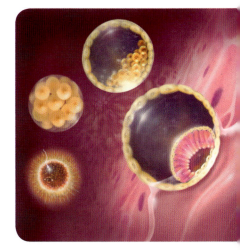

This illustration depicts the initial stages of fetal development, beginning at the lower left corner and moving clockwise. In the first stage, the spermatozoon penetrates the ovum. Their nuclei unite and the cells grow rapidly. At approximately day two, the cells form a solid cluster called a morula. As the number of cells increases, the mass descends down the uterine tube, forming a blastocyst, a hollow sphere of cells. In the last stage shown, the blastocyst implants itself in the uterine wall, approximately five days after fertilization.

Table 5.3 Fates of the Primary Germ Layers	
Ectoderm	Epidermis of skin, nervous system, sense organs
Mesoderm	Skeleton, muscles, blood vessels, heart, blood, gonads, kidneys, dermis of skin
Endoderm	Lining of digestive and respiratory tracts, liver, pancreas, thymus, thyroid

Table 5.4 Major Hormones of the Endocrine System			
Gland	*Hormone*	*Solubility*	*Effect*
Anterior pituitary	hGH	Water soluble	Growth of nearly all cells
	ACTH	Water soluble	Stimulates adrenal cortex
	FSH	Water soluble	Growth of follicles in female; Sperm production in male
	LH	Water soluble	Causes ovulation; stimulates estrogen and testosterone secretion
	TSH	Water soluble	Stimulates release of T_3 and T_4 in the thyroid
	Prolactin	Water soluble	Promotes milk production
Posterior pituitary	Oxytocin	Water soluble	Milk ejection and uterine contraction
	ADH	Water soluble	Water absorption by the kidney; increase blood pressure
Adrenal cortex	Aldosterone	Lipid soluble	Reduces Na^+ excretion; increases K^+ excretion; raises blood pressure
	Cortisol	Lipid soluble	Increase blood levels of carbohydrates, proteins, and fats
Adrenal medulla	Epinephrine	Water soluble	Stimulates sympathetic actions
	Norepinephrine	Water soluble	Stimulates sympathetic actions
Thyroid	T_3, T_4	Lipid soluble	Increases basal metabolic rate
	Calcitonin	Water soluble	Lowers blood calcium
Parathyroid	PH	Water soluble	Raises blood calcium
Pancreas	Insulin	Water soluble	Promotes entry of glucose into cells, decreasing glucose blood level
	Glucagon	Water soluble	Increases gluconeogenesis, increasing glucose blood levels
Ovaries	Estrogens	Lipid soluble	Growth of female sex organs; causes LH surge
	Progesterone	Lipid soluble	Prepares and maintains uterus for pregnancy
Testes	Testosterone	Lipid soluble	Secondary sex characteristics; closing of epiphyseal plate
Placenta	HCG	Water soluble	Stimulates corpus luteum to grow and release estrogen and progesterone
	Estrogens	Lipid soluble	Growth of mother sex organs; causes LH surge
	Progesterone	Lipid soluble	Prepares and maintains uterus for pregnancy

You are expected to know these hormones and their functions.

Terms		
Adrenal Cortex	Exocrine Glands	Ovulation
Adrenal Glands	Fallopian (Uterine)	Ovum
Adrenaline	Tube	Parathyroid Glands
Adrenocorticotropic	FSH	Parathyroid Hormone
Hormone (ACTH)	Gastrula	(PTH)
Aldosterone	Gastrulation	Peptide Hormones
Androgen	Glucagon	Placenta
Anterior Pituitary	Glucocorticoids	Posterior Pituitary
Antidiuretic Hormone	Gluconeogenesis	Prolactin
(ADH)	Gonadotropin (HCG)	Prostrate
Apoptosis	Gonads	Semen
Basal Metabolic Rate	HCG	Seminal Vesicles
Blastocyst	Human Growth	Seminiferous Tubules
Calcitonin	Hormone	Spermatogonia
Cleavage	(hGH)	Steroid Hormones
Corpus Albicans	Hypothalamus	T_3 and T_4
Corpus Luteum	Implantation	Testes
Cortisol	Insulin	Testosterone
Cowper's Glands	Intracellular Second	Thyroid-Stimulating
Determination	Messenger	Hormone (TSH)
Differentiation	LH	Tyrosine Derivatives
Ectoderm	Luteal Surge	Urethra
Effector	Mesoderm	Vas Deferens
Endocrine Glands	Mineral Corticoids	Vasopressin
Endoderm	Morula	Zona Pellucida
Epididymus	Neurula	Zygote
Epinephrine	Norepinephrine	
Estradiol	Notochord	
Estrogen	Oviduct	

113. A drug that causes increased secretion of testosterone from the interstitial cells of a physically mature male would most likely:

    A. cause the testes to descend prematurely.
    B. delay the onset of puberty.
    C. cause enhanced secondary sex characteristics.
    D. decrease core body temperature.

114. During the female menstrual cycle, increasing levels of estrogen cause:

    A. a positive feedback response, stimulating LH secretion by the anterior pituitary.
    B. a positive feedback response, stimulating FSH secretion by the anterior pituitary.
    C. a negative feedback response, stimulating a sloughing-off of the uterine lining.
    D. a negative feedback response, stimulating decreased progesterone secretion by the anterior pituitary.

115. The function of the epididymis is to:

    A. store sperm until they are released during ejaculation.
    B. produce and secrete testosterone.
    C. conduct the ovum from the ovary into the uterus.
    D. secrete FSH and LH to begin the menstrual cycle.

116. Decreasing progesterone levels during the luteal phase of the menstrual cycle are associated with:

    A. thickening of the endometrial lining in preparation for implantation of the zygote.
    B. increased secretion of LH, leading to the luteal surge and ovulation.
    C. degeneration of the corpus luteum in the ovary.
    D. increased secretion of estrogen in the follicle, leading to the flow phase of the menstrual cycle.

117. The inner linings of the Fallopian tubes are covered with a layer of cilia. The purpose of this layer is to:

    A. remove particulate matter that becomes trapped in the mucus layer covering the Fallopian tubes.
    B. maintain a layer of warm air close to the inner lining, protecting the ovum from temperature changes occurring in the external environment.
    C. kill incoming sperm, thus preventing fertilization
    D. facilitate movement of the ovum towards the uterus.

118. Which of the following endocrine glands produce testosterone?

    A. The anterior pituitary
    B. The pancreas
    C. The adrenal cortex
    D. The adrenal medulla

119. Which of the following does NOT describe cleavage in human embryos?

    A. The solid ball of cells produced during cleavage is called a morula.
    B. The size of the embryo remains constant throughout the cell divisions of cleavage.
    C. Cell division occurs in one portion of the egg in meroblastic cleavage.
    D. Daughter cells are genetically identical to parent cells.

120. The heart, bone and skeletal muscle most likely arise from which of the following primary germ layers?

    A. The ectoderm
    B. The endoderm
    C. The gastrula
    D. The mesoderm

THE DIGESTIVE SYSTEM; THE EXCRETORY SYSTEM

6.1 Anatomy

Digestion is the break down of ingested foods before they are absorbed into the body. The major reaction involved in the digestion of all macromolecules is hydrolysis.

You should know the basic anatomy of the digestive tract (Figure 6.1), which goes as follows: **mouth; esophagus; stomach; small intestine** (*duodenum, ileum, jejunum*); **large intestine** (*ascending colon, transverse colon, descending colon, sigmoid colon*); **rectum** and; **anus**.

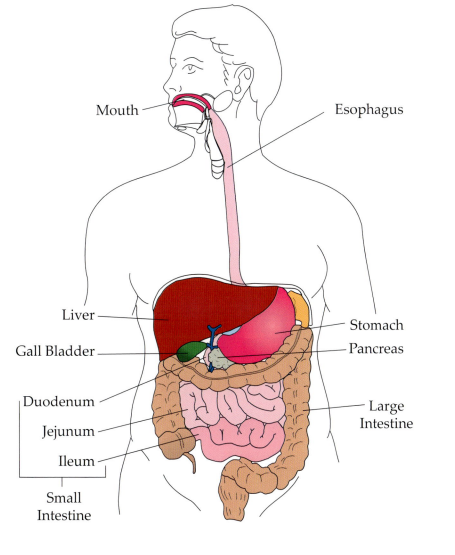

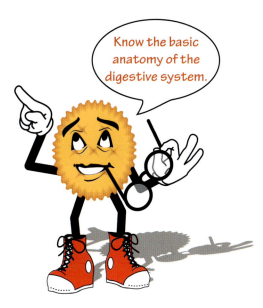

Know the basic anatomy of the digestive system.

Figure 6.1   Anatomy of the Digestive System

6.2 The Mouth and Esophagus

Digestion begins in the mouth with **α-amylase** contained in saliva. Starch is the major carbohydrate in the human diet. α-amylase begins breaking down the long straight chains of starch into polysaccharides. Chewing also increases the surface area of food, which enables more enzymes to act on the food at any one time. Chewed food forms a clump in the mouth called a *bolus*. The bolus is pushed into the **esophagus** by swallowing, and then moved down the esophagus via **peristaltic action**. (Technically, swallowing includes the movement of the bolus from the esophagus into the stomach, and is composed of a voluntary and involuntary stage.) Peristaltic action is a wave motion, similar to squeezing a tube of toothpaste at the bottom and sliding your fingers toward the top to expel the toothpaste. The peristaltic movement is performed by smooth muscle. Saliva acts to lubricate the food helping it to move down the esophagus. No digestion occurs in the esophagus.

6.3 The Stomach

The bolus moves into the stomach through the *lower esophageal sphincter* (or *cardiac sphincter*). (A sphincter is a ring of muscle that is normally contracted so that there is no opening at its center.) The stomach is a very flexible pouch that both mixes and stores food, reducing it to a semifluid mass called **chyme**. The stomach contains **exocrine glands** (two types that are very similar) whose *gastric pits* are shown in Figure 6.2. Another important function of the stomach is to begin protein digestion with the enzyme pepsin. The low pH of the stomach assists this process by denaturing the proteins. A full stomach has a pH of 2. The low pH also helps to kill ingested bacteria.

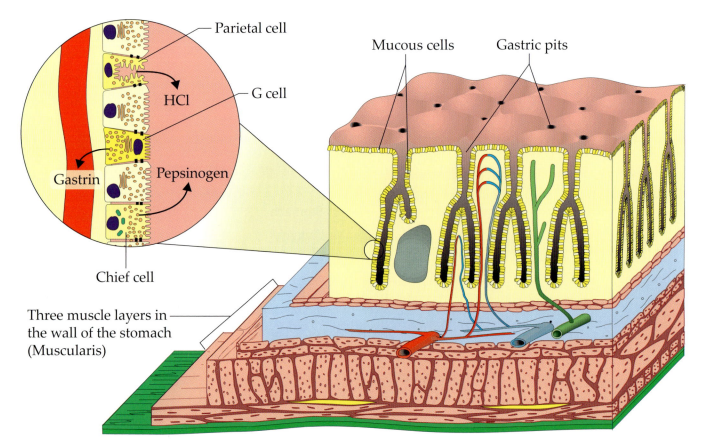

Figure 6.2 A Sectional View of the Stomach

There are four major cell types in the stomach (Figure 6.3):

1. **mucous cells**;
2. **chief (peptic) cells**;
3. **parietal (oxyntic) cells** and;
4. **G cells**.

There are different types of **mucous cells**, but all of them perform the same basic function, secreting mucus. The mucous cells line the stomach wall and the necks of the exocrine glands. Mucus, composed mainly of a sticky glycoprotein and electrolytes, lubricates the stomach wall so that food can slide along its surface without causing damage, and mucus protects the **epithelial lining** from the acidic environment of the stomach. Some mucous cells also secrete a small amount of pepsinogen.

Chief cells are found deep in the exocrine glands. They secrete **pepsinogen**, the zymogen precursor to **pepsin**. Pepsinogen is activated to pepsin by the low pH in the stomach. Once activated, pepsin begins protein digestion.

Parietal cells are also found in the exocrine glands of the stomach. Parietal cells secrete **hydrochloric acid (HCl)**, which diffuses to the lumen. The exact method used by the parietal cells to manufacture HCl has not been agreed upon, but the amount of energy necessary to produce the concentrated acid is great. Carbon dioxide is involved in the process, making carbonic acid inside the cell. The hydrogen from the carbonic acid is expelled to the lumen side of the cell, while the bicarbonate ion is expelled to the interstitial fluid side. The net result is to lower the pH of the stomach and raise the pH of the blood. Parietal cells also secrete *intrinsic factor*, which helps the ileum absorb B_{12}.

G cells secrete **gastrin** into the interstitium. Gastrin, a large peptide hormone, is absorbed into the blood and stimulates parietal cells to secrete HCl.

The major hormones that affect the secretion of the stomach juices are acetylcholine, gastrin, and *histamine*. Acetylcholine increases the secretion of all cell types. Gastrin and histamine mainly increase HCl secretion.

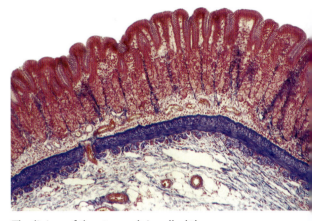

The lining of the stomach is called the mucosa.

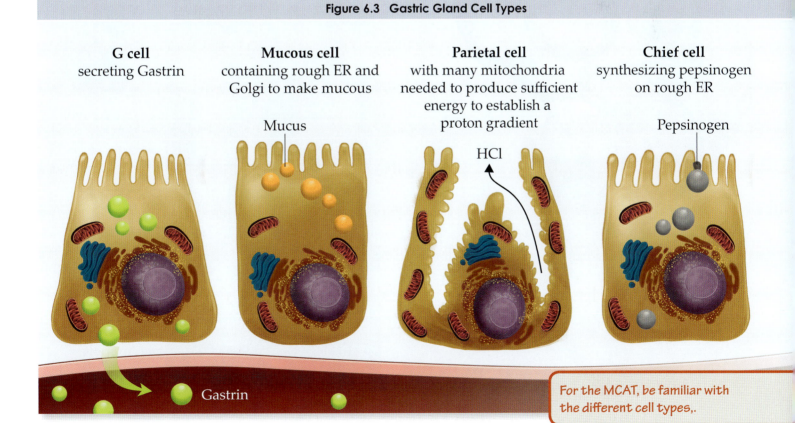

Figure 6.3 Gastric Gland Cell Types

G cell
secreting Gastrin

Mucous cell
containing rough ER and
Golgi to make mucous

Mucus

Parietal cell
with many mitochondria
needed to produce sufficient
energy to establish a
proton gradient

HCl

Chief cell
synthesizing pepsinogen
on rough ER

Pepsinogen

Gastrin

For the MCAT, be familiar with
the different cell types,.

6.4 The Small Intestine

About 90% of digestion and absorption occurs in the small intestine. In a living human the small intestine is about 3 m in length. (In a cadaver the length increases to about 6 m due to loss of smooth muscle tone.) The small intestine is divided into three parts. From smallest to largest they are the *duodenum, jejunum,* and *ileum.* Most of digestion occurs in the duodenum, and most of the absorption occurs in the jejunum and ileum. The wall of the small intestine is similar to the wall of the stomach except that the outermost layer contains finger-like projections called villi (Figure 6.4). The villi increase the surface area of the intestinal wall allowing for greater digestion and absorption. Within each villus are a capillary network and a lymph vessel, called a **lacteal**. Nutrients absorbed through the wall of the small intestine pass into the capillary network and the lacteal.

On the apical (lumen side) surface of the cells of each villus (cells called *enterocytes*) are much smaller finger-like projections called **microvilli**. The microvilli increase the surface area of the intestinal wall still further. Under a light microscope the microvilli appear as a fuzzy covering. This fuzzy covering is called the **brush border**. The brush border contains membrane bound digestive enzymes, such as carbohydrate-digesting enzymes (*dextrinase, maltase, sucrase,* and *lactase*) protein-digesting enzymes called peptidases; and nucleotide-digesting enzymes called *nucleosidases.* Some of the epithelial cells are **goblet cells** that secrete mucus to lubricate the intestine and help protect the brush border from mechanical and chemical damage. Dead cells regularly slough off into the lumen of the intestine and are replaced by new cells.

Located deep between the villi are the intestinal exocrine glands, the crypts of *Lieberkuhn.* These glands secrete an intestinal juice with a pH of 7.6 and *lysozyme.* Lysozyme helps to regulate the bacteria within the intestine.

False-colour scanning electron micrograph of a section through the wall of the human duodenum, showing the villi, which project 0.5 to 1 mm out into the intestinal lumen. They greatly increase the effective absorptive and secretory surface of the mucosa (mucus membrane) which lines the small intestine. Each villus contains a central core of connective tissue (yellowish orange), known as the lamina propria. This contains large blood vessels, capillaries, some smooth muscle cells and a blind-ended lymph vessel known as a lacteal.

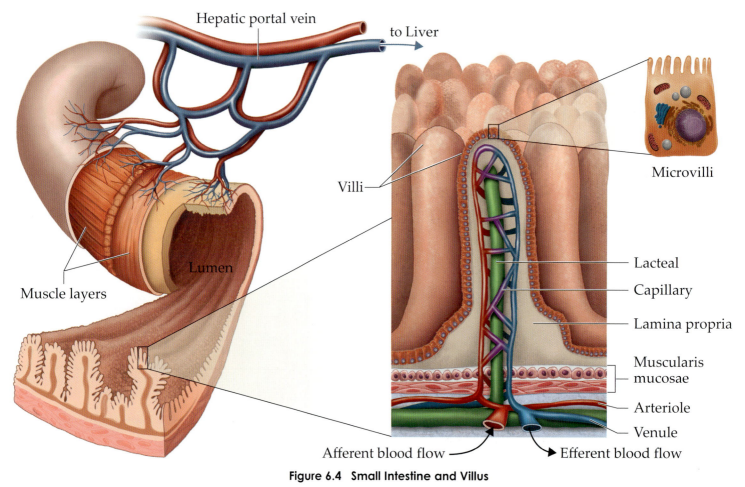

Figure 6.4 Small Intestine and Villus

The semifluid **chyme** is squeezed out of the stomach through the *pyloric sphincter* and into the duodenum. The fluid inside the duodenum has a **pH of 6** due mainly to **bicarbonate ion** secreted by the pancreas. The pancreas also acts as an exocrine gland, releasing enzymes from the *acinar cells* through the *pancreatic duct* into the duodenum. The major enzymes released by the pancreas are **trypsin, chymotrypsin, pancreatic amylase, lipase, ribonuclease, and deoxyribonuclease**. All enzymes are released as zymogens. Trypsin is activated by the enzyme *enterokinase* located in the brush border. Activated trypsin then activates the other enzymes.

Trypsin and **chymotrypsin** degrade proteins into small polypeptides. Another pancreatic enzyme, *carboxypolypeptidase*, cleaves amino acids from the sides of these peptides. Most proteins reach the brush border as small polypeptides. Here they are reduced to amino acids, dipeptides, and tripeptides before they are absorbed into the enterocytes. Enzymes within the *enterocytes* (the cells of the brush border) reduce the dipeptides and tripeptides to amino acids.

Like salivary amylase, **pancreatic amylase** hydrolyzes polysaccharides to disaccharides and trisaccharides; however, pancreatic amylase is much more powerful. Pancreatic amylase degrades nearly all the carbohydrates from the chyme into oligosaccharides. The brush border enzymes finish degrading these polymers to their respective monosaccharides before they are absorbed.

Lipase degrades fat, specifically **triglycerides**. However, since the intestinal fluid is an aqueous solution, the fat clumps together, reducing its surface area. This problem is solved by the addition of bile. Bile is produced in the **liver** and stored in the **gall bladder**. The gall bladder releases bile through the *cystic duct*, which empties into the *common bile duct* shared with the liver. The common bile duct empties into the *pancreatic duct* before connecting to the duodenum at the *ampulla of Vater*. Bile emulsifies the fat, which means it breaks it up into small particles without changing it chemically. This increases the surface area of the fat, allowing the lipase to degrade it into mainly fatty acids and monoglycerides. These products are shuttled to the brush border in bile micelles, and then absorbed by the enterocytes. Bile also contains *bilirubin*, an end product of hemoglobin degradation. Much of the bile is reabsorbed by the small intestine and transported back to the liver.

Chyme is moved through the intestines by **peristalsis**. A second type of intestinal motion, *segmentation*, mixes the chyme with the digestive juices.

Several lobes of the pancreas are seen here, separated by fissures. The smaller sections seen on each lobe are clusters of acini cells. These are exocrine cells, secreting digestive enzymes. The enzymes drain into a highly branched system of ducts of increasing size, that terminates in the main pancreatic duct, which feeds into the duodenum (small intestine). The other function of the pancreas is the endocrinal secretion of hormones, in particular insulin. Fragments of connective tissue and blood vessels are also seen.

> You must know the pancreatic enzymes trypsin, chymotrypsin, amylase, and lipase, and know their functions in the small intestine.

> Understand that bile is necessary to increase the surface area of fat, but that it does not digest the fat. In other words, bile physically separates fat molecules, but does not break them down chemically.

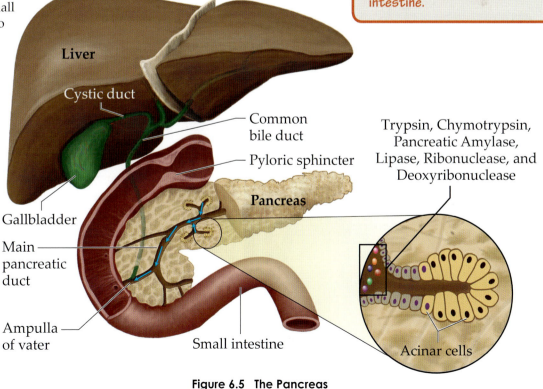

Figure 6.5 The Pancreas

The large intestine, or colon, has four parts:

1. *ascending colon;*
2. *transverse colon;*
3. *descending colon* and;
4. *sigmoid colon.*

The major functions of the large intestine are **water absorption** and **electrolyte absorption**. When this function fails, diarrhea results. The large intestine also contains the bacteria *E. coli*. The bacteria produce vitamin K, B$_{12}$, thiamin, and riboflavin.

Healthy feces are composed of 75% water. The remaining solid mass is 30% dead bacteria, 10-20% fat (mainly from bacteria and sloughed enterocytes), 10-20% inorganic matter, 2-3% protein, and 30% roughage (i.e. cellulose) and undigested matter (i.e. sloughed cells).

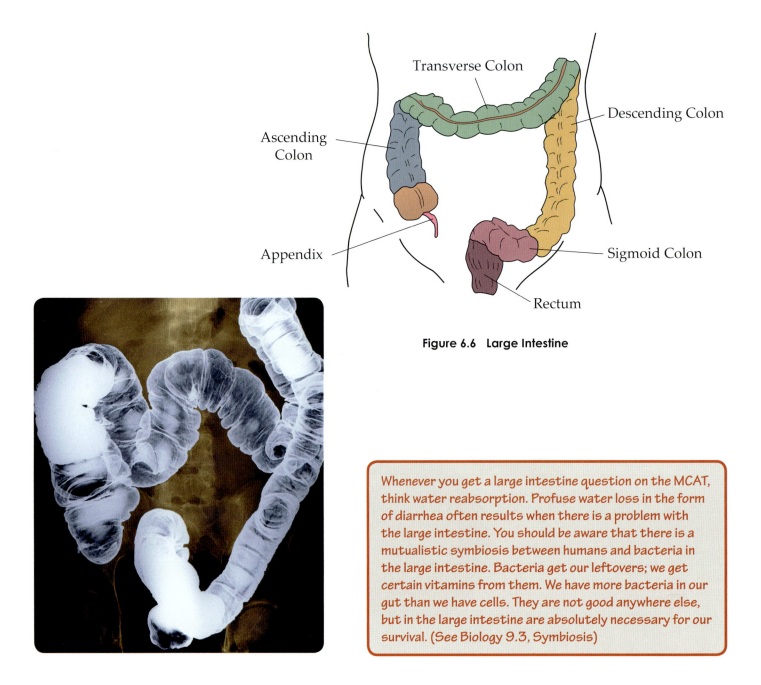

Figure 6.6   Large Intestine

Whenever you get a large intestine question on the MCAT, think water reabsorption. Profuse water loss in the form of diarrhea often results when there is a problem with the large intestine. You should be aware that there is a mutualistic symbiosis between humans and bacteria in the large intestine. Bacteria get our leftovers; we get certain vitamins from them. We have more bacteria in our gut than we have cells. They are not good anywhere else, but in the large intestine are absolutely necessary for our survival. (See Biology 9.3, Symbiosis)

Secretin, cholecystokinin, and *gastric inhibitory peptide* are local peptide hormones secreted by the small intestine after a meal. Each of these hormones increases blood insulin levels especially in the presence of glucose.

Gastric inhibitory peptide is released in response to fat and protein digestates in the duodenum, and to a lesser extent, in response to carbohydrates. It has a mild effect in decreasing the motor activity of the stomach.

Hydrochloric acid in the duodenum causes secretin release. Secretin stimulates sodium bicarbonate secretion by the pancreas.

Food in the upper duodenum, especially fat digestates, causes the release of cholecystokinin. Cholecystokinin causes gallbladder contraction and pancreatic enzyme secretion. It also decreases the motility of the stomach allowing the duodenum more time to digest fat.

You don't have to remember these gastrointestinal hormones, although they may appear in a passage. Instead, understand the ideas of digestion. The body eats to gain energy in the form of food. The digestive system breaks down the food so it can be absorbed into the body. One problem is that the food may move too fast through the digestive tract and come out undigested. The stomach stores food, and releases small amounts at a time to be digested and absorbed by the intestine. This way, the body can take in (eat) a large amount of food at a single time and take a long time to digest it. One of the jobs of the gastrointestinal hormones just described is to help regulate this process.

Table 6.1 Major Digestive Enzymes of Humans	
Source/Enzyme	*Action*
SALIVARY GLANDS	
Salivary amylase	Starch → Maltose
STOMACH	
Pepsin	Proteins → Peptides; autocatalysis
PANCREAS	
Pancreatic amylase	Starch → Maltose
Lipase	Fats → Fatty acid and glycerol
Nuclease	Nucleic acids → Nucleotides
Trypsin	Proteins →Peptides; Zymogen activation
Chymotrypsin	Proteins →Peptides
Carboxypeptidase	Peptides → Shorter peptides and amino acids
SMALL INTESTINE	
Aminopeptidase	Peptides → Shorter peptides and amino acids
Dipeptidase	Dipeptides → Amino acids
Enterokinase	Trypsinogen → Trypsin
Nuclease	Nucleic acids → Nucleotides
Maltase	Maltose → Glucose
Lactase	Lactose → Galactose and glucose
Sucrase	Sucrose → Fructose and glucose

We have looked at digestion, the breakdown of food. Next, we will look at absorption, the assimilation of the by-products of digestion.

121. As chyme is passed from the stomach to the small intestine, the catalytic activity of pepsin:

    A. increases because pepsin works synergistically with trypsin.
    B. increases because pepsin is activated from its zymogen form.
    C. decreases in response to the change in pH.
    D. decreases because pepsin is digested by pancreatic amylase in the small intestine.

122. Which of the following is the best explanation for why pancreatic enzymes are secreted in zymogen form?

    A. A delay in digestion is required in order for bile to increase the surface area chyme.
    B. Enzymes are most active in zymogen form.
    C. Zymogens will not digest bile in the pancreatic duct.
    D. Pancreatic cells are not as easily replaced as intestinal epithelium.

123. Omeprazole is used to treat duodenal ulcers that result from gastric acid hypersecretion. Omeprazole blocks the secretion of HCl from the parietal cells of the stomach. Which of the following is LEAST likely to occur in a patient taking omeprazole?

    A. an increase in microbial activity in the stomach
    B. a decrease in the activity of pepsin
    C. an increase in stomach pH
    D. a decrease in carbohydrate digestion in the stomach

124. Which of the following reaction types is common to the digestion of all macronutrients?

    A. hydrolysis
    B. reduction
    C. glycolysis
    D. phosphorylation

125. One function of the large intestine is:

    A. to absorb water.
    B. to secrete excess water.
    C. to digest fat.
    D. to secrete urea.

126. Salivary α–amylase begins the digestion of:

    A. lipids.
    B. nucleic acids.
    C. proteins.
    D. carbohydrates.

127. All of the following enzymes are part of pancreatic exocrine function EXCEPT:

    A. bile
    B. chymotrypsin
    C. pancreatic amylase
    D. lipase

128. In humans, most chemical digestion of food occurs in the:

    A. mouth.
    B. stomach.
    C. duodenum.
    D. ileum.

6.8 Absorption and Storage

The function of the entire digestive tract described in the previous section is to convert ingested food into basic nutrients that the small intestine is able to absorb. Once absorbed into the enterocytes, nutrients are processed and carried to the individual cells for use. The following section describes the process of absorption and the post-absorptive fates of the major nutrients: carbohydrates, proteins, and fats. This section is provided mainly as background knowledge; very little of this information will be tested directly on the MCAT.

6.9 Carbohydrates

By far the major carbohydrates in a human diet are sucrose, lactose, and starch. Cellulose (the polysaccharide making up the cell wall of plants) cannot be digested by humans, and is considered *roughage*. Sucrose and lactose are disaccharides made from glucose and fructose, and from glucose and galactose, respectively. Starch is a straight chain of glucose molecules. Typically, 80% of the end product of carbohydrate digestion is glucose. 95% of the carbohydrates in the blood are glucose.

Carbohydrate absorption is shown in Figure 6.7. Glucose is absorbed by a secondary active transport mechanism down the concentration gradient of sodium. Sodium is actively pumped out of the enterocyte on the basolateral side. The resulting low sodium concentration inside the enterocyte drags sodium from the intestinal lumen into the cell through a transport protein, but only after glucose has also attached itself to the protein. Thus glucose is dragged into the enterocyte by sodium. As the concentration of glucose inside the cell builds, it moves out of the cell on the basolateral side via facilitated transport. At high concentrations of lumenal glucose, glucose builds up in the paracellular space and raises the osmotic pressure there. The aqueous solution of the lumen is dragged into the paracellular space pulling glucose along with it. Glucose is absorbed by this second method only when present in high concentrations.

Galactose follows a similar absorption path to glucose. Fructose is absorbed via facilitated diffusion, and much of it is converted to glucose while inside the enterocyte.

All carbohydrates are absorbed into the bloodstream and carried by the portal vein to the liver. One of the jobs of the liver is to maintain a fairly constant blood glucose level (90 mg/dl between meals to 140 mg/dl after a meal). The liver absorbs the carbohydrates and converts nearly all the galactose and fructose into glucose, and then into glycogen for storage. The formation of glycogen is called **glycogenesis**. When the blood glucose level decreases, **glycogenolysis** takes place in the liver, and glucose is returned to the blood.

In all cells except enterocytes and the cells of the renal tubule, glucose is transported from high concentration to low concentration via facilitated diffusion. Nearly all cells are capable of producing and storing some glycogen; however, only muscle cells and especially liver cells store large amounts. When the cells have reached their saturation point with glycogen, carbohydrates are converted to fatty acids and then triglycerides in a process requiring a small amount of energy.

You can ignore most of the details here and concentrate on the big picture of carbohydrate digestion, absorption, and metabolism. Relate this information to glycolysis and the Krebs cycle in Biology Lecture 1 for a complete picture. Notice that most of the glucose is stored for later use. When the glycogen stores are full, the glucose is converted to fat, a long-term form of energy storage. The conversion of glucose to fat takes place in the liver and adipocytes and is stored in the adipocytes. Keep in mind the role of the liver in processing carbohydrates. We will talk more about this later.

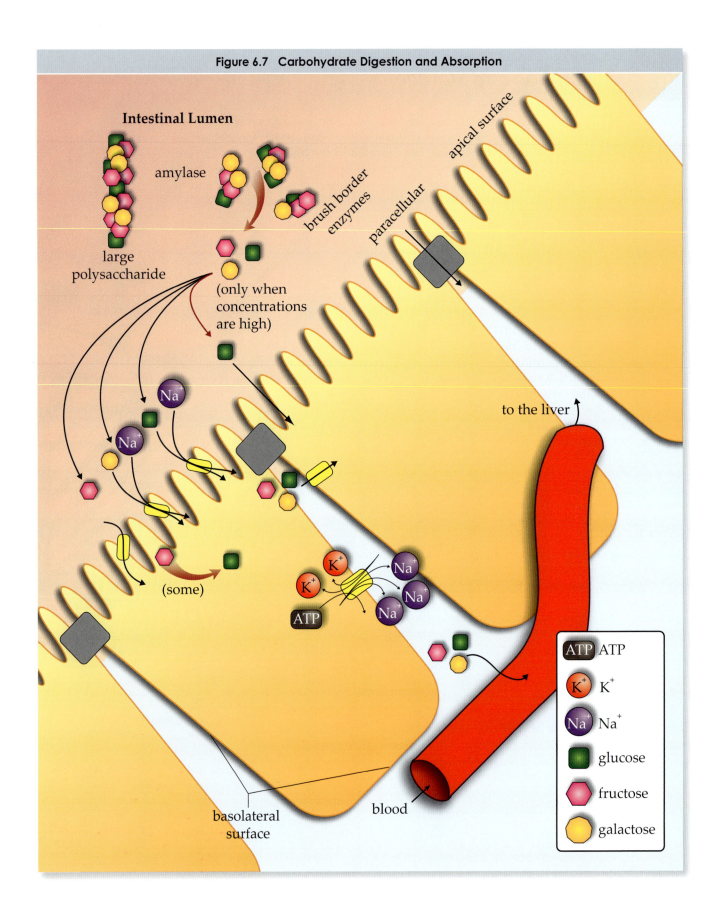

Figure 6.7 Carbohydrate Digestion and Absorption

6.10 Proteins

Protein digestion results in amino acids, dipeptides, and tripeptides. Absorption of many of these products occurs via a cotransport mechanism down the concentration gradient of sodium, similar to the mechanism used by glucose. A few amino acids are transported by facilitated diffusion. Because the chemistry of amino acids varies greatly, each transport mechanism is specific to a few amino acids or polypeptides.

This is the normal surface pattern of the jejunal mucosa.

Nearly all polypeptides absorbed into an enterocyte are hydrolyzed to their amino acid constituents by enzymes within the enterocytes. From the enterocytes, amino acids are absorbed directly into the blood and then quickly taken up by all cells of the body, especially the liver. Transport into the cells may be facilitated or active, but is never passive, since amino acids are too large and polar to diffuse through the membrane. The cells immediately create proteins from the amino acids so that the intracellular amino acid concentration remains low. However, most proteins are easily broken down and returned to the blood when needed. When the cells reach their upper limit for protein storage, amino acids can be burned for energy (see Biology Lecture 1) or converted to fat for storage. The energy that can be gained from burning protein is about 4 Calories per gram of protein. This can be compared to carbohydrates, which produce about 4.5 Calories per gram, and fat, which produces about 9 Calories per gram. Ammonia, a nitrogen containing compound, is a by-product of gluconeogenesis from proteins. Nearly all ammonia is converted to **urea** by the liver and then excreted in the urine by the kidney.

This is the jejunal mucosa of a person with Celiac Disease. Celiac Disease is an uncommon condition caused by hypersensitivity to a component of gluten, a protein in wheat flour.

Figure 6.8 Protein Digestion and Absorption

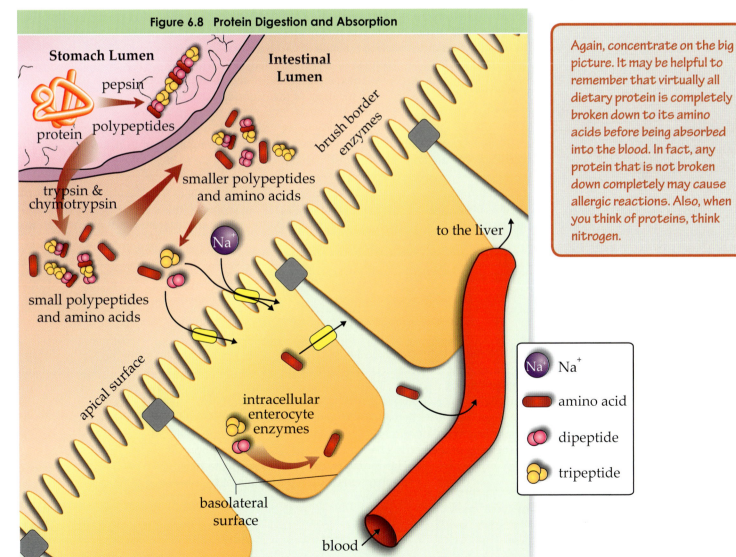

Stomach Lumen

Intestinal Lumen

pepsin

protein polypeptides

trypsin & chymotrypsin

smaller polypeptides and amino acids

brush border enzymes

Na⁺

small polypeptides and amino acids

to the liver

apical surface

intracellular enterocyte enzymes

basolateral surface

blood

Na⁺   Na⁺

amino acid

dipeptide

tripeptide

Again, concentrate on the big picture. It may be helpful to remember that virtually all dietary protein is completely broken down to its amino acids before being absorbed into the blood. In fact, any protein that is not broken down completely may cause allergic reactions. Also, when you think of proteins, think nitrogen.

6.11 Fats

Most of dietary fat consists of triglycerides, which are broken down to monoglycerides, and fatty acids, before they are shuttled to the brush border by bile micelles, and diffuse through the enterocyte membrane (Figure 6.9). After delivering their cargo, the micelles shuttle back to the chyme to pick up more fat digestates. Micelles also carry other fat digestates such as small amounts of hydrolyzed phospholipids and cholesterol, which also diffuse through the enterocyte membrane.

Once inside the enterocyte, monoglycerides, and fatty acids, are turned back into triglycerides at the smooth endoplasmic reticulum. The newly synthesized triglycerides aggregate within the smooth endoplasmic reticular lumen along with some cholesterol and phospholipids. These amphipathic molecules orient themselves like a micelle with their charged ends pointing outward toward the aqueous solution of the lumen. Apoproteins attach to the outside of these globules. (See Biology Lecture 1 for more on apoproteins.) The globules move to the Golgi apparatus and are released from the cell into the interstitial fluid via exocytosis. Most of these globules, now called *chylomicrons*, move into the lacteals of the lymph system. 80-90% of ingested fat that is absorbed by this process, moves through the lymph system, and is emptied into the large veins of the neck at the *thoracic duct*. Small amounts of more water soluble fatty acids (short chain fatty acids) are absorbed directly into the blood of the villi.

The chylomicron concentration in the blood peaks about 1-2 hours after a meal, but falls rapidly (chylomicrons have a half life of about 1 hour) as the fat digestates are absorbed into the cells of the body. The major absorption of fat occurs in the liver and adipose tissue. Chylomicrons stick to the side of capillary walls where *lipoprotein lipase* hydrolyzes the triglycerides, the products of which immediately diffuse into the fat and liver cells. Inside the fat and liver cells, the triglycerides are reconstituted at the smooth endoplasmic reticulum. Thus, the first stop for most of the digested fat is the liver.

From adipose tissue, most fatty acids are transported in the form of *free fatty acid*, which combines immediately in the blood with **albumin**. A single albumin molecule typically carries 3 fatty acid molecules, but is capable of carrying up to 30.

Between meals (called the *postabsorptive state*) 95% of lipids in the plasma are in the form of *lipoproteins*. Lipoproteins look like small chylomicrons, or, more precisely, chylomicrons are large lipoproteins. Besides chylomicrons, there are four different types of lipoproteins:

1. *very low-density lipoproteins*;
2. *intermediate-density lipoproteins*;
3. *low-density lipoproteins* and;
4. *high-density lipoproteins*.

All are made from triglycerides, cholesterol, phospholipids, and protein. As the density increases, first the amount of triglycerides decrease, and then the amount of cholesterol and phospholipids decrease. Thus, very low-density lipoproteins have a lot of triglycerides, and high-density lipoproteins have very few triglycerides. Most lipoproteins are made in the liver. Very-low density lipoproteins transport triglycerides from the liver to adipose tissue. Intermediate and low-density lipoproteins transport cholesterol and phospholipids to the cells of the body. The function of high-density lipoproteins is less well understood. Hardening of the arteries seems to be induced by the lower density lipoproteins, but impeded by high-density lipoproteins.

Although lipoproteins are a hot topic, they are not a required topic for the MCAT. Once again, look at the big picture here. Keep in mind that fat is insoluble in water, so typically requires a carrier (i.e. a lipoprotein, or albumin). For the MCAT, you should associate fat with efficient long-term energy storage; lots of calories (energy) with little weight.

Take a look at Orgo Lecture 4 and relate the chemistry to this biology section.

Figure 6.9 Fat Digestion and Absorption

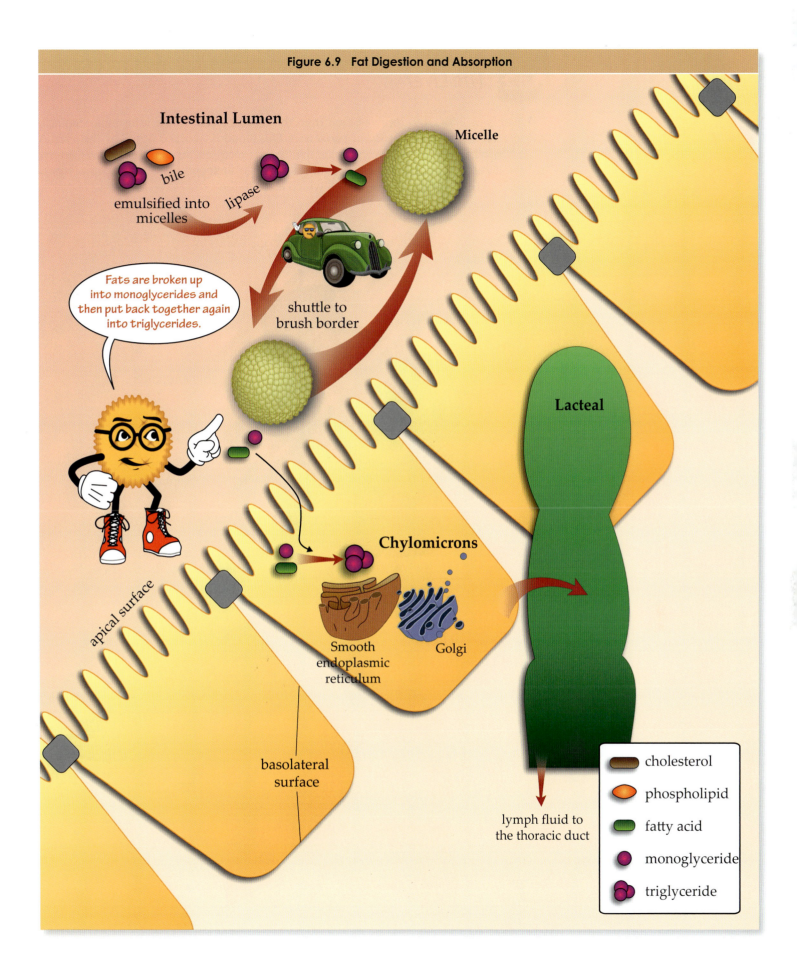

The liver is positioned to receive blood from the capillary beds of the intestines, stomach, spleen, and pancreas via the *hepatic portal vein*. This blood is 'worked upon' by the liver. A second blood supply, used to oxygenate the liver, is received through the hepatic artery. All blood received by the liver moves through large flattened spaces called the *hepatic sinusoids* and collects in the *hepatic vein*, which leads to the **vena cava**.

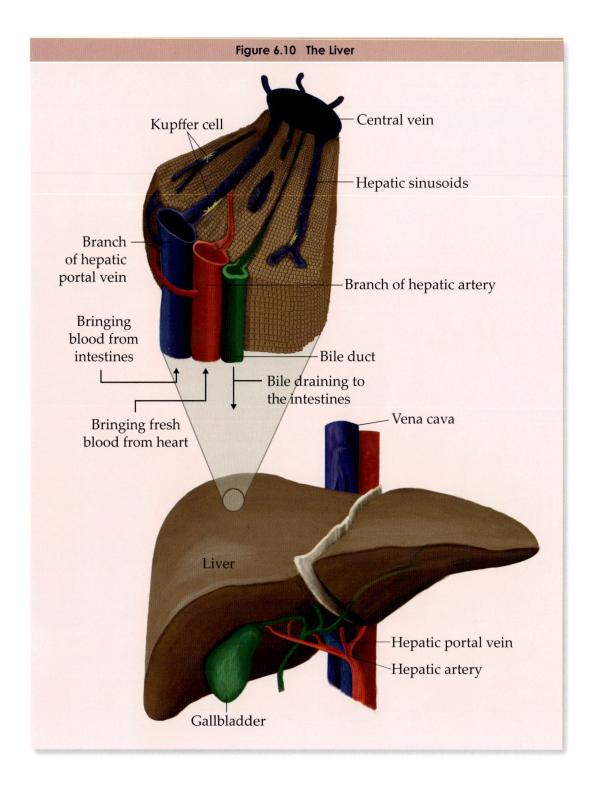

Figure 6.10   The Liver

The liver has the following interrelated functions (Figure 6.11).

- **Blood storage**: the liver can expand to act as a blood reservoir for the body.

- **Blood filtration**: *Kupffer cells* phagocytize bacteria picked up from the intestines.

- **Carbohydrate metabolism**: The liver maintains normal blood glucose levels through **gluconeogenesis** (the production of glycogen and glucose from noncarbohydrate precursors), glycogenesis, and storage of glycogen.

- **Fat metabolism**: The liver synthesizes bile from cholesterol and converts carbohydrates and proteins into fat. It oxidizes fatty acids for energy, and forms most lipoproteins.

- **Protein metabolism**: The liver deaminates amino acids, forms **urea** from ammonia in the blood, synthesizes plasma proteins such as fibrinogen, prothrombin, albumin, and most globulins, and synthesizes nonessential amino acids.

- **Detoxification**: Detoxified chemicals are excreted by the liver as part of bile or polarized so they may be excreted by the kidney.

- **Erythrocyte destruction**: *Kupffer cells* also destroy irregular erythrocytes, but most irregular erythrocytes are destroyed by the spleen.

- **Vitamin storage**: The liver stores vitamins such as vitamins A, D, and B12. The liver also stores iron combining it with the protein *apoferritin* to form of *ferritin*.

Prothrombin and fibrinogen are two important clotting factors. Albumin is the major osmoregulatory protein in the blood. Globulins are a group of proteins that include antibodies. Antibodies, however, are not made in the liver. They are made by plasma cells.

When the liver mobilizes fat for energy, it produces acids called *ketone bodies*. This often results in a condition called *ketosis* or *acidosis*. For the MCAT, you should know that when the liver mobilizes fat or protein for energy, the blood acidity increases.

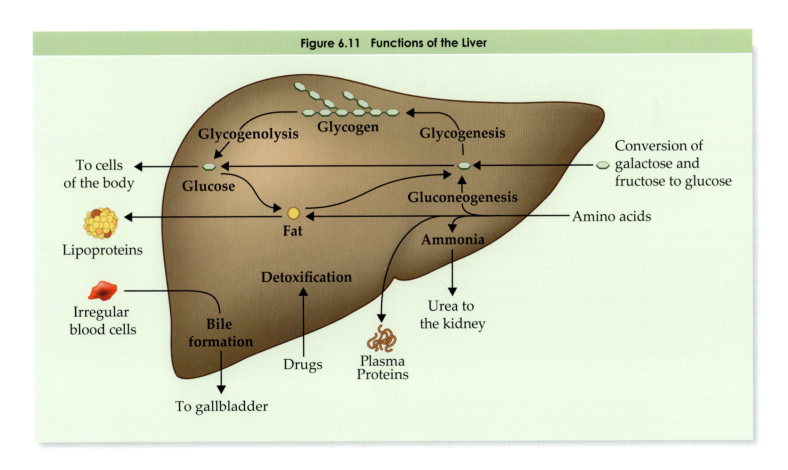

Figure 6.11 Functions of the Liver

Questions 129 through 136 are **NOT** based on a descriptive passage.

129. A stomach ulcer may increase the acidity of the stomach. The stomach cells most affected by a stomach ulcer are:

    A. goblet cells.
    B. parietal cells.
    C. chief cells.
    D. G cells.

130. Which of the following occurs mainly in the liver?

    A. fat storage
    B. protein degradation
    C. glycolysis
    D. gluconeogenesis

131. Dietary fat consists mostly of neutral fats called triglycerides. Most digestive products of fat:

    A. enter intestinal epithelial cells as chylomicrons.
    B. are absorbed directly into the capillaries of the intestines.
    C. are degraded to fatty acids by the smooth endoplasmic reticulum of enterocytes.
    D. enter the lymph system before entering the blood stream.

132. Which of the following is not true concerning the digestive products of dietary protein?

    A. They are used to synthesize essential amino acids in the liver.
    B. Some of the products are absorbed into the intestines by facilitated diffusion.
    C. Energy is required for the intestinal absorption of at least some of these products.
    D. Deamination of these products in the liver leads to urea in the blood.

133. Cholera is an intestinal infection that can lead to severe diarrhea causing profuse secretion of water and electrolytes. A glucose-electrolyte solution may be administered orally to patients suffering from cholera. What is the most likely reason for mixing glucose with the electrolyte solution?

    A. When digested, glucose increases the strength of the patient.
    B. The absorption of glucose increases the uptake of electrolytes.
    C. Glucose is an electrolyte.
    D. Glucose stimulates secretion of the pancreatic enzyme, amylase.

134. Most of the glycogen in the human body is stored in the liver and the skeletal muscles. Which of the following hormones inhibits glycogenolysis?

    A. Cortisol
    B. Insulin
    C. Glucagon
    D. Aldosterone

135. Free fatty acids do not dissolve in the blood, so they must be transported within the body bound to protein carriers. The most likely explanation for this is:

    A. Blood is an aqueous solution and only hydrophobic compounds are easily dissolved.
    B. Blood is an aqueous solution and only hydrophilic compounds are easily dissolved.
    C. Blood serum contains chylomicrons which do not bind to fatty acids.
    D. Blood serum is lipid based and the polar region of a fatty acid will not be dissolved.

136. Essential amino acids must be ingested because they cannot be synthesized by the body. In what form are these amino acids likely to enter the blood stream?

    A. single amino acids
    B. dipeptides
    C. polypeptides
    D. proteins

6.13 The Kidney

The function of the **kidney** is:

    1. to excrete waste products, such as urea, uric acid, ammonia, and phosphate;

    2. to maintain homeostasis of the body fluid volume and solute composition and;

    3. to help control plasma pH.

There are two kidneys. Each kidney is a fist-sized organ made up of an outer **cortex** and an inner **medulla**. Urine is created by the kidney and emptied into the **renal pelvis**. The renal pelvis is emptied by the **ureter**, which carries urine to the **bladder**. The bladder is drained by the **urethra**.

The functional unit of the kidney is the **nephron** (Figures 6.12, 6.13, and 6.14). Blood flows into the first capillary bed of the nephron called the **glomerulus**. Together, **Bowman's capsule** and the glomerulus make up the **renal corpuscle**. **Hydrostatic pressure** forces some plasma through **fenestrations** of the glomerular endothelium and into Bowman's capsule. Like a sieve, the fenestrations screen out blood cells and large proteins from entering Bowman's capsule. The fluid that finds its way into Bowman's capsule is called *filtrate* or *primary urine*. Filtrate moves from Bowman's capsule to the **proximal tubule**. The proximal tubule is where most **reabsorption** takes place. Secondary active transport proteins in the apical membranes of the proximal tubule cells are responsible for the reabsorption of nearly all glucose, most proteins, and other solutes. These transport proteins can become saturated. The concentration of a solute that saturates its transport proteins is called the *transport maximum*. Once a solute has reached its transport maximum, any more solute is washed into the urine. Some solutes that are not actively reabsorbed are reabsorbed by passive or facilitated diffusion. Water is reabsorbed into the renal interstitium of the proximal tubules across relatively permeable tight junctions due to the favorable osmotic gradient.

Drugs, toxins, and other solutes are **secreted** into the filtrate by the cells of the proximal tubule. Hydrogen ions are secreted through an **antiport** system with sodium, which is driven by the sodium concentration gradient. This antiport system is similar to the transport system of glucose with sodium, except the proton crosses the membrane in the opposite direction to sodium. Uric acid, bile pigments, antibiotics and other drugs are also secreted into the proximal tubule.

The net result of the proximal tubule is to reduce the amount of filtrate in the nephron while changing the solute composition without changing the osmolarity.

From the proximal tubule, the filtrate flows into the **loop of Henle**. The loop of Henle dips into the medulla. The function of the loop of Henle is to increase the solute concentration, and thus the osmotic pressure, of the medulla. As filtrate descends into the medulla, water passively diffuses out of the loop of Henle and into the medulla. The descending loop of Henle has low permeability to salt, so filtrate osmolarity goes up. As the filtrate rises out of the medulla, salt diffuses out of the ascending loop of Henle, passively at first, then actively. The

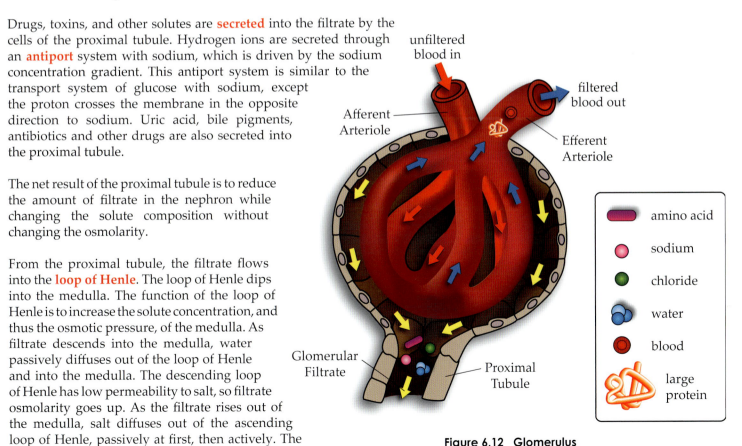

Figure 6.12   Glomerulus

ascending loop of Henle is nearly impermeable to water. A second capillary bed, called the *vasa recta*, surrounds the loop of Henle and helps to maintain the concentration of the medulla.

The **distal tubule** reabsorbs Na^+, and Ca^{2+} while secreting K^+, H^+, and HCO_3^-. Aldosterone acts on the distal tubule cells to increase sodium and potassium membrane transport proteins. The net effect of the distal tubule is to lower the filtrate osmolarity. At the end of the distal tubule, called the *collecting tubule* (The collecting tubule is a portion of the distal tubule, not to be confused with the collecting duct), ADH acts to increase the permeability of the cells to water. Therefore, in the presence of ADH, water flows from the tubule, concentrating the filtrate.

The distal tubule empties into the **collecting duct**. The collecting duct carries the filtrate into the highly osmotic medulla. The collecting duct is impermeable to water, but is also sensitive to ADH. In the presence of ADH, the collecting duct becomes permeable to water allowing it to passively diffuse into the medulla, concentrating the urine. Many collecting ducts line up side by side in the medulla to make the *renal pyramids*.

The collecting ducts lead to a *renal calyx*, which empties into the renal pelvis.

6.14 The Juxtaglomerular Apparatus

The **juxtaglomerular apparatus** monitors filtrate pressure in the distal tubule. Specialized cells, called *granular cells*, in the juxtaglomerular apparatus secrete the enzyme renin. Renin initiates a regulatory cascade producing angiotensin I, II, and III, which ultimately stimulates the adrenal cortex to secrete aldosterone. Aldosterone acts on the distal tubule, stimulating the formation of membrane proteins that absorb sodium and secrete potassium.

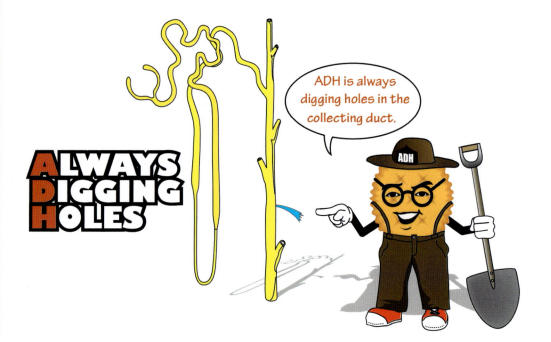

There are many details about the kidney that you must be able to recall for the MCAT. You should know the function of each section of the nephron: filtration occurs in the renal corpuscle; reabsorption and secretion mostly in the proximal tubule; the loop of Henle concentrates solute in the medulla; the distal tubule empties into the collecting duct; the collecting duct concentrates the urine. Understand that the amount of filtrate is related to the hydrostatic pressure of the glomerulus. You should know that the descending loop of Henle is permeable to water, and that the ascending loop of Henle is impermeable to water and actively transports sodium into the kidney.

Don't lose sight of the big picture; the function of the kidney is homeostasis.

Figure 6.13 The Nephron

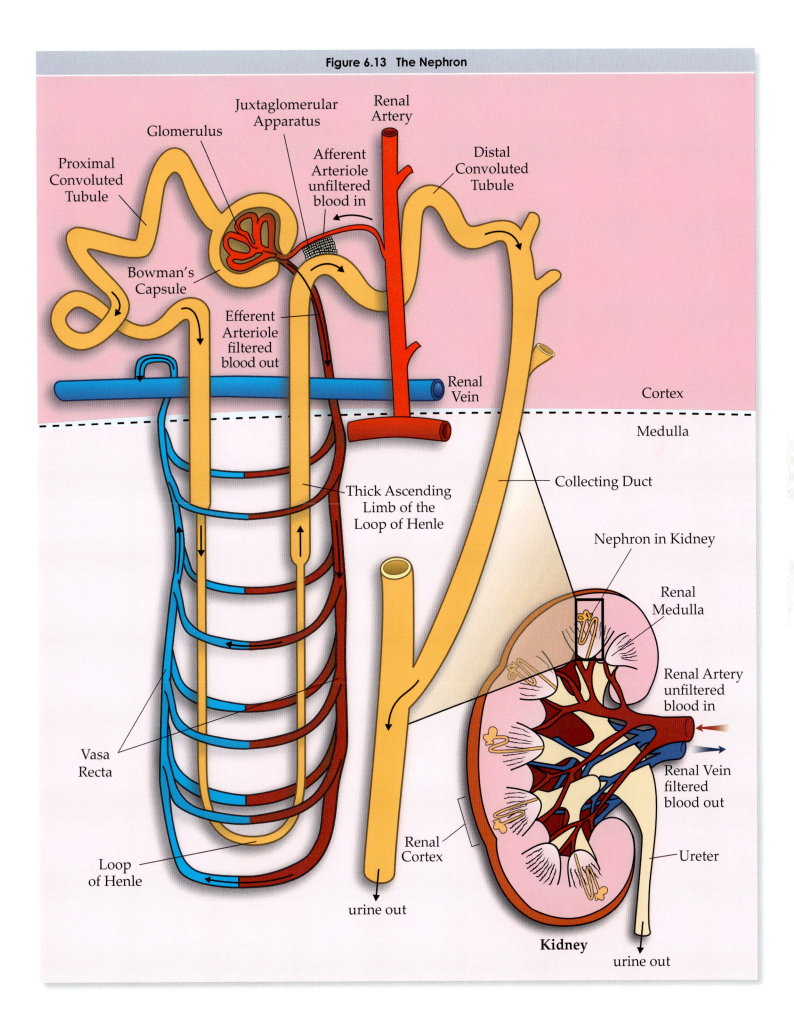

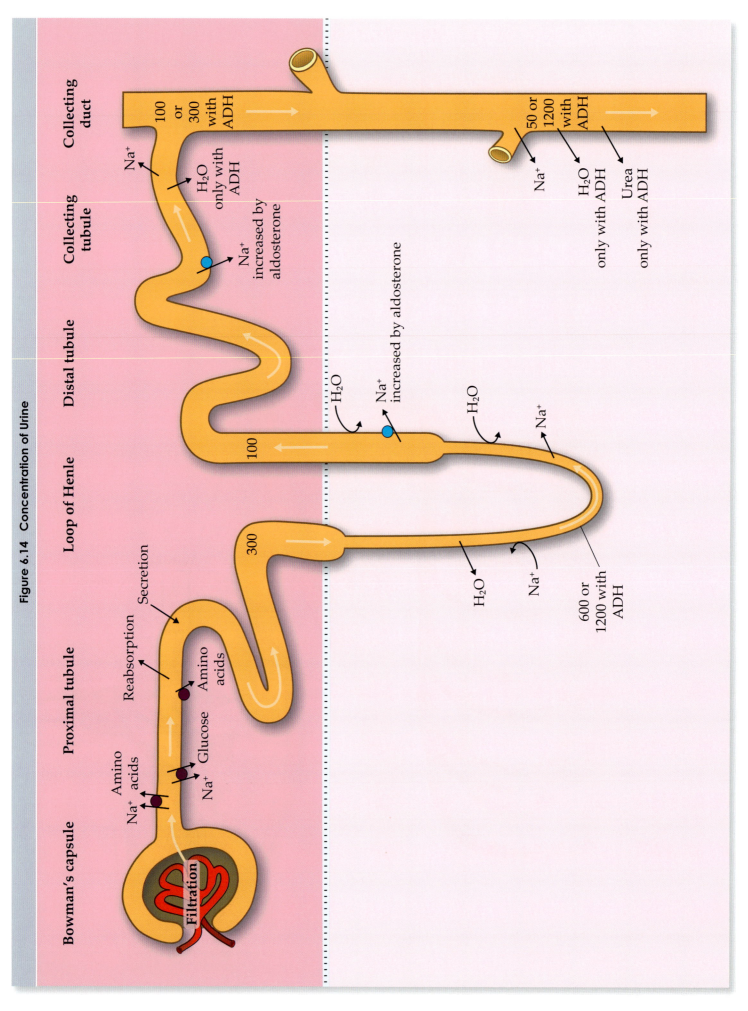

Figure 6.14 Concentration of Urine

Terms		
Antiport	Glycogenesis	Pancreatic Amylase
Bicarbonate Ion	Glycogenolysis	Parietal (Oxyntic) Cells
Bladder	Goblet Cells	Pepsinogen
Bowman's Capsule	Hydrochloric Acid	Peristalsis
Brush Border	(HCl)	Proximal Tubule
Chief (Peptic) Cells	Hydrostatic Pressure	Reabsorption
Chyme	Juxtaglomerular	Rectum
Chymotrypsin	Apparatus	Renal Corpuscle
Collecting Duct	Kidney	Renal Pelvis
Cortex	Lacteal	Ribonuclease
Deoxyribonuclease	Large Intestine	Salivary Amylase
Distal Tubule	Lipase	Small Intestine
Esophagus	Liver	Stomach
Exocrine Glands	Loop of Henle	Triglycerides
Fenestrations	Medulla	Trypsin
G Cells	Microvilli	Urea
Gallbladder	Mouth	Ureter
Gastrin	Mucous Cells	Urethra
Glomerulus	Nephron	

Notes:

Questions 137 through 144 are **NOT** based on a descriptive passage.

137. Bowman's capsule assists in clearing urea from the blood by:

    A. actively transporting urea into the filtrate using ATP-driven pumps.
    B. exchanging urea for glucose in an antiport mechanism.
    C. allowing urea to diffuse into the filtrate under filtration pressure.
    D. converting urea to amino acids.

138. Tests reveal the presence of glucose in a patient's urine. This is an indication that:

    A. glucose transporters in the loop of Henle are not functioning properly.
    B. the patient is healthy, as glucose normally appears in the urine.
    C. the proximal tubule is over-secreting glucose.
    D. glucose influx into the filtrate is occurring faster than it can be reabsorbed.

139. The epithelial cells of the proximal convoluted tubule contain a brush border similar to the brush border of the small intestine. The most likely function of the brush border in the proximal convoluted tubule is to:

    A. increase the amount of filtrate that reaches the loop of Henle.
    B. increase the surface area available for the absorption.
    C. slow the rate of at which the filtrate moves through the nephron.
    D. move the filtrate through the nephron with cilia like action.

140. If a patient were administered a drug that selectively bound and inactivated renin, which of the following would most likely result?

    A. The patient's blood pressure would increase.
    B. Platelets would be found in the urine.
    C. The amount of filtrate entering Bowman's capsule would increase.
    D. Sodium reabsorption by the distal tubule would decrease.

141. Which of the following correctly orders the structures through which urine flows as it leaves the body?

    A. urethra, urinary bladder, ureter, collecting duct
    B. collecting duct, urinary bladder, urethra, ureter
    C. collecting duct, ureter, urinary bladder, urethra
    D. ureter, collecting duct, urethra, urinary bladder

142. How are the blood levels of vasopressin and aldosterone in a dehydrated individual likely to compare with those of a healthy individual?

    A. Vasopressin and aldosterone levels are likely to be lower in a dehydrated individual.
    B. Vasopressin and aldosterone levels are likely to be higher in a dehydrated individual.
    C. Vasopressin levels are likely to be higher while aldosterone levels are likely to be lower in a dehydrated individual.
    D. Vasopressin levels are likely to be lower while aldosterone levels are likely to be higher in a dehydrated individual.

143. An afferent arteriole in a glomerular tuft contains microscopic fenestrations which increase fluid flow. In a hypertensive patient (a patient with high blood pressure):

    A. these fenestrations would constrict resulting in decreased urinary output.
    B. filtrate volume would be expected to be larger due to increased fluid pressure.
    C. filtrate volume would be expected to be smaller due to increased fluid pressure.
    D. urinary output will most likely be diminished due to increased solute concentration.

144. Long loops of Henle on juxtamedullary nephrons allow for greater concentration of urine. For an individual with highly concentrated urine, filtrate entering the loop of Henle is likely to be:

    A. more concentrated than filtrate exiting the loop Henle.
    B. less concentrated than filtrate exiting the loop of Henle.
    C. more voluminous than filtrate exiting the loop of Henle.
    D. less voluminous than filtrate exiting the loop of Henle.

THE CARDIOVASCULAR SYSTEM; THE RESPIRATORY SYSTEM

7.1 Cardiovascular Anatomy

The cardiovascular system consists of the heart, blood, and blood vessels (Figure 7.1). For the MCAT, you must be able to trace the circulatory path of the blood. Beginning with the **left ventricle**, blood is pumped through the **aorta**. From the aorta, branch many smaller **arteries**, which themselves branch into still smaller **arterioles**, which branch into still smaller **capillaries**. Blood from the capillaries is collected into **venules**, which themselves collect into larger **veins**, which collect again into the **superior and inferior vena cava**. The vena cava empty into

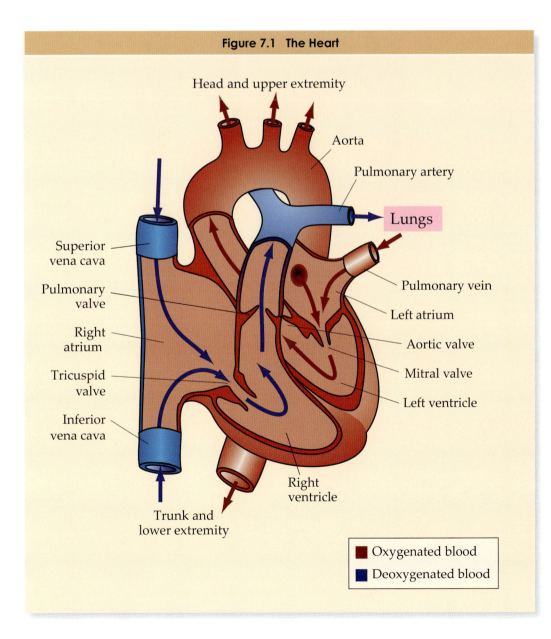

Figure 7.1 The Heart

Head and upper extremity

Aorta

Pulmonary artery

Lungs

Superior vena cava

Pulmonary valve

Right atrium

Tricuspid valve

Inferior vena cava

Pulmonary vein

Left atrium

Aortic valve

Mitral valve

Left ventricle

Right ventricle

Trunk and lower extremity

■ Oxygenated blood
■ Deoxygenated blood

the **right atrium** of the heart. This first half of the circulation as just described is called the **systemic circulation**. From the right atrium, blood is squeezed into the **right ventricle**. The right ventricle pumps blood through the **pulmonary arteries**, to arterioles, to the capillaries of the lungs. From the capillaries of the lungs, blood collects in venules, then in veins, and finally in the **pulmonary veins** leading to the heart. (True capillaries branch off arterioles, and do not represent the only route between an arteriole and venule.) The pulmonary veins empty into the **left atrium**, which fills the left ventricle. This second half of the circulation is called the **pulmonary circulation**. Since there are no openings for the blood to leave the vessels, the entire system is said to be a **closed circulatory system**.

The heart itself is a large muscle. Unlike skeletal muscle, it is not attached to bone. Instead, its fibers form a net and the net contracts upon itself squeezing blood into the arteries. Systole occurs when the ventricles contract; diastole occurs during relaxation of the entire heart and then contraction of the atria.

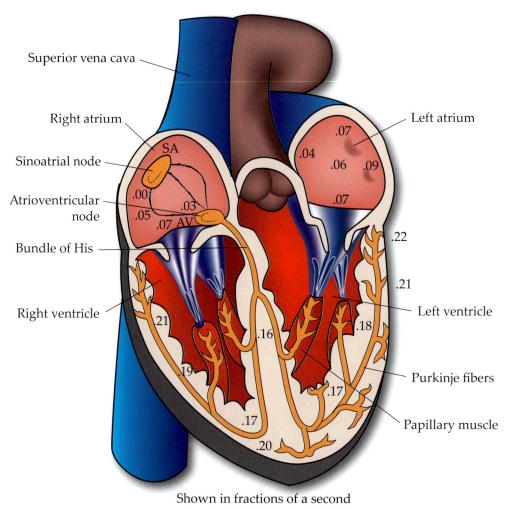

Shown in fractions of a second

Figure 7.2 Transmission of the Cardiac Impulse

The blood is propelled by the hydrostatic pressure created by the contraction of the heart. The rate of these contractions is controlled by the autonomic nervous system, but the autonomic nervous system does not initiate the contractions. The heart contracts automatically, paced by a group of specialized cardiac muscle cells called the **sinoatrial node (SA node)** located in the right atrium. The SA node is *autorhythmic* (contracts by itself at regular intervals), spreading its contractions to the surrounding cardiac muscles via **electrical synapses** made from **gap junctions**. The pace of the SA node is faster than normal heartbeats but the parasympathetic **vagus nerve** innervates the SA node, slowing the contractions. The action potential generated by the SA node spreads around both atria causing them to contract, and, at the same time, spreads to the **atrioventricular node (AV node)** located in the interatrial septa (the wall of cardiac muscle between the atria). The AV node is slower to contract, creating a delay which allows the atria to finish their contraction, and to squeeze their contents into the ventricles before the ventricles begin to contract. From the AV node, the action potential moves down conductive fibers called the **bundle of His**. The bundle of His is located in the wall separating the ventricles. The action potential branches out through the ventricular walls via conductive fibers called Purkinje fibers. From the **Purkinje fibers**, the action potential is spread through gap junctions from one cardiac muscle to the next. The Purkinje fibers in the ventricles allow for a more unified, and stronger, contraction.

Arteries are elastic, and stretch as they fill with blood. When the ventricles finish their contraction, the stretched arteries recoil, keeping the blood moving more smoothly. Arteries are wrapped in smooth muscle that is typically innervated by the sympathetic nervous system. Epinephrine is a powerful vasoconstrictor causing arteries to narrow. Larger arteries have less smooth muscle per volume than medium size arteries, and are less affected by sympathetic innervation. Medium sized arteries, on the other hand, constrict enough under sympathetic stimulation to reroute blood. **Arterioles** are very small. They are wrapped by smooth muscle. Constriction and dilation of arterioles can be used to regulate blood pressure as well as rerouting blood.

Capillaries are microscopic blood vessels (Figure 7.3). Capillary walls are only one cell thick, and the diameter of a capillary is roughly equal to that of a single red blood cell. Nutrient and gas exchange with any tissue other than vascular tissue takes place only across capillary walls, and not across arterioles or venules. There are four methods for materials to cross capillary walls:

1. pinocytosis;
2. diffusion or transport through capillary cell membranes;
3. movement through pores in the cells called fenestrations;
4. movement through the space between the cells.

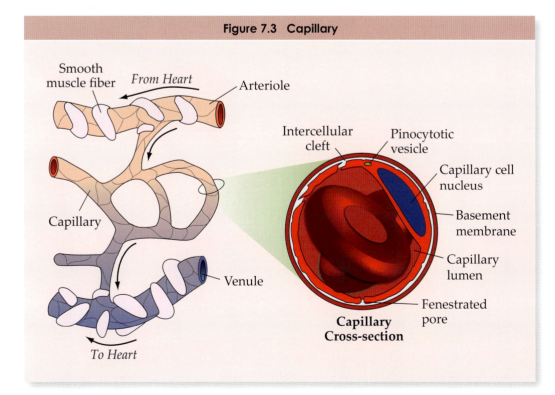

Figure 7.3 Capillary

Smooth muscle fiber — *From Heart* — Arteriole
Capillary
Venule
To Heart

Intercellular cleft — Pinocytotic vesicle — Capillary cell nucleus — Basement membrane — Capillary lumen — Fenestrated pore

Capillary Cross-section

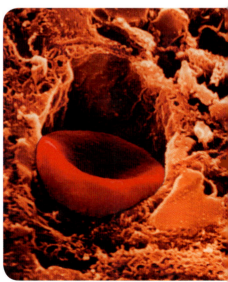

A red blood cell enters a capillary.

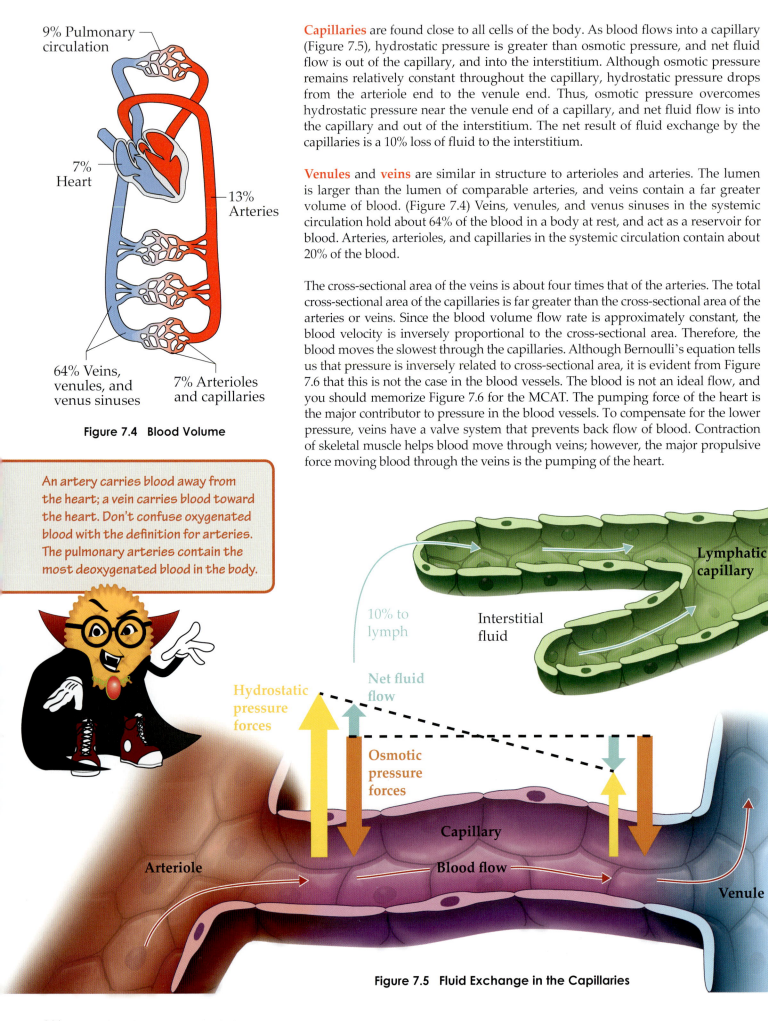

9% Pulmonary circulation

7% Heart

13% Arteries

64% Veins, venules, and venus sinuses

7% Arterioles and capillaries

Figure 7.4 Blood Volume

An artery carries blood away from the heart; a vein carries blood toward the heart. Don't confuse oxygenated blood with the definition for arteries. The pulmonary arteries contain the most deoxygenated blood in the body.

Capillaries are found close to all cells of the body. As blood flows into a capillary (Figure 7.5), hydrostatic pressure is greater than osmotic pressure, and net fluid flow is out of the capillary, and into the interstitium. Although osmotic pressure remains relatively constant throughout the capillary, hydrostatic pressure drops from the arteriole end to the venule end. Thus, osmotic pressure overcomes hydrostatic pressure near the venule end of a capillary, and net fluid flow is into the capillary and out of the interstitium. The net result of fluid exchange by the capillaries is a 10% loss of fluid to the interstitium.

Venules and **veins** are similar in structure to arterioles and arteries. The lumen is larger than the lumen of comparable arteries, and veins contain a far greater volume of blood. (Figure 7.4) Veins, venules, and venus sinuses in the systemic circulation hold about 64% of the blood in a body at rest, and act as a reservoir for blood. Arteries, arterioles, and capillaries in the systemic circulation contain about 20% of the blood.

The cross-sectional area of the veins is about four times that of the arteries. The total cross-sectional area of the capillaries is far greater than the cross-sectional area of the arteries or veins. Since the blood volume flow rate is approximately constant, the blood velocity is inversely proportional to the cross-sectional area. Therefore, the blood moves the slowest through the capillaries. Although Bernoulli's equation tells us that pressure is inversely related to cross-sectional area, it is evident from Figure 7.6 that this is not the case in the blood vessels. The blood is not an ideal flow, and you should memorize Figure 7.6 for the MCAT. The pumping force of the heart is the major contributor to pressure in the blood vessels. To compensate for the lower pressure, veins have a valve system that prevents back flow of blood. Contraction of skeletal muscle helps blood move through veins; however, the major propulsive force moving blood through the veins is the pumping of the heart.

Lymphatic capillary

10% to lymph

Interstitial fluid

Net fluid flow

Hydrostatic pressure forces

Osmotic pressure forces

Capillary

Arteriole

Blood flow

Venule

Figure 7.5 Fluid Exchange in the Capillaries

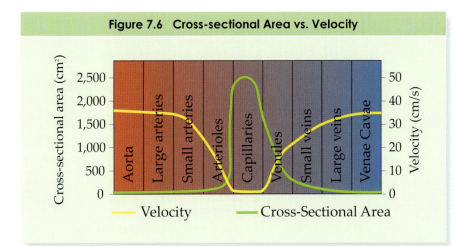

Figure 7.6 Cross-sectional Area vs. Velocity

— Velocity   — Cross-Sectional Area

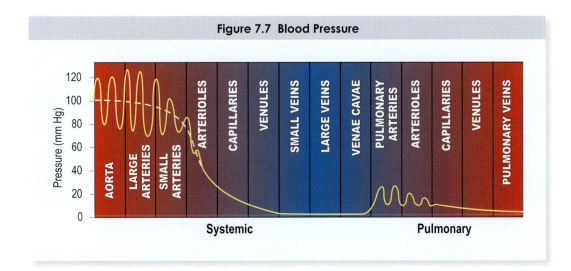

Figure 7.7 Blood Pressure

Systemic   Pulmonary

Information on blood pressure, volume, velocity, and cross-sectional area of vessels is more likely to be found in a physics passage. Don't memorize anything, but take a moment here to consider the relationships.

Pressure: Blood pressure increases near the heart and decreases to its lowest in the capillaries.

Velocity: A single artery is much bigger than a capillary, but there are far more capillaries than arteries. The total cross-sectional area of all those capillaries put together is much greater than the cross sectional area of a single aorta or a few arteries. Blood flow follows the Continuity Equation, Q = Av, reasonably well, so velocity is greatest in the arteries where cross-sectional area is smallest, and velocity is lowest where cross-sectional area is the greatest.

145. The atrioventricular node:

A. is a parasympathetic ganglion located in the right atrium of the heart.

B. conducts an action potential from the vagus nerve to the heart.

C. sets the rhythm of cardiac contractions.

D. delays the contraction of the ventricles of the heart.

146. Cardiac output, which is the product of the heart rate and the stroke volume (the amount of blood pumped per contraction by either the left or the right ventricle) would most likely be:

A. greater if measured using the stroke volume of the left ventricle.

B. greater if measured using the stroke volume of the right ventricle.

C. the same regardless of which stroke volume is used.

D. dependent on the viscosity of the blood.

147. Which of the following is responsible for the spread of the cardiac action potential from one cardiac muscle cell to the next?

A. gap junctions

B. desmosomes

C. tight junctions

D. acetylcholine

148. In the congenital heart defect known as patent ductus arteriosus, the ductus arteriosus, which connects the aorta and the pulmonary arteries during fetal development, fails to close at birth. This will likely lead to all of the following EXCEPT:

A. equal, or increased, oxygen concentration in the blood that reaches the systemic tissues.

B. increased oxygen concentration in the blood that reaches the lungs.

C. increased work load imposed on the left ventricle.

D. increased work load imposed on the right ventricle.

149. Which chambers of the heart pump oxygenated blood?

A. The right and left atria

B. The right and left ventricles

C. The right atria and the left ventricle

D. The left atria and the left ventricle

150. Hypovolemic shock represents a set of symptoms that occur when a patient's blood volume falls abruptly. Hypovolemic shock is most likely to occur during:

A. arterial bleeding

B. venous bleeding

C. low oxygen intake

D. excess sodium consumption

151. The capillary network comprises the greatest cross sectional area of blood vessels in the body with the highest resistance to blood flow. In a healthy individual, the highest blood pressure would most likely be found in:

A. the aorta

B. the vena cavae

C. the systemic capillaries

D. the pulmonary capillaries

152. Gas exchange between the blood and tissues occurs:

A. throughout the circulatory system.

B. in the arteries, arterioles and capillaries.

C. in the systemic arteries only.

D. in the capillaries only.

STOP.

7.2 The Respiratory System

The respiratory system provides a path for gas exchange between the external environment and the blood (Figure 7.8). Air enters through the nose, moves through the pharynx, larynx, trachea, bronchi, bronchioles, and into the alveoli where oxygen is exchanged for carbon dioxide with the blood. Inspiration occurs when the *medulla oblongata* of the midbrain signals the diaphragm to contract. The **diaphragm** is skeletal muscle, and innervated by the phrenic nerve. When relaxed, the diaphragm is dome-shaped. It flattens upon contraction, expanding the chest cavity and creating negative gauge pressure. (Gauge pressure is measured relative to local atmospheric conditions. See Physics Lecture 5.) *Intercostal muscles* (rib muscles) also help to expand the chest cavity. Atmospheric pressure forces air into the lungs. Upon relaxation of the diaphragm, the chest cavity shrinks (aided by different intercostal muscles and abdominal muscles), and the elasticity of the lungs along with the increased pressure in the chest cavity forces air out of the body.

Figure 7.8 Respiratory System

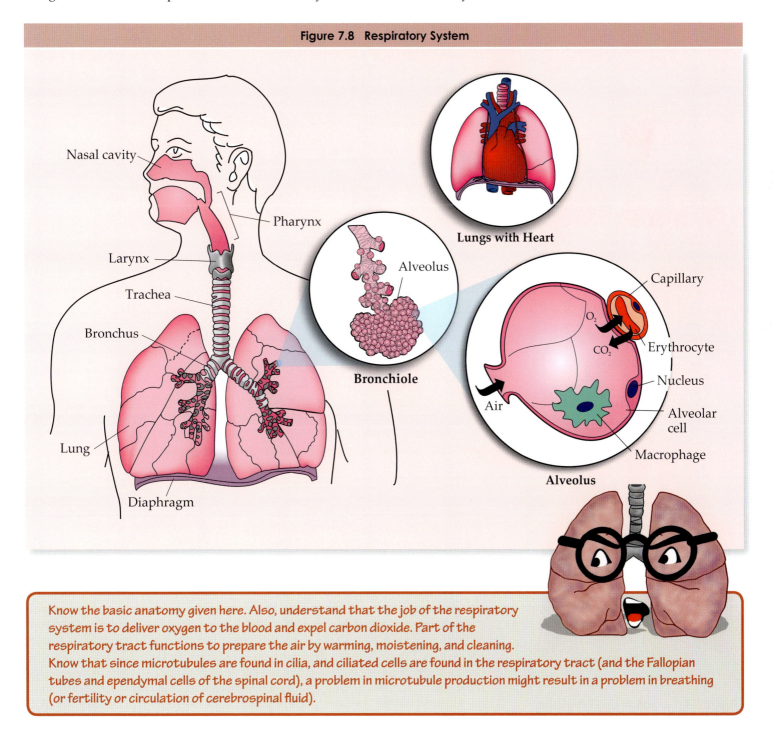

Know the basic anatomy given here. Also, understand that the job of the respiratory system is to deliver oxygen to the blood and expel carbon dioxide. Part of the respiratory tract functions to prepare the air by warming, moistening, and cleaning.
Know that since microtubules are found in cilia, and ciliated cells are found in the respiratory tract (and the Fallopian tubes and ependymal cells of the spinal cord), a problem in microtubule production might result in a problem in breathing (or fertility or circulation of cerebrospinal fluid).

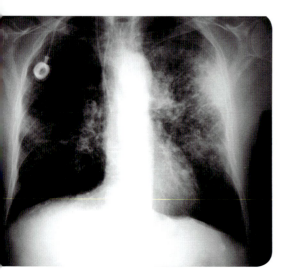

X-ray of a lung cancer.

The **nasal cavity** is the space inside the nose. It **filters, moistens,** and **warms** incoming air. **Coarse hair** at the front of the cavity traps large dust particles. **Mucus** secreted by goblet cells traps smaller dust particles and moistens the air. Capillaries within the nasal cavity warm the air. **Cilia** moves the mucus and dust back toward the pharynx, so that it may be removed by spitting or swallowing.

The **pharynx** (or throat) functions as a passageway for food and air.

The **larynx** is the voice box. It sits behind the **epiglottis**, which is the cartilaginous member that prevents food from entering the trachea during swallowing. When nongaseous material enters the larynx, a coughing reflex is triggered forcing the material back out. The larynx contains the vocal cords.

The **trachea** (or windpipe) lies in front of the esophagus. It is composed of ringed cartilage covered by ciliated mucous cells. Like the nasal cavity, the mucus and cilia in the trachea collect dust and usher it toward the pharynx. Before entering the lungs the trachea splits into the right and left **bronchi**. Each bronchus branches many more times to become tiny **bronchioles**. Bronchioles terminate in grape-like clusters called *alveolar sacs* composed of tiny **alveoli**. From each alveolus, oxygen diffuses into a capillary where it is picked up by red blood cells. The red blood cells release carbon dioxide, which diffuses into the alveolus, and is expelled upon exhalation.

7.3 The Chemistry of Gas Exchange

Typically, the air we inspire is 79% nitrogen and 21% oxygen, with negligible amounts of other trace gases. Exhaled air is 79% nitrogen, 16% oxygen, and 5% carbon dioxide and trace gases. Inside the lungs, the partial pressure of oxygen is approximately 110 mm Hg, and carbon dioxide is approximately 40 mm Hg. Under these pressures, oxygen diffuses into the capillaries, and carbon dioxide diffuses into the alveoli.

98% of the oxygen in the blood binds rapidly and reversibly with the protein **hemoglobin** inside the erythrocytes forming **oxyhemoglobin**. Hemoglobin is composed of four polypeptide subunits, each with a single heme cofactor. The *heme cofactor* is an organic molecule with an atom of **iron** at its center. Each of the four iron atoms in hemoglobin can bind with one O_2 molecule. When one **O_2 molecule** binds with an iron atom in hemoglobin, oxygenation of the other heme groups is accelerated. Similarly, release of an O_2 molecule by any of the heme groups, accelerates release by the others. This phenomenon is called *cooperativity*.

As O_2 pressure increases, the O_2 saturation of hemoglobin increases sigmoidally. The *oxyhemoglobin* (HbO_2) *dissociation curve* (Figure 7.10) shows the percent of hemoglobin that is bound with oxygen at various partial pressures of oxygen. In the arteries of a normal person breathing room air, the oxygen saturation is 97%. The flat portion of the curve in this region shows that small fluctuations in oxygen pressure have little effect.

The oxygen saturation of hemoglobin also depends upon carbon dioxide pressure, pH, and temperature of the blood. The **oxygen dissociation curve is shifted** to the right by an increase in **carbon dioxide pressure, hydrogen ion concentration, or temperature**. A shift to the right indicates a lowering of hemoglobin's affinity for oxygen. The shift due to pH change is called the *Bohr shift*. 2,3-DPG, a chemical found in red blood cells, also shifts the curve to the right. Carbon monoxide has more than 200 times greater affinity for hemoglobin than does oxygen but shifts the curve to the left. In cases of carbon monoxide poisoning, pure oxygen can be

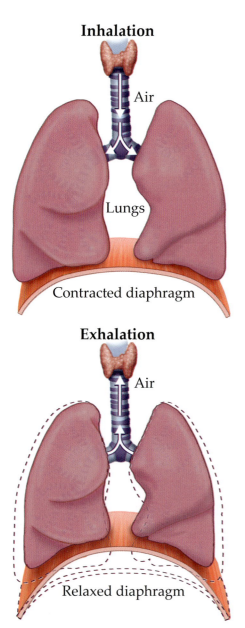

Inhalation

Air

Lungs

Contracted diaphragm

Exhalation

Air

Relaxed diaphragm

Figure 7.9 Inhalation and Exhalation

administered to displace the CO from hemoglobin.

Oxygen pressure is typically 40 mm Hg in body tissues. As the blood moves through the systemic capillaries, oxygen diffuses to the tissues, and carbon dioxide diffuses to the blood. Carbon dioxide is carried by the blood in three forms:

1. in physical solution;
2. as bicarbonate ion and;
3. in carbamino compounds (combined with hemoglobin and other proteins).

Ten times as much is carried as bicarbonate than as either of the other forms. The bicarbonate ion formation is governed by the enzyme **carbonic anhydrase** in the reversible reaction:

$$CO_2 + H_2O \rightarrow HCO_3^- + H^+$$

Because carbonic anhydrase is inside the red blood cell and not in the plasma, when carbon dioxide is absorbed in the lungs, bicarbonate ion diffuses into the cell. To balance the electrostatic forces, chlorine moves out of the cell in a phenomenon called the *chloride shift* (Figure 7.11).

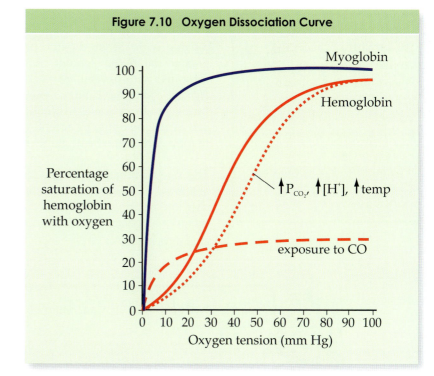

Figure 7.10 Oxygen Dissociation Curve

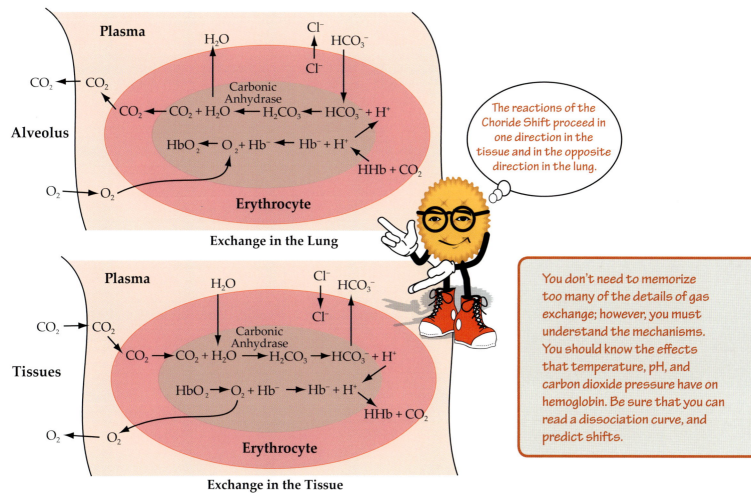

The reactions of the Choride Shift proceed in one direction in the tissue and in the opposite direction in the lung.

Exchange in the Lung

Exchange in the Tissue

Figure 7.11 Chloride Shift

You don't need to memorize too many of the details of gas exchange; however, you must understand the mechanisms. You should know the effects that temperature, pH, and carbon dioxide pressure have on hemoglobin. Be sure that you can read a dissociation curve, and predict shifts.

Carbon dioxide has its own dissociation curve, which relates blood content of carbon dioxide with carbon dioxide pressure. The greater the pressure of carbon dioxide, the greater the blood content of carbon dioxide. However, when hemoglobin becomes saturated with oxygen, its capacity to hold carbon dioxide is reduced. This is called the *Haldane effect*. The Haldane effect facilitates the transfer of carbon dioxide from blood to lungs, and from tissues to blood. Reduced hemoglobin (Hb) (hemoglobin without oxygen) acts as a blood buffer by accepting protons. It is the greater capacity of reduced hemoglobin to form carbamino hemoglobin that explains the Haldane effect.

The rate of breathing is affected by *central chemoreceptors* located in the medulla, and *peripheral chemoreceptors* located in the carotid arteries and aorta. Central and peripheral chemoreceptors monitor carbon dioxide concentration in the blood and increase breathing when levels get too high. Oxygen concentration and pH are monitored mainly by peripheral chemoreceptors.

What about all that nitrogen? What effect does nitrogen have on the body?

Remember your chemistry. Nitrogen is extremely stable due to its strong triple bond. Thus, nitrogen diffuses into the blood, but doesn't react with the chemicals in the blood. However, people that go diving must be careful. As the pressure increases with depth, more nitrogen diffuses into the blood. When divers come back up, the pressure decreases and the gas volume increases. If they don't allow enough time for the nitrogen to diffuse out of the blood and into the lungs, the nitrogen will form bubbles. Among other problems, these bubbles may occlude (block) vessels causing decompression sickness also known as 'the bends.'

Notes:

153. Alkalosis is increased blood pH resulting in a leftward shift of the oxy hemoglobin dissociation curve. Which of the following might cause alkalosis?

- **A.** hypoventilation
- **B.** hyperventilation
- **C.** breathing into a paper bag
- **D.** adrenal steroid insufficiency

154. Which of the following would most likely occur in the presence of a carbonic anhydrase inhibitor?

- **A.** The blood pH would increase.
- **B.** The carbamino hemoglobin concentration inside erythrocytes would decrease.
- **C.** The rate of gas exchange in the lungs would decrease.
- **D.** The oxy hemoglobin concentration inside erythrocytes would increase.

155. Which of the following will most likely occur during heavy exercise?

- **A.** Blood pH will decrease in the active tissues.
- **B.** Less oxygen will be delivered to the tissues due to increased cardiac contractions resulting in increased blood velocity.
- **C.** Capillaries surrounding contracting skeletal muscles will constrict to allow increased freedom of movement.
- **D.** The respiratory system will deliver less nitrogen to the blood.

156. An athlete can engage in blood doping by having blood drawn several weeks before an event, removing the blood cells, and having them reinjected into her body a few days before an athletic activity. Blood doping is most likely an advantage to athletes because:

- **A.** the increased concentration of immune cells in the blood after reinjection can decrease the chances of becoming ill just before the competition.
- **B.** the increased red blood cell count in the blood after reinjection can facilitate greater gas exchange with the tissues.
- **C.** the increased blood volume after reinjection can ensure that the athlete maintains adequate hydration during the event.
- **D.** the decreased red blood cell count of the blood in the weeks before the competition can facilitate training by decreasing the viscosity of the blood.

157. Carbon dioxide partial pressure:

- **A.** increases in the blood as it travels from the systemic venules to the inferior vena cava.
- **B.** increases in the blood as it travels from the pulmonary arteries to the pulmonary veins.
- **C.** is greater in the blood in the systemic capillary beds than in the alveoli of the lungs.
- **D.** is greater in the blood in the systemic capillary beds than in the systemic tissues.

158. At high altitude, water vapor pressure in the lungs remains the same and carbon dioxide pressure falls slightly. Oxygen pressure falls. The body of a person remaining at high altitudes for days, weeks, and even years will acclimatize. All of the following changes assist the body in coping with low oxygen EXCEPT:

- **A.** increased red blood cells.
- **B.** decreased vascularity of the tissues.
- **C.** increased pulmonary ventilation.
- **D.** increased diffusing capacity of the lungs.

159. In an asthma attack, a patient suffers from difficulty breathing due to constricted air passages. The major causative agent is a mixture of leukotrienes called slow reacting substance of anaphylaxis. During an asthma attack, slow reacting substance of anaphylaxis most likely causes:

- **A.** smooth muscle spasms of the bronchioles.
- **B.** cartilaginous constriction of the trachea.
- **C.** edema in the alveoli.
- **D.** skeletal muscle spasms in the thorax.

160. Sustained heavy exercise results in all of the following changes to blood chemistry EXCEPT:

- **A.** lowered pH.
- **B.** raised CO_2 tension.
- **C.** increased temperature.
- **D.** decreased carboxyhemoglobin.

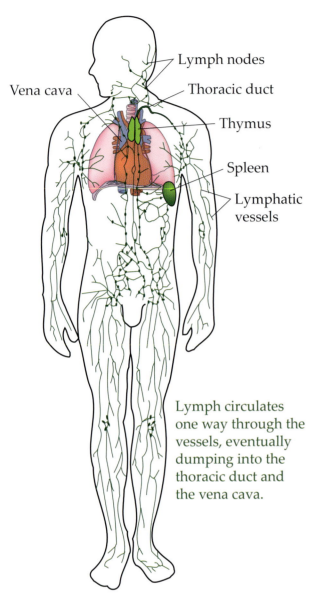

Lymph nodes

Vena cava

Thoracic duct

Thymus

Spleen

Lymphatic vessels

Lymph circulates one way through the vessels, eventually dumping into the thoracic duct and the vena cava.

Figure 7.12 Lymphatic System

7.4   *The Lymphatic System*

The lymphatic system collects excess interstitial fluid and returns it to the blood. Proteins and large particles that cannot be taken up by the capillaries are removed by the lymph system. The pathway to the blood takes the excess fluid through lymph nodes, which are well prepared to elicit an immune response if necessary. Thus, the lymph system recycles the interstitial fluid and monitors the blood for infection. In addition, the lymph system reroutes low soluble fat digestates around the small capillaries of the intestine and into the large veins of the neck. Most tissues are drained by lymphatic channels. A notable exception is the central nervous system.

The lymph system is an **open system**. In other words, fluid enters at one end and leaves at the other. Lymph capillaries are like tiny fingers protruding into the tissues. To enter the lymph system, interstitial fluid flows between overlapping endothelial cells (Figure 7.13). Large particles literally push their way between the cells into the lymph. The cells overlap in such a fashion that, once inside, large particles cannot push their way out.

Typically, interstitial fluid pressure is slightly negative. (Of course, we mean gauge pressure. See Physics Lecture 5.) As the interstitial pressure rises toward zero, lymph flow increases. Factors that affect interstitial pressure include: blood pressure; plasma osmotic pressure; interstitial osmotic pressure (e.g. from proteins, infection response, etc.); permeability of capillaries. Like veins, lymph vessels are constructed with intermittent valves, which allow fluid to flow in only one direction. Fluid is propelled through these valves in two ways. First, smooth muscle in the walls of larger lymph vessels contracts when stretched. Second, the lymph vessels may be squeezed by adjacent skeletal muscles, body movements, arterial pulsations, and compression from objects outside the body. Lymph flow in an active individual is considerably greater than in an individual at rest.

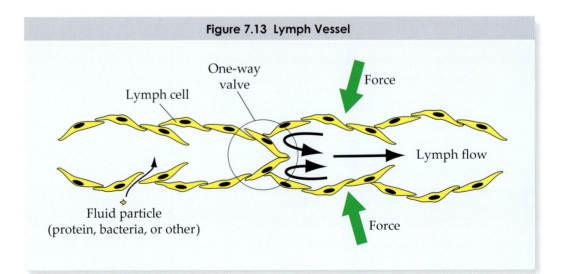

Figure 7.13 Lymph Vessel

One-way valve

Force

Lymph cell

Lymph flow

Fluid particle (protein, bacteria, or other)

Force

The lymph system empties into large veins at the *thoracic duct* and the *right lymphatic* duct. Lymph from the right arm and head enters the blood through the right lymphatic duct. The rest of the body is drained by the thoracic duct.

Throughout the lymphatic system there are many lymph nodes, containing large quantities of lymphocytes.

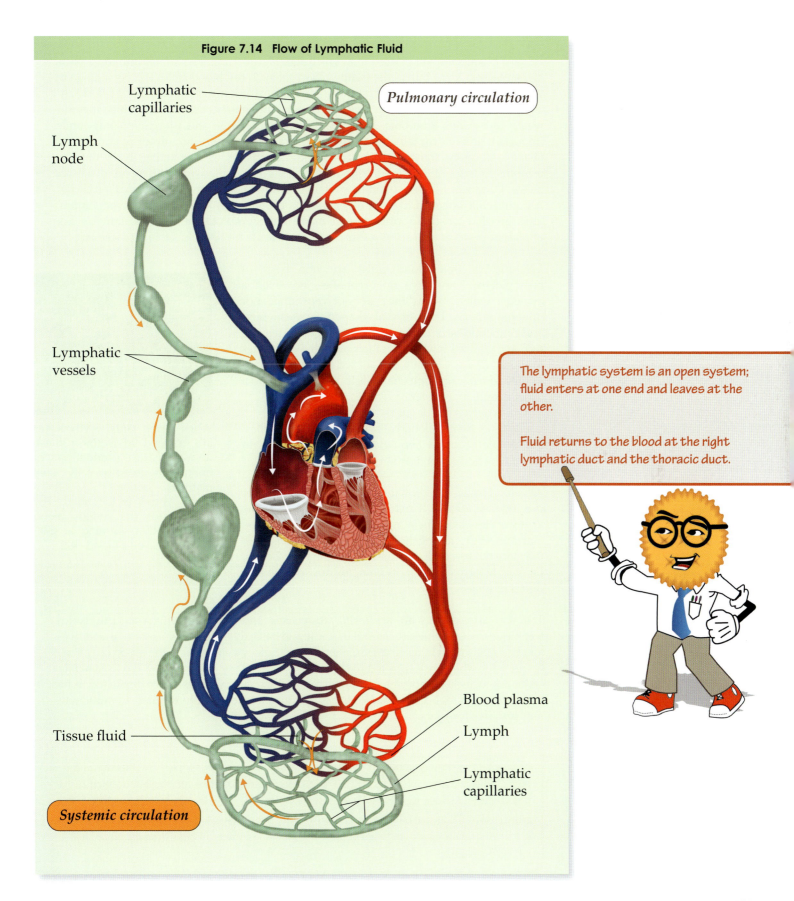

Figure 7.14   Flow of Lymphatic Fluid

Lymphatic capillaries

Lymph node

Pulmonary circulation

Lymphatic vessels

The lymphatic system is an open system; fluid enters at one end and leaves at the other.

Fluid returns to the blood at the right lymphatic duct and the thoracic duct.

Blood plasma

Lymph

Tissue fluid

Lymphatic capillaries

Systemic circulation

Figure 7.15 Blood Composition

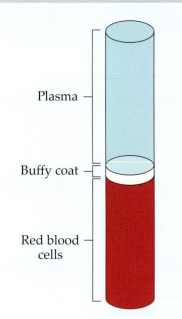

Plasma

Buffy coat

Red blood cells

7.5 The Blood

The blood is connective tissue. Like any **connective tissue**, it contains cells and a matrix. Blood regulates the extracellular environment of the body by transporting nutrients, waste products, hormones, and even heat. Blood also protects the body from injury and foreign invaders.

When a blood sample is placed in a centrifuge (Figure 7.15), it separates into three parts: 1. the plasma; 2. the buffy coat (white blood cells); 3. red blood cells. The percentage by volume of red blood cells is called the *hematocrit*. Hematocrit is normally 35-50%, and is greater in men than women. Plasma contains the matrix of the blood, which includes water, ions, urea, ammonia, proteins, and other organic and inorganic compounds. Important proteins contained in the plasma are **albumin**, **immunoglobulins**, and clotting factors. Albumins transport fatty acids and steroids, as well as acting to regulate the osmotic pressure of the blood. Immunoglobulins (also called **antibodies**) are discussed below. Plasma in which the clotting protein **fibrinogen** has been removed is called **serum**. Albumin, fibrinogen, and most other plasma proteins are formed in the liver. Gamma globulins that constitute antibodies are made in the lymph tissue. An important function of plasma proteins is to act as a source of amino acids for tissue protein replacement.

> Know that the job of erythrocytes is to deliver oxygen and remove carbon dioxide.

Erythrocytes (red blood cells) are like bags of hemoglobin. They have no organelles, not even a nucleus, which means they do not reproduce nor undergo mitosis. They are disk-shaped vesicles whose main function is to transport O_2 and CO_2. Squeezing through capillaries wears out their plasma membranes in about 120 days. Most worn out red blood cells burst as they squeeze through channels in the spleen or, to a lesser extent, in the liver.

Leukocytes (white blood cells) do contain organelles, but do not contain hemoglobin. They function to protect the body from foreign invaders.

All blood cells differentiate from the same type of precursor, a **stem cell** residing in the bone marrow. Erythrocytes lose their nucleus while still in the marrow. After entering the blood stream as *reticulocytes*, they lose the rest of their organelles within 1 or 2 days. Leukocyte formation is more complex due to the many different types.

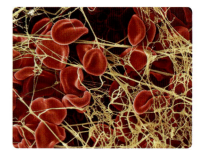

When blood clots, erythrocytes are trapped in a fibrin mesh (brown). The production of fibrin is triggered by cells called platelets, activated when a blood vessel is damaged. The fibrin binds the various blood cells together, forming a solid structure called a blood clot.

The *granular leukocytes* are *neutrophils, eosinophils,* and *basophils*. With respect to dyeing techniques, neutrophils are neutral to acidic and basic dyes, eosinophils stain in acid dyes, and basophils stain in basic dyes. Generally, granulocytes remain in the blood only 4 to 8 hours before they are deposited in the tissues, where they live for 4 to 5 days. *Agranular leukocytes* include monocytes, lymphocytes, and megakaryocytes. Once deposited in the tissues, monocytes become macrophages and may live for months to years. Lymphocytes may also live for years.

Figure 7.16 Granulocytes

Neutrophil

Eosinophil

Basophil

> Notice that granulocytes live a very short time, whereas agranulocytes-other white blood cells-live a very long time. This is because granulocytes function nonspecifically against all infective agents, whereas most agranulocytes work against specific agents of infection. Thus, agranulocytes need to hang around in case the same infective agent returns; granulocytes multiply quickly against any infection, and then die once the infection is gone.

Platelets are small portions of membrane-bound cytoplasm torn from megakaryocytes. *Megakaryocytes* remain mainly in the bone marrow. Platelets are similar to tiny cells without a nucleus. They contain actin and myosin, residuals of the Golgi and the ER, mitochondria, and are capable of making protaglandins and some important enzymes. Its membrane is designed to avoid adherence to healthy endothelium while adhering to injured endothelium. When platelets come into conact with injured endothelium, they become sticky and begin to swell releasing various chemicals and activating other platelets. The platelets stick to the endothelium and to each other forming a loose *platelet plug*. Healthy individuals have many platelets in their blood. Platelets The platelet has a half-life of 8-12 days in the blood.

Coagulation occurs in three steps: 1) A dozen or so coagulation factors form a comlex called *protrombin activator*. 2) Protrombin activator catalyzes the conversion of *prothrombin*, a plasma protein, into *thrombin*. 3) Thrombin is an enzyme that governs the polymerization of the plasma protein fibrinogen to fibrin threads that attach to the platelets and form a tight plug. This *blood cot* formation (or coagulation) begins to appear in seconds in small injuries and 1 to 2 minutes in larger injuries.

> Just know that the coagulation process involves many factors starting wth platelets and include the plasma proteins prothrombin and fibrin.

The following is the leukocyte composition in the blood:

Neutrophils 62%
Lymphocytes................................30%
Monocytes 5.3%
Eosinophils..................................2.3%
Basophils 0.4%

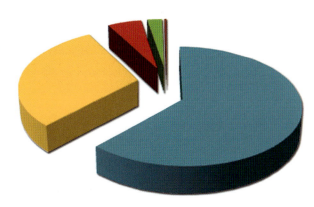

7.6 The Immune System

The human body protects itself from infectious microbes and toxins in two ways: *innate immunity* and *acquired immunity*. Innate immunity involves a generalized protection from most intruding organisms and toxins. Acquired immunity is protection against specific organisms or toxins. Acquired immunity develops after the body is first attacked. Innate immunity includes:

1. the skin as a barrier to organisms and toxins;

2. stomach acid and digestive enzymes to destroy ingested organisms and toxins;

3. phagocytotic cells; and

4. chemicals in the blood.

Injury to tissue results in **inflammation**, which includes dilation of blood vessels, increased permeability of capillaries, swelling of tissue cells, and migration of granulocytes and macrophages to the inflamed area (Figure 7.17). *Histamine, prostaglandins,* and *lymphokines* are just some of the causative agents of inflammation that are released by the tissues. Part of the effect of inflammation is to 'wall-off' the affected tissue and local lymph vessels from the rest of the body, impeding the spread of the infection.

Infectious agents that are able to pass through the skin or the digestive defenses and enter the body are first attacked by local *macrophages*. These phagocytotic giants can engulf as many as 100 bacteria. *Neutrophils* are next on the scene. Most neutrophils are stored in the bone marrow until needed, but some are found circulating in the blood or in the tissues. Neutrophils move toward infected or injured areas, drawn by chemicals (a process called *chemotaxis*) released from damaged tissue or by the infectious agents themselves. To enter the tissues, neutrophils slip between endothelial cells of the capillary walls, using an amaeboid-like process called *diapedesis*. A single neutrophil can phagocytize from 5 to 20 bacteria.

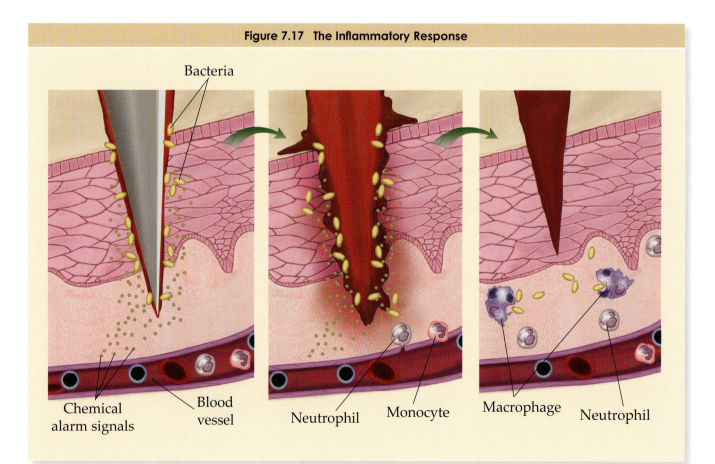

Figure 7.17   The Inflammatory Response

Bacteria

Chemical alarm signals

Blood vessel

Neutrophil   Monocyte

Macrophage   Neutrophil

Monocytes circulate in the blood until they, too, move into the tissues by diapedesis. Once inside the tissues, monocytes mature to become macrophages.

When the neutrophils and macrophages engulf necrotic tissue and bacteria, they die. These dead leukocytes, along with tissue fluid and necrotic tissue, make up what is known as *pus*.

Eosinophils work mainly against parasitic infections. *Basophils* release many of the chemicals of the inflammation reaction.

There are two types of acquired immunity: **humoral** or **B-cell immunity**; **cell-mediated** or **T-cell immunity**. Humoral immunity is promoted by **B lymphocytes**. B lymphocytes differentiate and mature in the bone marrow and the liver. Each B lymphocyte is capable of making a single type of **antibody** or (**immunoglobulin**), which it displays on its membrane. An antibody recognizes a foreign particle, called an **antigen**. The portion of the antibody that binds to an antigen is highly specific for that antigen. The portion of the antigen that binds to the antibody is called an *antigenic determinant*. An antigenic determinant that is removed from an antigen is called a *hapten*. Haptens can only stimulate an immune response if the individual has been previously exposed to the full antigen. Macrophages present the antigenic determinants of engulfed microbes on their surfaces. If the B lymphocyte antibody contacts a matching antigen (presented by a macrophage), the B lymphocyte, assisted by a **helper T cell**, differentiates into **plasma cells** and **memory B cells**. Plasma cells begin synthesizing free antibodies, and releasing them into the blood. Free antibodies may attach their base to *mast cells*. When an antibody whose base is bound to a mast cell also binds to an antigen, the mast cell releases histamine and other chemicals. When other free antibodies contact the specific antigen, they bind to it. Once bound, the antibodies may begin a cascade of reactions involving blood proteins (called *complement*) that cause the antigen bearing cell to be perforated. The antibodies may mark the antigen for phagocytosis by macrophages and *natural killer cells*. The antibodies may cause the antigenic substances to **agglutinate** or even precipitate, or, in the case of a toxin, the antibodies may block its chemically active portion. The first time the immune system is exposed to an antigen is known as the **primary response**. The primary immune response requires 20 days to reach its full potential.

Memory B cells proliferate, and remain in the body. In the case of re-infection, each of these cells can be called upon to synthesize antibodies, resulting in a faster acting and more potent effect called the **secondary response**. The secondary response requires approximately 5 days to reach its full potential.

Humoral immunity is effective against bacteria, fungi, parasitic protozoans, viruses, and blood toxins.

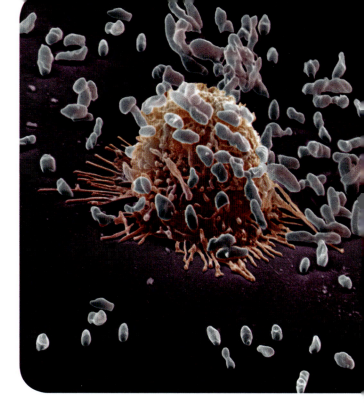

Macrophages ingest bacteria as part of the immune response to infection.

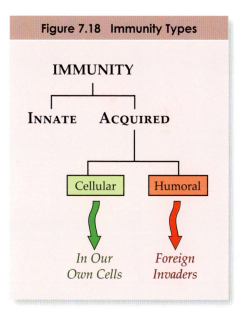

Figure 7.18 Immunity Types

IMMUNITY

INNATE ACQUIRED

Cellular Humoral

In Our Own Cells *Foreign Invaders*

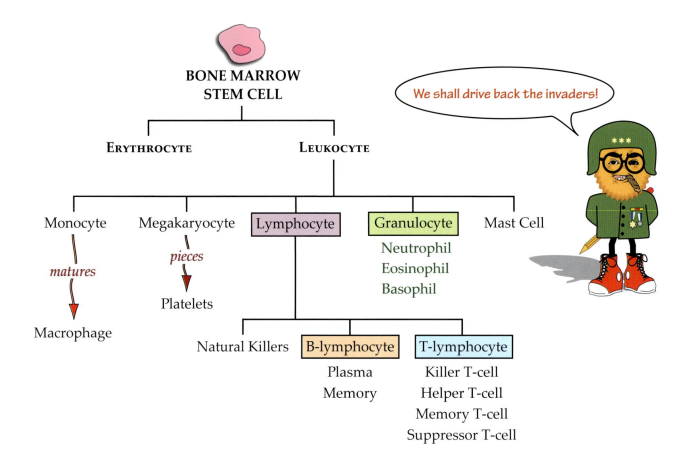

BONE MARROW
STEM CELL

We shall drive back the invaders!

ERYTHROCYTE — LEUKOCYTE

Monocyte Megakaryocyte Lymphocyte Granulocyte Mast Cell

Neutrophil
Eosinophil
Basophil

matures *pieces*

Platelets

Macrophage

Natural Killers B-lymphocyte T-lymphocyte

Plasma Killer T-cell
Memory Helper T-cell
Memory T-cell
Suppressor T-cell

Cell-mediated immunity involves **T-lymphocytes**. T-lymphocytes mature in the thymus. Similar to B lymphocytes, T lymphocytes have an antibody-like protein at their surface that recognizes antigens. However, T-lymphocytes never make free antibodies. In the thymus, T-lymphocytes are tested against *self-antigens* (antigens expressed by normal cells of the body). If the T-lymphocyte binds to a self-antigen, that T lymphocyte is destroyed. If it does not, it is released to lodge in lymphoid tissue or circulate between the blood and the lymph fluid. T lymphocytes that are not destroyed differentiate into **helper T cells**, **memory T cells**, **suppressor T cells**, and **killer T cells** (also called *cytotoxic* T cells). As discussed above, T helper cells assist in activating B lymphocytes as well as killer and suppressor T cells. Helper T cells are the cells attacked by HIV. Memory T cells have a similar function to Memory B cells. Suppressor T cells play a negative feedback role in the immune system. Killer T cells bind to the antigen-carrying cell and release *perforin*, a protein which punctures the antigen-carrying cell. Killer T cells can attack many cells because they do not phagocytize their victims. Killer T cells are responsible for fighting some forms of cancer, and for attacking transplanted tissue.

Cell-mediated immunity is effective against infected cells.

Let's imagine a bacterial infection. First we have inflammation. Macrophages, then neutrophils, engulf the bacteria. Interstitial fluid is flushed into the lymphatic system where lymphocytes wait in the lymph nodes. Macrophages process and present the bacterial antigens to B lymphocytes. With the help of Helper T cells, B lymphocytes differentiate into memory cells and plasma cells. The memory cells are preparation in the event that the same bacteria ever attack again (the secondary response). The plasma cells produce antibodies, which are released into the blood to attack the bacteria.

You must know that a single antibody is specific for a single antigen, and that a single B lymphocyte produces only one antibody type.

Figure 7.19 Summary Diagram of Adaptive Immunity

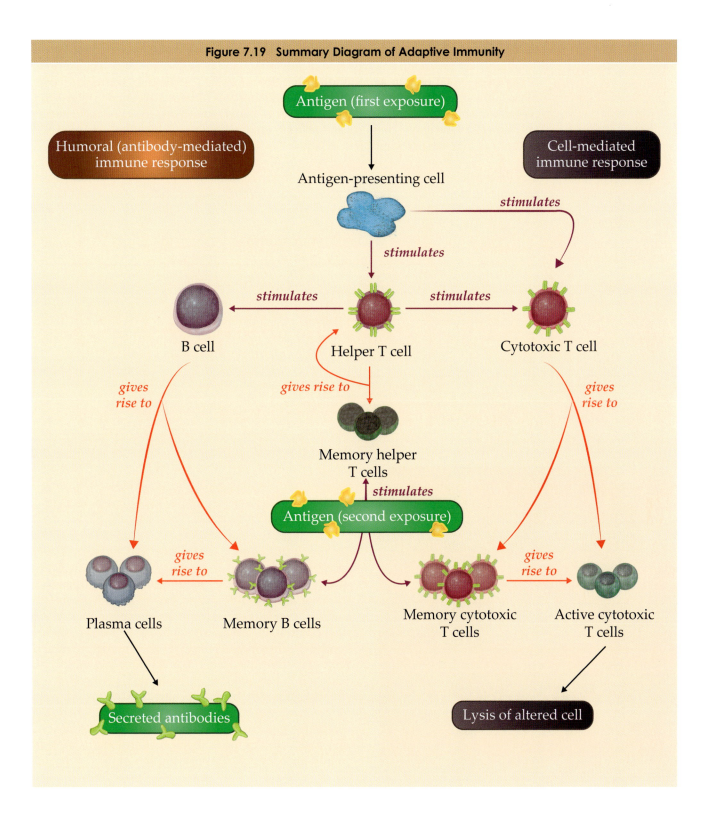

Blood types are identified by the A and B surface antigens. For instance, type A blood means that the red blood cell membrane has A antigens and does not have B antigens. Of course, if the erythrocytes have A antigens, the immune system does not make A antibodies. Type O blood has neither A nor B antigens, and makes both A and B antibodies. Thus, a blood donor may donate blood only to an individual that does not make antibodies against the donor blood. Table 7.1 shows a '+' sign when blood agglutinates (is rejected), and a '–' sign when no agglutination occurs. Notice that an individual with type O blood may donate to anyone (all minuses in the donor column), and an individual with type AB blood may receive from anyone (all minuses in the recipient column).

The genes which produce the A and B antigens are co-dominant. Thus, an individual having type A or B blood may be heterozygous or homozygous. An individual with type O blood has two recessive alleles.

Table 7.1   Blood Types

		Donor			
		A	**B**	**AB**	**O**
Recipient	A	–	+	+	–
	B	+	–	+	–
	AB	–	–	–	–
	O	+	+	+	–

Blood type	Genotype
A	$I^A I^A$ or $I^A i$
B	$I^B I^B$ or $I^B i$
AB	$I^A I^B$
O	$i\,i$

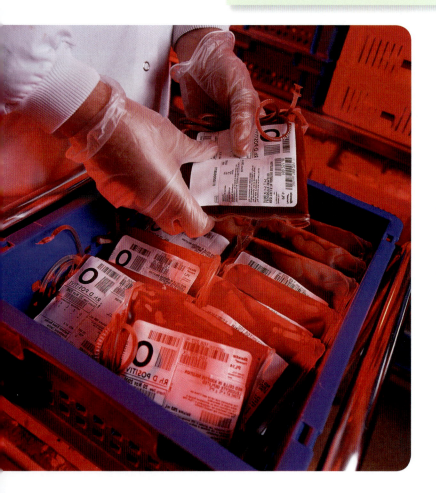

Rh factors are surface proteins on red blood cells first identified in Rhesus monkeys. Individuals having genotypes that code for nonfunctional products of the Rh gene are said to be Rh-negative. All others are Rh-positive. Transfusion reactions involving the Rh factor, if they occur at all, are usually mild. Rh factor is more of a concern during the pregnancy of an Rh-negative mother with an Rh-positive fetus. For the first pregnancy, the mother is not exposed to fetal blood until giving birth and problems are rare. Upon exposure, the mother develops an immune response against the Rh-positive blood. In a second pregnancy, the second fetus that is Rh-positive may be attacked by the antibodies of the mother, which are small enough to pass the placental barrier. The problem is life threatening, and treatment usually involves complete replacement of the fetal blood with Rh-negative blood for the first few weeks of life.

Terms	
Agglutination	Larynx
Albumin	Left Atrium
Alveoli	Left Ventricle
Antibodies	Leukocytes
Antibody	Memory B Cells
Antigen	Memory T Cells
Aorta	Nasal Cavity
Arteries	Oxyhemoglobin
Arterioles	Pharynx
Atrioventricular Node (AV Node)	Plasma cells
B lymphocytes	Platelets
Bronchi	Primary Immune Response
Bronchioles	Pulmonary Circulation
Bundle of His	Purkinje Fibers
Capillaries	Right Atrium
Carbonic Anhydrase	Right Ventricle
Cell-mediated Immunity	Secondary Immune Response
Circulatory System	Serum
Diaphragm	Sinoatrial Node (SA Node)
Epiglottis	Stem Cell
Erythrocytes	Superior Vena Cava
Fibrinogen	Suppressor T Cells
Helper T Cell	Systemic Circulation
Hemoglobin	T Lymphocytes
Humoral Immunity	Trachea
Immunoglobulins	Vagus Nerve
Inferior Vena Cava	Veins
Inflammation	Ventricle
Killer T Cells	Venules

161. Anemia (decreased red blood cell count) can be caused by over activity of which of the following organs?

    A. Thymus
    B. Thyroid
    C. Spleen
    D. Lymph nodes

162. "Swollen glands" are often observed in the neck of a person with a cold. The most likely explanation for this is that:

    A. blood pools in the neck in an attempt to keep it warm.
    B. lymph nodes swell as white blood cells proliferate within them to fight the infection.
    C. the infection sets off an inflammatory response in the neck, causing fluid to be drained from the area.
    D. fever causes a general expansion of the tissues of the head and neck.

163. Humoral immunity involves the action of:

    A. cytotoxic T lymphocytes.
    B. stomach acid.
    C. pancreatic enzymes.
    D. immunoglobulins.

164. Antibodies function by:

    A. phagocytizing invading antigens.
    B. adhering to circulating plasma cells and marking them for destruction by phagocytizing cells.
    C. preventing the production of stem cells in the bone marrow.
    D. attaching to antigens via their variable portions.

165. Lymphatic vessels absorb fluid from the interstitial spaces and carry it to the:

    A. kidneys, where it is excreted.
    B. large intestine, where it is absorbed and returned to the bloodstream.
    C. lungs, where the fluid is vaporized and exhaled.
    D. lymphatic ducts, which return it to the circulation.

166. Which of the following is true concerning type B negative blood?

    A. Type B negative blood will make antibodies that attack type A antigens but not type B antigens.
    B. Type B negative blood will make antibodies that attack type B antigens but not type A antigens.
    C. Type B negative blood will make antibodies that attack type O antigens only.
    D. Type B negative blood will make antibodies that attack both type A and type O antigens.

167. Which of the following would you not expect to find in a lymph node?

    A. B lymphocytes
    B. proteins discarded by tissue cells
    C. invading bacteria
    D. old erythrocytes

168. An individual exposed to a pathogen for the first time will exhibit an innate immune response involving:

    A. B lymphocytes.
    B. T lymphocytes.
    C. granulocytes.
    D. An individual exposed to a pathogen for the first time must acquire immunity before it can respond.

MUSCLE, BONE AND SKIN

8.1 Muscle

There are three types of muscle tissue:

1. skeletal muscle;
2. cardiac muscle;
3. smooth muscle.

Any muscle tissue generates a force only by contracting its cells. The mechanisms by which muscle cells contract differ between the three types of tissue, and are described below. Muscle contraction has four possible functions:

1. body movement;
2. stabilization of body position;
3. movement of substances through the body;
4. generating heat to maintain body temperature.

> Know the three types of muscle and their four possible functions.

8.2 Skeletal Muscle

Skeletal muscle is **voluntary muscle tissue**. It can be consciously controlled.

Skeletal muscle connects one bone to another. The muscle does not attach directly to the bone, but instead is attached via a **tendon**. (A tendon connects muscle to bone; a **ligament** connects bone to bone.) Typically, a muscle stretches across a joint. The muscle *origin* is on the larger bone, which remains relatively stationary, and its *insertion* is on the smaller bone, which moves relative to the larger bone upon contraction of the muscle. Muscles work in groups. The **agonist** (the muscle responsible for the movement) contracts, while a second muscle, the **antagonist**, stretches. When the antagonist contracts, the bone moves

> A muscle uses leverage by applying a force to a bone at its insertion point and rotating the bone in some fashion about the joint. This is a likely MCAT topic because it applies the physics concept of leverage to a biological system. It may seem strange, but most lever systems of the body typically act to increase the required force of a muscle contraction. In other words, a greater force than mg is required to lift a mass m. This is done in order to reduce the bulk of the body and increase the range of movement. If the muscle has a shorter lever arm, it is closer to the body, and thus creates less bulk.

in the opposite direction, stretching the agonist. An example of antagonistic muscles is the upper arm muscles, the biceps and the triceps. In addition to antagonistic muscles, there are usually **synergistic** muscles. Synergistic muscles assist the agonist by stabilizing the origin bone or by positioning the insertion bone during the movement. In this way, skeletal muscle allows for movement and posture.

Contraction of skeletal muscle may squeeze blood and lymph vessels aiding circulation.

Contraction of skeletal muscle produces large amounts of heat. Shivering, controlled by the hypothalamus upon stimulation by receptors in the skin and spinal cord, is the rapid contraction of skeletal muscle to warm the body.

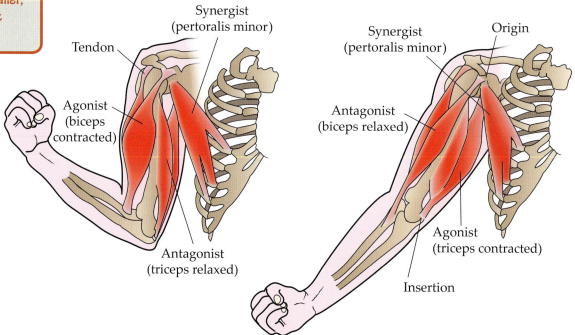

Figure 8.1 Agonist, Antagonist, and Synergist Muscles

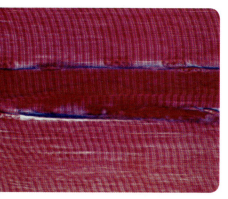

The striated (striped) banding-pattern of skeletal muscle fibers can be seen in this photo. The cross-striations are the arrangement of proteins (actin and myosin) which cause the fibers to contract. Along the junction of the fibers (blue) can be seen multiple nuclei (fuzzy pink); each cylindrical fiber is a single muscle cell.

8.3 Physiology of Skeletal Muscle Contraction

The smallest functional unit of skeletal muscle is the **sarcomere** (Figure 8.2). A sarcomere is composed of many strands of two protein filaments, the **thick and the thin filament**, laid side by side to form a cylindrical segment. Sarcomeres are positioned end to end to form a *myofibril*. Each myofibril is surrounded by the specialized endoplasmic reticulum of the muscle cell called the sarcoplasmic reticulum. The lumen of the sarcoplasmic reticulum is filled with Ca^{2+} ions for reasons that shall become clear shortly. Lodged between the myofibrils are mitochondria and many nuclei. Skeletal muscle is **multinucleate**. A modified membrane called the **sarcolemma** wraps several myofibrils together to form a muscle cell or muscle fiber. Many muscle fibers are further bound into a fasciculus, and many fasciculae make up a single muscle.

The **thick filament** of a sarcomere is made of the protein **myosin**. Several long myosin molecules wrap around each other to form one thick filament. Globular heads protrude along both ends of the thick filament. **The thin filament** is composed mainly of a polymer of the globular protein **actin**. Attached to the actin are the proteins *troponin* and *tropomyosin*.

Figure 8.2 Structure of Skeletal Muscle

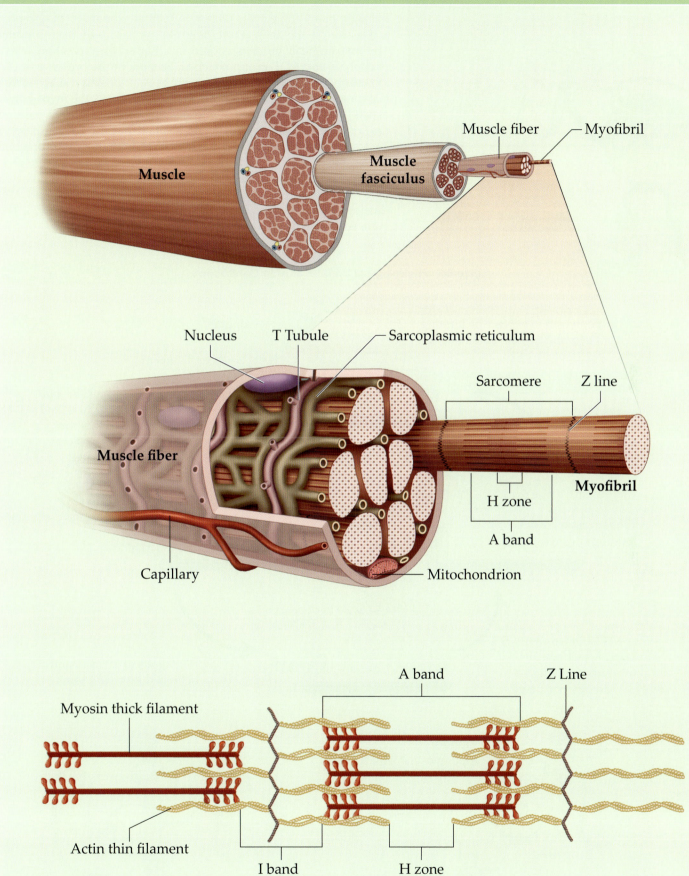

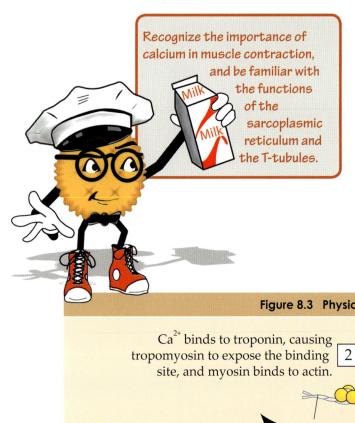
Myosin and actin work together sliding alongside each other to create the contractile force of skeletal muscle. Each myosin head crawls along the actin in a **5 stage cycle** (Figure 8.3). First, tropomyosin covers an active site on the actin preventing the myosin head from binding. The myosin head remains cocked in a high-energy position with a phosphate and ADP group attached. Second, in the presence of Ca^{2+} ions, troponin pulls the tropomyosin back, exposing the active site, allowing the myosin head to bind to the actin. Third, the myosin head expels a phosphate and ADP and bends into a low energy position, dragging the actin along with it. This is called the power stroke because it causes the shortening of the sarcomere and the muscle contraction. In the fourth stage, ATP attaches to the myosin head. This releases the myosin head from the active site, which is covered immediately by tropomyosin. Fifth, ATP splits to inorganic phosphate and ADP causing the myosin head to cock into the high-energy position. This cycle is repeated many times to form a contraction.

Figure 8.3 Physiology of Skeletal Muscle Contraction

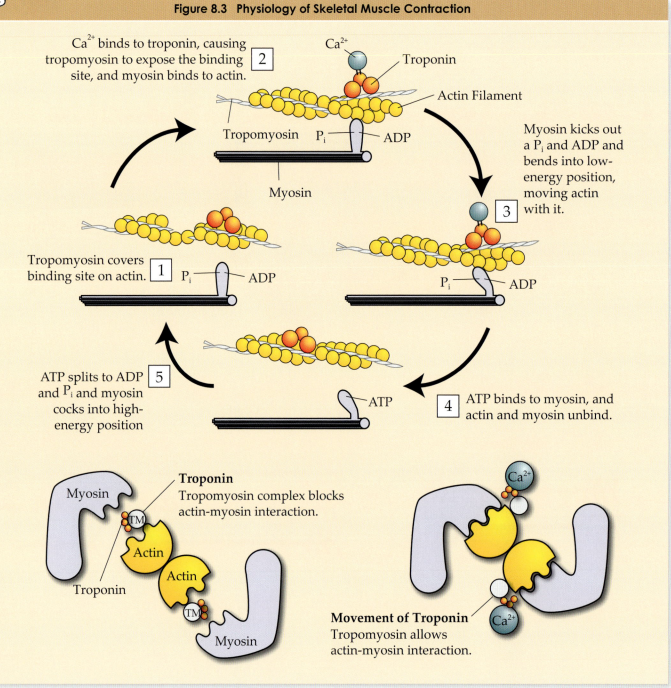

A muscle contraction begins with an action potential. A neuron attaches to a muscle cell forming a **neuromuscular synapse**. The action potential of the neuron releases **acetylcholine** into the synaptic cleft. The acetylcholine activates ion channels in the sarcolemma of the muscle cell creating an action potential. The action potential moves deep into the muscle cell via small tunnels in the membrane called **T-tubules**. T-tubules allow for a uniform contraction of the muscle by allowing the action potential to spread through the muscle cell more rapidly. The action potential is transferred to the sarcoplasmic reticulum, which suddenly becomes permeable to Ca^{2+} ions. The Ca^{2+} ions begin the 5 stage cycle described above. At the end of each cycle, Ca^{2+} is actively pumped back into the sarcoplasmic reticulum.

8.4   A Motor Unit

The muscle fibers of a single muscle do not all contract at once. Instead, from 2 to 2000 fibers spread throughout the muscle are innervated by a single neuron. The neuron and the muscle fibers that it innervates are called a *motor unit*. Motor units are independent of each other. The force of a contracting muscle depends upon the number and size of the active motor units, and the frequency of action potentials in each neuron of the motor unit. Typically, smaller motor units are the first to be activated, and larger motor units are recruited as needed. This results in a smooth increase in the force generated by the muscle. Another important point concerning motor units is that muscles requiring intricate movements, like those in the finger, have smaller motor units, whereas muscles requiring greater force, such as those in the back, have larger motor units.

> A motor unit consists of a nerve and all the muscle fibers it synapses with.

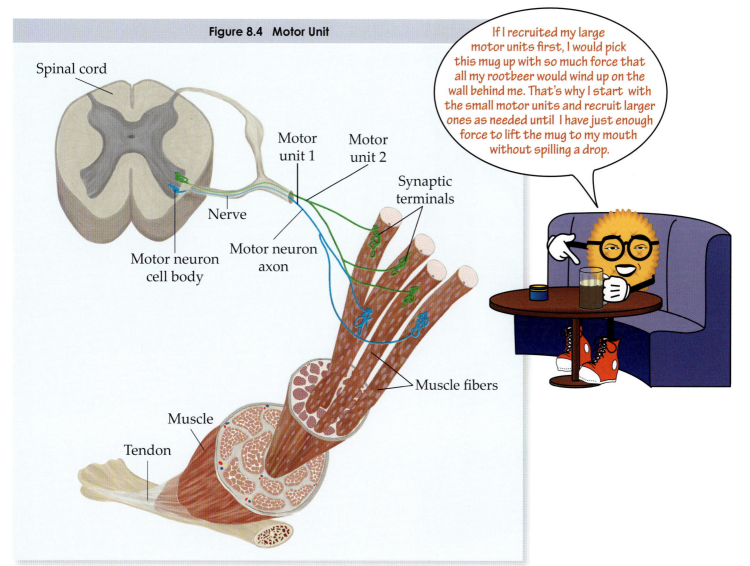

Figure 8.4   Motor Unit

8.5 Skeletal Muscle Type

There are three types of skeletal muscle fibers: 1) *slow oxidative (type I) fibers*; 2) *fast oxidative (type II A) fibers*; and *fast glycolytic (type II B) fibers*. Type I or slow-twitch muscle fibers are red from large amounts of **myoglobin**. Myoglobin is an oxygen storing protein similar to hemoglobin, but which has only one protein subunit. Type I fibers also contain large amounts of mitochondria. They split ATP at a slow rate. As a result, they are slow to fatigue, but also have a slow contraction velocity. Type II A or fast-twitch A fibers are also red, but they split ATP at a high rate. Type II A fibers contract rapidly. Type II A fibers are resistant to fatigue, but not as resistant as type I fibers. Type II B or fast-twitch B fibers have a low myoglobin content, appear white under the light microscope, and contract very rapidly. They contain large amounts of glycogen.

Most muscles in the body have a mixture of fiber types. The ratio of the mixture depends upon the contraction requirements of the muscle and upon the genetics of the individual. Large amounts of type I fibers are found in the postural muscles. Large amounts of type II A fibers are found in the upper legs. Large amounts of type II B fibers are found in the upper arms.

Adult human skeletal muscle does not generally undergo mitosis to create new muscle cells (*hyperplasia*). Instead, a number of changes occur over time when the muscles are exposed to forceful, repetitive contractions. These changes include: the diameter of the muscle fibers increases, the number of sarcomeres and mitochondria increases, and sarcomeres lengthen. This increase in muscle cell diameter and change in muscle conformation is called *hypertrophy*.

Myoglobin stores oxygen inside muscle cells. A molecule of myoglobin looks like one subunit of hemoglobin. It is capable of storing only one molecule of oxygen.

Like many cell types, human muscle cells are so specialized that they have lost the ability to undergo mitosis. Only in rare cases does one muscle cell split to form two cells.

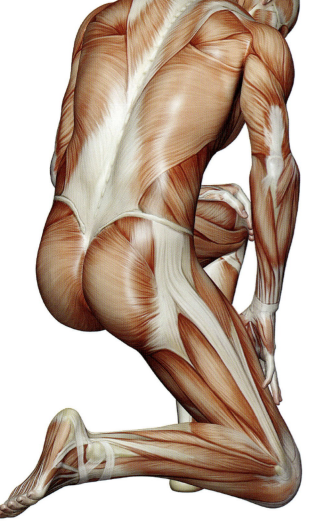

Notes:

169. During a muscular contraction:

    A. both the thin and thick filaments contract.
    B. the thin filament contracts, but the thick filament does not.
    C. the thick filament contracts, but the thin filament does not.
    D. neither the thin nor the thick filament contract.

170. Irreversible sequestering of calcium in the sarcoplasmic reticulum would most likely:

    A. result in permanent contraction of the muscle fibers, similar to what is seen in rigor mortis.
    B. create a sharp increase in bone density as calcium is resorbed from bones to replace the sequestered calcium.
    C. prevent myosin from binding to actin.
    D. depolymerize actin filaments in the sarcomere.

171. Shivering increases body temperature by:

    A. serving as a warning that body temperature is too low, prompting the person to seek warmer locations.
    B. causing bones to rub together, creating heat through friction.
    C. increasing the activity of muscles.
    D. convincing the hypothalamus that body temperature is higher than it actually is.

172. Muscles cause movement at joints by:

    A. inciting neurons to initiate an electrical "twitch" in tendons.
    B. increasing in length, thereby pushing the muscle's origin and insertion farther apart.
    C. filling with blood, thereby expanding and increasing the distance between the ends of a muscle.
    D. decreasing in length, thereby bringing the muscle's origin and insertion closer together.

173. When one of a pair of antagonistic muscles contracts, what usually happens to the other muscle to produce movement?

    A. It acts synergistically by contracting to stabilize the moving bone.
    B. It relaxes to allow movement.
    C. It contracts in an isometric action.
    D. Its insertion slides down the bone to allow a larger range of movement.

174. When undergoing physical exercise, healthy adult skeletal muscle is likely to respond with an increase in all of the following except:

    A. glycolysis.
    B. the Citric Acid Cycle.
    C. mitosis.
    D. protein production.

175. The biceps muscle is connected to the radius bone by:

    A. biceps tendon.
    B. annular ligament of the radius.
    C. articular cartilage.
    D. the triceps muscle.

176. Skeletal muscle contraction may assist in all of the following EXCEPT:

    A. movement of fluid through the body.
    B. body temperature regulation.
    C. posture.
    D. peristalsis.

As seen through a light microscope cardiac muscle cells form a net-like structure. The junctions between individual cells are the intercalated discs (dark lines).

8.6 Cardiac Muscle

The human heart is composed mainly of **cardiac muscle** (Figure 8.5). Like skeletal muscle, cardiac muscle is **striated**, which means that it is composed of **sarcomeres**. However, each cardiac muscle cell contains only *one nucleus*, and is separated from its neighbor by an **intercalated disc**. The intercalated discs contain gap junctions which allow an action potential to spread from one cardiac cell to the next via electrical synapses. The mitochondria of cardiac muscle are larger and more numerous. Skeletal muscle connects bone to bone via tendons; cardiac muscle, on the other hand, is not connected to bone. Instead, cardiac muscle forms a net which contracts in upon itself like a squeezing fist. Cardiac muscle is *involuntary*. Like skeletal muscle, cardiac muscle grows by hypertrophy.

The action potential of cardiac muscle exhibits a plateau after depolarization. The plateau is created by slow voltage-gated calcium channels which allow calcium to enter and hold the inside of the membrane at a positive potential difference. The plateau lengthens the time of contraction.

> Be sure to remember the importance of calcium in the cardiac action potential. Without it, the heart would beat far too quickly to serve as a functional pump, and we would die.

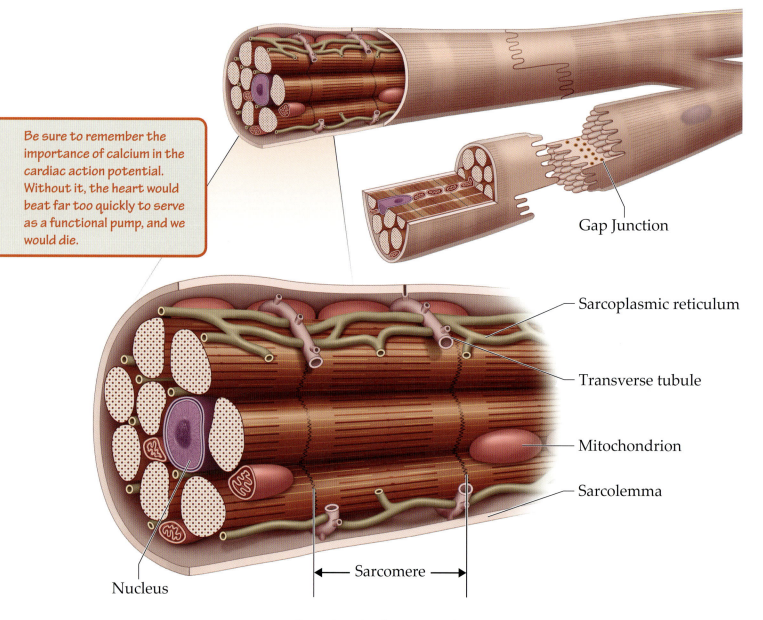

Figure 8.5 Cardiac Muscle Tissue

8.7 Smooth Muscle

Smooth muscle is mainly *involuntary*, so it is innervated by the **autonomic nervous system** (Figure 8.6). Like cardiac muscle, smooth muscle cells contain only one nucleus. Smooth muscles also contain thick and thin filaments, but they are not organized into sarcomeres. In addition, smooth muscle cells contain **intermediate filaments**, which are attached to **dense bodies** spread throughout the cell. The thick and thin filaments are attached to the intermediate filaments, and, when they contract, they cause the intermediate filaments to pull the dense bodies together. Upon contraction, the smooth muscle cell shrinks length-wise.

There are two types of smooth muscle: 1. *single-unit* and 2. *multiunit*. Single unit smooth muscle, also called *visceral*, is the most common. Single-unit smooth muscle cells are connected by gap junctions spreading the action potential from a single neuron through a large group of cells, and allowing the cells to contract as a single unit. Single-unit smooth muscle is found in small arteries and veins, the stomach, intestines, uterus, and urinary bladder.

Each multiunit smooth muscle fiber is attached directly to a neuron. A group of multiunit smooth muscle fibers can contract independently of other muscle fibers in the same location. Multiunit smooth muscle is found in the large arteries, bronchioles, pili muscles attached to hair follicles, and the iris.

In addition to responding to neural stimulus, smooth muscle also contracts or relaxes in the presence of hormones, or to changes in pH, O_2 and CO_2 levels, temperature, and ion concentrations.

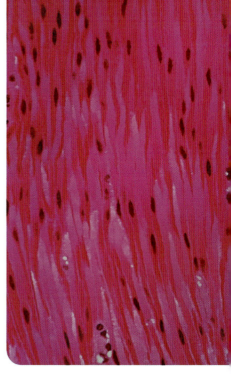

Smooth muscle is composed of spindle-shaped cells grouped in irregular bundles. Each cell contains one nucleus, seen here as a dark stained spot.

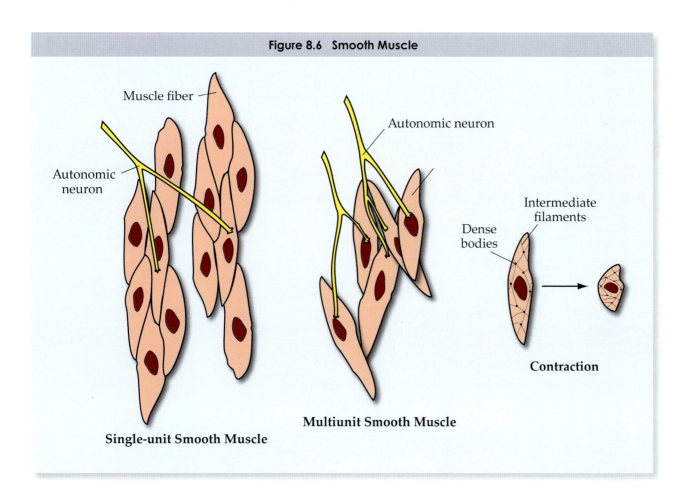

Figure 8.6 Smooth Muscle

Muscle fiber

Autonomic neuron

Autonomic neuron

Dense bodies

Intermediate filaments

Contraction

Single-unit Smooth Muscle

Multiunit Smooth Muscle

177. The function of gap junctions in the intercalated discs of cardiac muscle is to:

    A. anchor the muscle fibers together.
    B. insure that an action potential is spread to all fibers in the muscle network.
    C. control blood flow by selectively opening and closing capillaries.
    D. release calcium into the sarcoplasmic reticulum.

178. Which of the following muscular actions is controlled by the autonomic nervous system?

    A. the knee-jerk reflex
    B. conduction of cardiac muscle action potential from cell to cell
    C. peristalsis of the gastrointestinal tract
    D. contraction of the diaphragm

179. During an action potential, a cardiac muscle cell remains depolarized much longer than a neuron. This is most likely to:

    A. prevent the initiation of another action potential during contraction of the heart.
    B. ensure that adjacent cardiac muscle cells will contract at different times.
    C. keep the neuron from firing twice in rapid succession.
    D. allow sodium voltage-gated channels to remain open long enough for all sodium to exit the cell.

180. All of the following are true concerning smooth muscle EXCEPT:

    A. Smooth muscle contractions are longer and slower than skeletal muscle contractions.
    B. A chemical change in the environment around smooth muscle may create a contraction.
    C. Smooth muscle does not require calcium to contract.
    D. Smooth muscle is usually involuntary.

181. Which of the following muscles is under voluntary control?

    A. the diaphragm
    B. the heart
    C. the smooth muscle of the large intestines
    D. the iris

182. Cardiac muscle is excited by:

    A. parasympathetic nervous excitation.
    B. constriction of T-tubules.
    C. increased cytosolic sodium concentration .
    D. increased cytosolic calcium concentration.

183. When left alone, certain specialized cardiac muscle cells have the capacity for self-excitation. The SA node is a collection of such cells. The SA node is innervated by the vagus nerve. The frequency of self excitation of the cardiac cells of the SA Node is likely to be:

    A. slower than a normal hearbeat because excitation by the vagus nerve decreases the heart rate.
    B. slower than a normal hearbeat because excitation by the vagus nerve increases the heart rate.
    C. faster than a normal hearbeat because excitation by the vagus nerve decreases the heart rate.
    D. faster than a normal hearbeat because excitation by the vagus nerve increases the heart rate.

184. In extreme cold, just before the onset of frostbite, sudden vasodilation occurs manifested by flushed skin. This vasodilation is most likely the result of:

    A. paralysis of smooth muscle in the vascular walls.
    B. paralysis of skeletal muscle surrounding the vascular walls.
    C. sudden tachycardia with a resultant increase in blood pressure.
    D. blood shunting due to smooth muscle sphincters.

8.8 Bone

Bone is living tissue. Its functions are *support* of soft tissue, *protection* of internal organs, assistance in *movement* of the body, **mineral storage**, **blood cell production**, and **energy storage** in the form of adipose cells in bone marrow.

Bone tissue contains four types of cells surrounded by an extensive matrix.

1. *Osteoprogenitor (or osteogenic) cells* differentiate into osteoblasts.

2. **Osteoblasts** secrete collagen and organic compounds upon which bone is formed. Osteoblasts are incapable of mitosis. As osteoblasts release matrix materials around themselves, they become enveloped by the matrix and differentiate into osteocytes.

3. **Osteocytes** are also incapable of mitosis. Osteocytes exchange nutrients and waste materials with the blood.

4. **Osteoclasts** resorb bone matrix, releasing minerals back into the blood. Osteoclasts are believed to develop from the white blood cells called monocytes.

A bone-making osteoblast cell is shown here surrounded by a dense network of collagen fibers.

An osteocyte is an osteoblast that has become trapped within a bone cavity (lacuna). Osteocytes are responsible for bone formation, but eventually become embedded in the bone matrix.

Know the functions of osteoblasts, osteocytes, and osteoclasts.

A typical long bone (Figure 8.9) has a long shaft, called the *diaphysis*, and two ends, each end composed of a *metaphysis* and *epiphysis*. A sheet of cartilage in the metaphysis, called the *epiphyseal plate*, is where long bones grow in length. **Spongy bone** contains **red bone marrow**, the site of *hemopoiesis* or red blood cell development. **Compact bone** surrounds the *medullary cavity*, which holds **yellow bone marrow**. Yellow bone marrow contains adipose cells for fat storage. Compact bone is highly organized. In a continuous remodeling process, osteoclasts burrow tunnels, called **Haversian (central) canals**, through compact bone. The osteoclasts are followed by osteoblasts, which lay down a new matrix onto the tunnel walls forming concentric rings called **lamellae**. Osteocytes trapped between the lamellae exchange nutrients via **canaliculi**. Haversian canals contain blood and lymph vessels, and are connected by crossing canals called **Volkmann's canals**. The entire system of lamellae and Haversian canal is called an **osteon** (**Haversian system**).

8.9 Bone Function in Mineral Homeostasis

Calcium salts are only slightly soluble, so most calcium in the blood is not in the form of free calcium ions, but is bound mainly by proteins and, to a much lesser extent, by phosphates (HPO_4^{2-}) and other anions. It is the concentration of free calcium ions (Ca^{2+}) in the blood that is important physiologically. Too much Ca^{2+} results in membranes becoming hypo-excitable producing lethargy, fatigue, and memory loss; too little produces cramps and convulsions.

Most of the Ca^{2+} in the body is stored in the bone matrix as **hydroxyapatite** [$Ca_{10}(PO_4)_6(OH)_2$]. Collagen fibers lie along the lines of tensile force of the bone, giving the bone great tensile strength. Hydroxyapatite crystals lie alongside collagen fibers, and give bone greater compressive strength than the best reinforced concrete. Some of the body's Ca^{2+} exists in bone in the form of slightly soluble calcium salts such as $CaHPO_4$. It is these salts that buffer the plasma Ca^{2+} levels. Thus bone acts as a storage site for Ca^{2+} and HPO_4^{2-}.

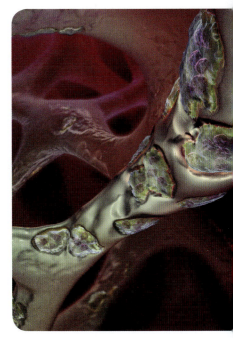

This computer-generated image shows multi-nucleated osteoclasts etching away 0trabecular bone in a process called bone resorption.

Figure 8.7 The Human Skeleton

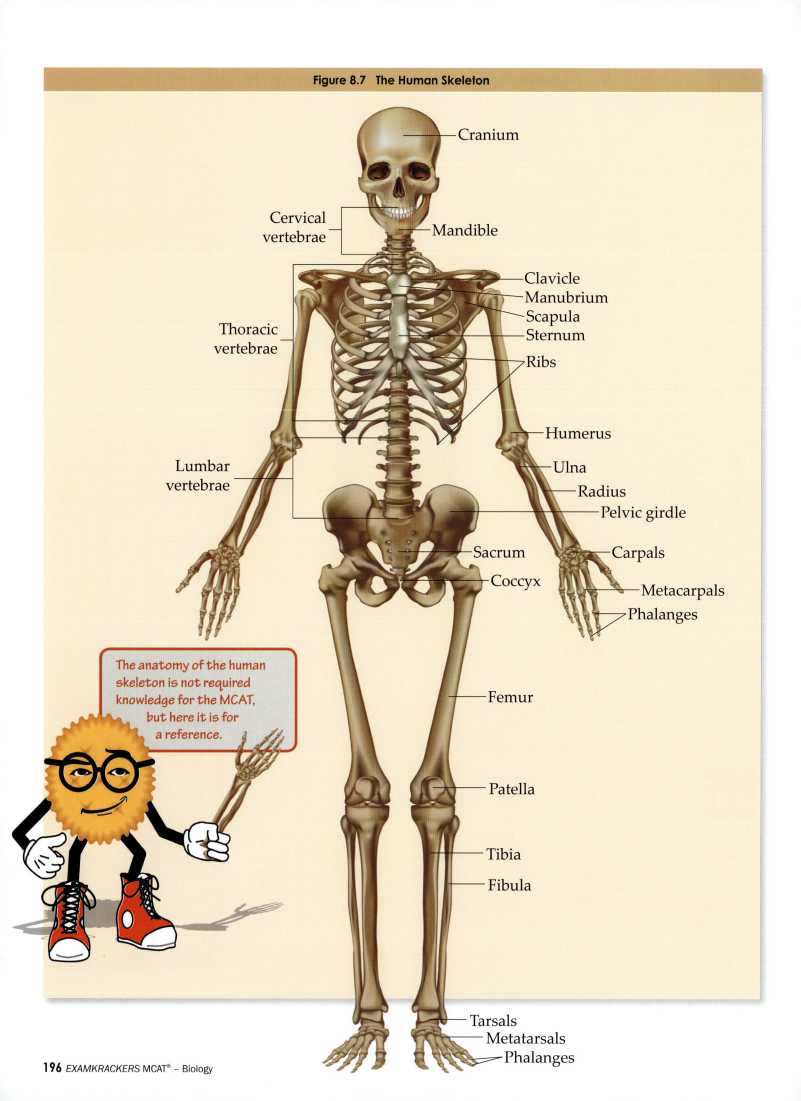

8.10 Bone Types and Structures

Most bones fall into one of four types: 1. long, 2. short, 3. flat, or 4. irregular. Long bones have a shaft that is curved for strength. They are composed of compact and spongy bone. Leg, arm, finger and toe bones are long bones. Short bones are cuboidal. They are the ankle and wrist bones. Flat bones are made from spongy bone surrounded by compact bone. They provide large areas for muscle attachment, and organ protection. The skull, sternum, ribs and shoulder blades are flat bones. Irregular bone has an irregular shape and variable amounts of compact and spongy bone. The oscicles of the ear are an example of irregular bones.

Remember that bone is not just for support, protection, and movement. Bone also stores calcium and phosphate, helping to maintain a homeostatic concentration of these ions in the blood. Bone stores energy in the form of fat. Bone is also the site of blood cell formation.

= Long Bone

= Cuboidal Bone

= Flat Bone

= Irregular Bone

Figure 8.8   Bone Types

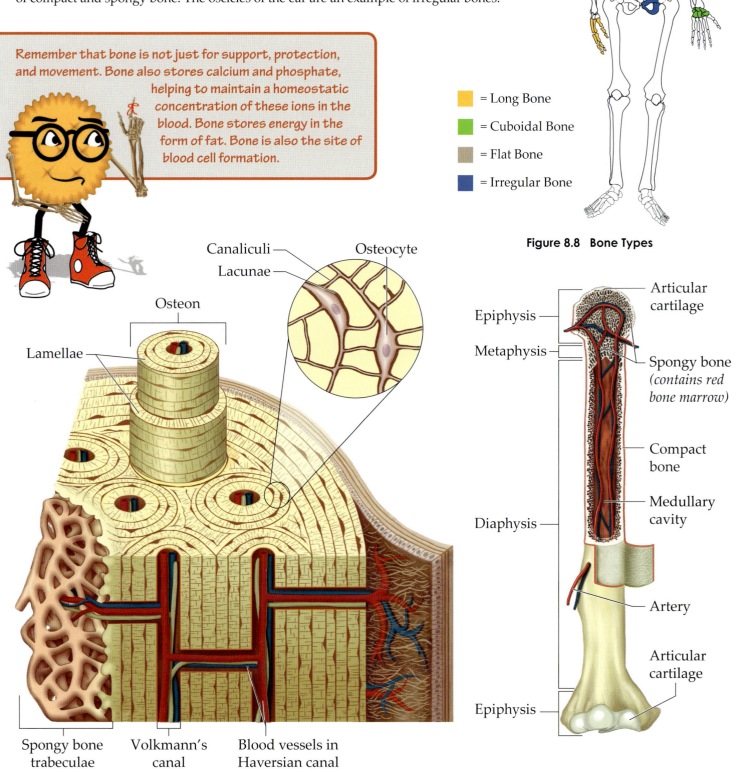

Figure 8.9   Bone Structure

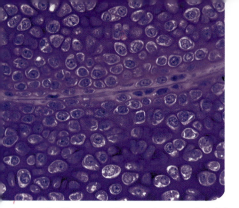

Hyaline cartilage is semi-rigid connective tissue composed of many chondrocytes (cartilage cells, pale purple). These synthesize an extracellular matrix (dark purple) of proteoglycans, collagen and water that keeps them apart from each other in spaces known as lacunae. Hyaline cartilage is strong but compressible due to its high water content. It reduces friction between the bones in the knee joint as they move against each other.

8.11 Cartilage

Cartilage is flexible, resilient connective tissue. It is composed primarily of collagen, and has great tensile strength. Cartilage contains no blood vessels or nerves except in its outside membrane called the *perichondrium*. There are three types of cartilage: 1. *hyaline*; 2. *fibrocartilage*; and 3. *elastic*. Hyaline cartilage is the most common. Hyaline cartilage reduces friction and absorbs shock in joints.

8.12 Joints

Joints can be classified by structure into three types:

1. *Fibrous joints* occur between two bones held closely and tightly together by fibrous tissue permitting little or no movement. Skull bones form fibrous joints with each other, and the teeth form fibrous joints with the mandible.

2. *Cartilaginous joints* also allow little or no movement. They occur between two bones tightly connected by cartilage, such as the ribs and the sternum, or the pubic symphysis in the pelvis.

3. *Synovial joints* (Figure 8.10) are not bound directly by the intervening cartilage. Instead, they are separated by a capsule filled with *synovial fluid*. Synovial fluid provides lubrication and nourishment to the cartilage. In addition, the synovial fluid contains phagocytotic cells that remove microbes and particles which result from wear and tear from joint movement. Synovial joints allow for a wide range of movement.

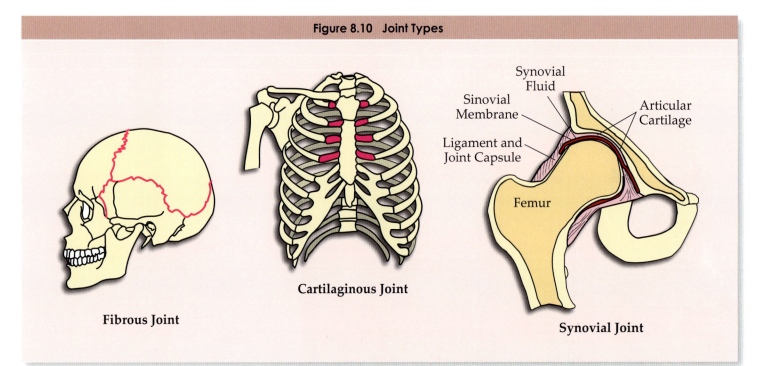

Figure 8.10   Joint Types

Fibrous Joint

Cartilaginous Joint

Synovial Joint

Skin

The **skin** (Figure 8.11) is an organ, which means that it is a group of tissues working together to perform a specific function. Some important functions of the skin are:

1. **Thermoregulation**: The skin helps to regulate the body temperature. Blood conducts heat from the core of the body to skin. Some of this heat can be dissipated by the endothermic evaporation of sweat, but most is dissipated by radiation. Of course, radiation is only effective if the body is higher than room temperature. Blood can also be shunted away from the capillaries of the skin to reduce heat loss, keeping the body warm. Hairs can be erected (*piloerection*) via sympathetic stimulation trapping insulating air next to the skin. Skin has both warmth and cold receptors.

2. **Protection**: The skin is a physical barrier to abrasion, bacteria, dehydration, many chemicals, and ultra violet radiation.

3. **Environmental sensory input**: The skin gathers information from the environment by sensing temperature, pressure, pain, and touch.

4. **Excretion**: Water and salts are excreted through the skin. This water loss occurs by diffusion through the skin and is independent of sweating. Adults lose one quarter to one half liter of water per day via this type of *insensible fluid loss*. Burning of the skin can increase this type of water loss dramatically.

5. **Immunity**: Besides being a physical barrier to bacteria, specialized cells of the epidermis are components of the immune system.

6. **Blood reservoir**: Vessels in the dermis hold up to 10% of the blood of a resting adult.

7. **Vitamin D synthesis**: Ultra violet radiation activates a molecule in the skin that is a precursor to vitamin D. The activated molecule is modified by enzymes in the liver and kidneys to produce vitamin D.

This is a sample of skin from the back of a human hand.

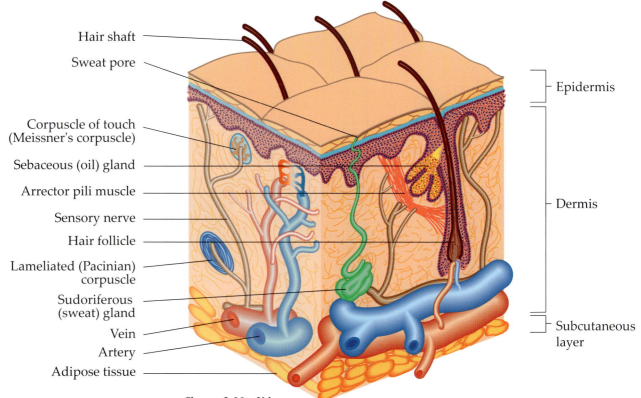

Figure 8.11   Skin

The skin has two principal parts: 1) the **epidermis** and 2) the **dermis**. Beneath the skin is a subcutaneous tissue called the *superficial fascia* or *hypodermis*. The fat of this subcutaneous layer is an important heat insulator for the body. The fat helps maintain normal core body temperatures on cold days while the skin approaches the temperature of the environment.

The epidermis is avascular (no blood vessels) epithelial tissue. It consists of four major cell types: 1) 90 % of the epidermis is composed of *Keratinocytes,* which produce the protein keratin that helps waterproof the skin. 2) *Melanocytes* transfer *melanin* (skin pigment) to keratinocytes. 3) *Langerhans cells* interact with the the helper T-cells of the immune system. 4) *Merkel cells* attach to sensory neurons and function in the sensation of touch.

There are five strata or layers of the epidermis. The deepest layer contains Merkel cells and stem cells. The stem cells continually divide to produce keratinocytes and other cells. Keratinocytes are pushed to the top layer. As they rise, they accumulate keratin and die, losing their cytoplasm, nucleus, and other organelles. When the cells reach the outermost layer of skin, they slough off the body. The process of keratinization from birth of a cell to sloughing off takes two to four weeks. The outermost layer of epidermis consists of 25 to 30 layers of flat, dead cells. Exposure to friction or pressure stimulates the epidermis to thicken froming a **callus**.

The **dermis** is connective tissue derived from mesodermal cells. The dermis is embedded by blood vessels, nerves, glands, and hair follicles. Collagen and elastic fibers in the dermis provide skin with strength, extensibility, and elasticity. The dermis is thick in the palms and soles.

The skin, hair, nails, glands, and some nerve endings make up the *integumentary system*. Hair, nails, and some glands are derivatives of embryonic epidermis. Hair is a column of keratinized cells held tightly together. As new cells are added to its base, the hair grows. Most hairs are associated with a *sebaceous (oil) gland* that empties oil directly into the follicle and onto the skin. When contracted, smooth muscle *(arrector pili)*, also associated with each hair, stands hair up pointing it perpendicular to the skin. *Nails* are also keratinized cells. *Sudoriferous (sweat) glands* are found in the skin separate from hair follicles. *Ceruminous glands* produce a wax-like material found in the ears.

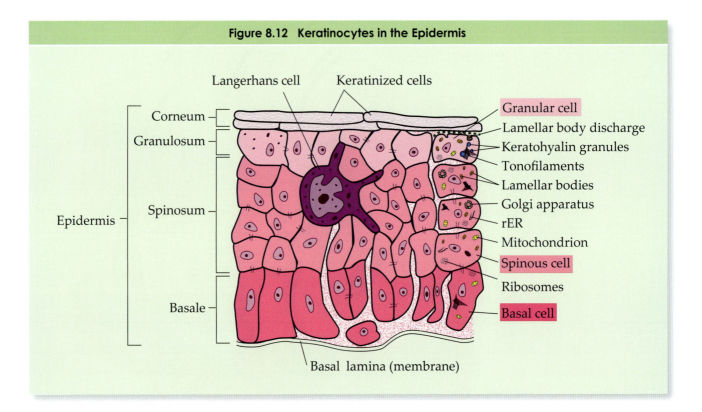

Figure 8.12 Keratinocytes in the Epidermis

Terms You Need To Know

Terms		
Acetylcholine	Lamellae	Skeletal Muscle
Actin	Ligament	Skin
Agonist	Mineral Storage	Smooth Muscle
Antagonist	Muscle Contraction	Spongy Bone
Bone	Cycle	Striated
Callus	Multinucleated	Synergistic Muscle
Canaliculi	Myoglobin	T-Tubules
Cartilage	Myosin	Tendon
Compact Bone	Neuromuscular	Thermoregulation
Cardiac Muscle	Synapse	Thick Filament
Dense Bodies	Osteoblasts	Thin Filament
Dermis	Osteoclasts	Tropomyosin
Epidermis	Osteocytes	Troponin
Haversian (Central)	Osteon (Haversian	Vitamin D Synthesis
Canals	System)	Volkmann's Canals
Hydroxyapatite	Red Bone Marrow	Yellow Bone Marrow
Intercalated Disc	Sarcolemma	
Intermediate Filaments	Sarcomere	

Questions 185 through 192 are **NOT** based on a descriptive passage.

185. The production of which of the following cells would most likely be increased by parathyroid hormone?

A. osteoprogenitor
B. osteocyte
C. osteoclast
D. osteoblast

186. In a synovial joint, the purpose of the synovial fluid is to:

A. reduce friction between bone ends.
B. keep bone cells adequately hydrated.
C. occupy space until the bones of the joint complete their growth.
D. maintain a rigid connection between two flat bones.

187. Surgical cutting of which of the following tissues would result in the LEAST amount of pain?

A. muscle
B. bone
C. cartilage
D. skin

188. In a synovial joint, the connective tissue holding the bones together are called:

A. ligaments.
B. tendons.
C. muscles.
D. osseous tissue.

189. All of the following are functions of bone EXCEPT:

A. mineral storage.
B. structural support.
C. blood temperature regulation.
D. fat storage.

190. All of the following are found in compact bone EXCEPT:

A. yellow marrow.
B. haversian canals.
C. canaliculi.
D. Volkmann's canals.

191. Hydroxyapatite is the mineral portion of bone. Hydroxyapatite contains all of the following elements except:

A. calcium.
B. sulfur.
C. phosphate.
D. hydrogen.

192. The spongy bone of the hips is most important in:

A. red blood storage.
B. red blood cell synthesis.
C. fat storage.
D. lymph fluid production.

_____

POPULATIONS

LECTURE

9

9.1 Mendelian Concepts

Gregor Mendel was a 19th century monk, who performed hybridization experiments with pea plants. The difference between Mendel and those who had come before him was that Mendel quantitated his results; he counted and recorded his findings. Mendel found that when he crossed purple flowered plants with white flowered plants, the **first filial**, or **F_1 generation**, produced purple flowers. He called the purple trait **dominant**, and the white trait **recessive**. Mendel examined seven traits in all, and each trait proved to have dominant and recessive alternatives. When Mendel self-pollinated the F_1 generation plants, the F_2 generation expressed both the dominant and recessive traits in a **3 to 1 ratio**, now referred to as the **Mendelian ratio** (Fig. 9.1). When the F_2 generation was self-pollinated, 33% of the dominants produced only dominants, and the rest of the dominants produced the Mendelian ratio. The white flowered plants produced only white flowered plants. Thus, half of the F_2 generation expressed the dominant trait with the recessive trait latent.

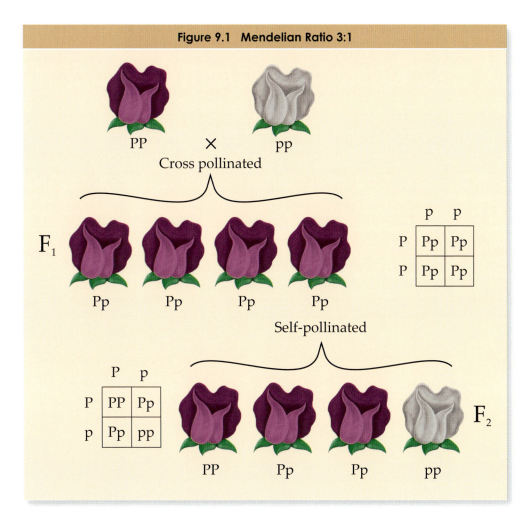

Figure 9.1   Mendelian Ratio 3:1

The expression of a trait is the **phenotype**, and an individual's genetic make up is the **genotype**. In Figure 9.1 the phenotypes are purple and white; the genotypes are PP, Pp, and pp. The phenotype is expressed through the action of enzymes and other structural proteins. Which are encoded by genes. In **complete dominance**, exhibited by the flowers in Mendel's experiment, for any one trait, a diploid individual will have two chromosomes each containing a separate gene that codes for that specific trait. These two chromosomes are homologous by definition. Their corresponding genes are located at the same **locus** or position on respective chromosomes. Each gene contributes an **allele** to the genotype. However, only one allele, the dominant allele, is expressed. If both alleles are dominant, then the dominant phenotype is expressed; if both alleles are recessive, then the recessive phenotype is expressed. An individual with a genotype having two dominant or two recessive alleles is said to be **homozygous** for that trait. An individual with a genotype having one dominant and one recessive allele is said to be **heterozygous** for the trait, and is called a **hybrid**.

Mendel's First Law of Heredity, the **Law of Segregation**, states that alleles segregate independently of each other when forming gametes. Any gamete is equally likely to posses any allele. Also, the phenotypic expression of the alleles is not a blend of the two, but an expression of the dominant allele (the principle of complete dominance).

When a heterozygous individual exhibits a phenotype that is intermediate between its homozygous counterparts, the alleles are referred to as partial, or incomplete dominants. Alleles showing partial dominance are represented with the same capital letter, and distinguished with a prime or superscript. For instance, a cross between red flowered sweet peas and white flowered sweet peas may produce pink flowers. The genotype for the pink flowered individual would be expressed as either CC' or Cr Cw. If the heterozygote exhibits both phenotypes, the alleles are codominant. Human blood type alleles are codominant because a heterozygote exhibits A and B antigens on the blood cell membranes.

Figure 9.2 shows a **Punnett square** for predicting genotypic ratios of offspring. The genotypes of all possible gametes of each parent are displayed in the first column and first row respectively. The alleles are then combined in the corresponding boxes to show the possible genotypes of the offspring. Since, according to the law of segregation, each gametic genotype is equally likely, each offspring genotype is also equally likely.

Mendel's Second Law of Heredity, the **Law of Independent Assortment**, states that genes located on different chromosomes assort independently of each other. In other words, genes that code for different traits (such as pea shape and pea color), when located on different chromosomes, do not affect each other during gamete formation. If two genes are located on the same chromosome, the likelihood that they will remain together during gamete formation is indirectly proportional to the distance separating them. Thus, the closer they are on the chromosome, the more likely they will remain together. In Figure 9.2, we use a Punnett square to predict the phenotypic ratio of a **dihybrid cross**. 'W' is the allele for a round pea shape, which is dominant, and 'w' is the allele for wrinkled pea shape, which is recessive. 'G' is the allele for yellow color, which is dominant, and 'g' is the allele for green color, which is recessive. We make the assumption that the genes for pea shape and pea color are on separate chromosomes, and will assort independently of each other. Notice the **phenotypic ratio of a dihybrid cross, 9:3:3:1**.

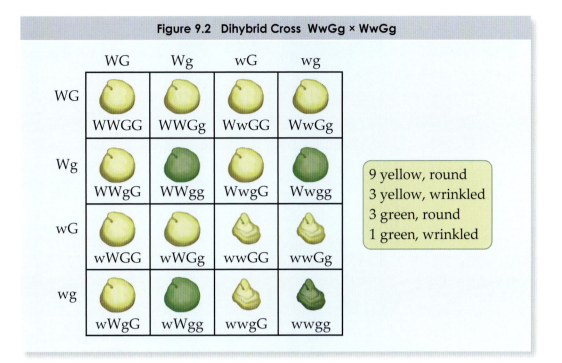

Figure 9.2 Dihybrid Cross WwGg × WwGg

	WG	Wg	wG	wg
WG	WWGG	WWGg	WwGG	WwGg
Wg	WWgG	WWgg	WwgG	Wwgg
wG	wWGG	wWGg	wwGG	wwGg
wg	wWgG	wWgg	wwgG	wwgg

9 yellow, round
3 yellow, wrinkled
3 green, round
1 green, wrinkled

The chromosomes of males and females differ. In humans, the 23rd pair of chromosomes establishes the sex of the individual, and each partner is called a **sex chromosome**. One of the 23rd chromosomes of a male is abbreviated. Instead of appearing as two Xs in a **karyotype** (a map of the chromosomes), the chromosome pair appears as an X and a Y. All other chromosomes appear as two Xs. When a gene is found on the sex chromosome it is called **sex-linked**. Generally, the Y chromosome does not carry the allele for the sex-linked trait; thus, the allele that is carried by the X chromosome in the male is expressed whether it is dominant or recessive. Since the female has two X chromosomes, her genotype is found through the normal rules of dominance. However, in most somatic cells, one of the X chromosomes will condense, and most of its genes will become inactive. The tiny dark object formed is called a **Barr body**. Barr bodies are formed at random, so the active allele is split about evenly among the cells. Nevertheless, in most cases, the recessive phenotype is only displayed in homozygous recessive individuals. Thus, the female may carry a recessive trait on her 23rd pair of chromosomes without expressing it. If she does, she is said to be a **carrier** for the trait. Such a recessive trait has a strong chance of being expressed in her male offspring regardless of the genotype of her mate.

Hemophilia is a sex-linked disease. The Punnett square shown in Figure 9.3 shows a cross between a female carrier for hemophilia and a healthy male. Since there are two possible phenotypes for the males, and one is the recessive phenotype, the male offspring from such a pairing have a 1 in 2 chance of having the disease.

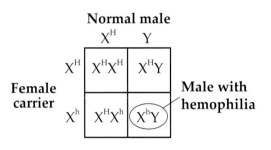

Normal male

	X^H	Y
X^H	$X^H X^H$	$X^H Y$
X^h	$X^H X^h$	$X^h Y$

Female carrier

Male with hemophilia

Figure 9.3 Sex-linked Traits

193. Color-blindness is a sex-linked recessive trait. A woman who is a carrier for the trait has two boys with a color-blind man. What is the probability that both boys are color-blind?

    A. 0%
    B. 25%
    C. 50%
    D. 100%

194. In the pedigree below, the darkened figures indicate an individual with hemophilia, a sex-linked recessive disease. The genotype of the female marked A is:

    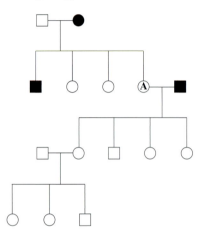

    A. $X^H X^h$
    B. $X^H X^H$
    C. $X^h X^h$
    D. $X^h Y$

195. What fraction of offspring are likely to display both dominant phenotypes in a dihybrid cross?

    A. 1/16
    B. 3/16
    C. 9/16
    D. 15/16

196. In a dihybrid cross, what fraction of the offspring are likely to be dihybrids?

    A. 1/16
    B. 1/4
    C. 1/2
    D. 9/16

197. Sickle cell anemia is an autosomal recessive disease. A male with the disease and a female that is not diseased, but carries the trait, produce two girls. What is the probability that neither girl carries a recessive allele?

    A. 0%
    B. 25%
    C. 50%
    D. 66%

198. The parents of a dihybrid cross:
    A. are genetic opposites at the genes of interest.
    B. are genetically identical at the genes of interest.
    C. are genetic opposites at one gene and genetically identical at the other.
    D. have no genetic relationship.

199. Sex-linked traits in men usually result due to genes located:
    A. on both chromosomes of a pair of homologous chromosomes.
    B. on one chromosome from a pair of homologous chromosomes
    C. on both chromosomes of a pair of nonhomologous chromosomes
    D. on one chromosome from a pair of nonhomologous chromosomes

200. Colorblindness is a sex-linked recessive trait. A woman is born colorblind. What can be said with certainty?

    A. Her father and mother are colorblind.
    B. Her mother and daughter are colorblind.
    C. Her father and son are colorblind.
    D. Her father is colorblind, but her son may or may not be.

9.2 Evolution

The **gene pool** is the total of all alleles in a population. **Evolution** is a change in the gene pool. Figure 9.4 shows the eye color alleles for a small population. Even if the ratio of blue eyed to brown eyed individuals temporarily changes, as long as the gene pool remains 30% b alleles and 70% B alleles, the population has not evolved.

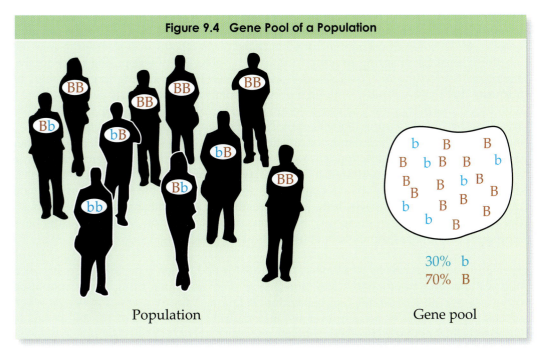

Figure 9.4   Gene Pool of a Population

30% b
70% B

Population   Gene pool

Most taxonomical classification systems are based upon genetic similarity. The classification system for animals that you must know for the MCAT contains ever more specific groupings in the following order: **Kingdom, Phylum, Class, Order, Family, Genus, Species**. (Plants and fungi use divisions instead of phyla.) Within each group are many subgroups, which are unimportant for the MCAT (except for the subphylum **Vertebrata**, which is in the phylum Chordata). Since organisms within the same group have similar genetic structures, they probably share similar phylogenies (evolutionary histories). For instance, all mammals belong to the class **Mammalia** and the phylum **Chordata**; thus, all mammals probably share a common ancestor that they do not share with birds, which are also in the phylum Chordata, but in the class Aves.

The taxonomy is changing, and for the MCAT you may want to be aware of the new superkingdoms called **domains**. There are three domains: Bacteria, Archaea, and Eukarya. This basically puts the kingdoms of Protista, Fungi, Plantae, and Animalia into the domain Eukarya. It makes the kingdom Monera obsolete dividing it into the domains of Bacteria and Archaea. The two domains of Bacteria and Archaea are divided into several kingdoms each. Archaea is more closely related to Eukarya than is Bacteria.

When naming an organism, the genus and species name are given in order. Typically, they are both written in italics, and the genus is capitalized while the species is not.

Species is loosely limited to, but not inclusive of, all organisms that can reproduce fertile offspring with each other. In other words, if two organisms can reproduce fertile offspring, they might be the same species; if their gametes are incompatible, they are definitely not the same species. Another guideline for species (but still an imperfect guideline) is all organisms which normally reproduce selectively fit offspring in the wild. Organisms of different species may be prevented from producing fit offspring by such things as *geographic isolation* (separated by geography), *habitat isolation* (live in the same location but have different habitats), *seasonal isolation* (mate in different seasons), *mechanical isolation* (physically impossible to mate), *gametic isolation* (gametes are incompatible), *developmental*

> You should be aware that ontogeny recapitulates phylogeny. In other words, the course of development of an organism from embryo to adult reflects its evolutionary history. For instance, the human fetus has pharyngeal pouches reflecting a gilled ancestor.

isolation (fertilized embryo develops improperly), *hybrid inviability or sterility* (hybrid malformed), *selective hybrid elimination* (hybrid is less fit), and *behavioral isolation* (different mating rituals).

In order to survive, the members of the same species will exploit their environment in a unique manner not shared by any other species. The way in which a species exploits its environment is called its **niche**. No two species can occupy the same niche indefinitely. The theory of **survival of the fittest** predicts that one species will exploit the environment more efficiently, eventually leading to the extinction of the other with the same niche. The definition of **the "fittest" organism** in this theory is the organism which can best survive to reproduce offspring which will, in turn reproduce offspring and so on generation after generation. This definition may include living beyond reproduction in order to provide a better chance for offspring to reproduce. In fact, there are two opposing reproductive strategies: *r*-selection and *K*-selection. *r*-selection involves producing large numbers of offspring that mature rapidly with little or no parental care. *r*-strategists generally have a high brood mortality rate. Their population growth curves are exponential. *r*-strategists are generally found in unpredictable, rapidly changing environments affected by density independent factors such as floods, or drastic temperature change. *K*-selection is the other side of the spectrum. *K*-selection involves small brood size with slow maturing offspring and strong parental care. *K*-strategists tend to have a sigmoidal growth curve which levels off at the carrying capacity. (The *K* comes from an equation variable representing carrying capacity.) The carrying capacity is the maximum number of organisms that an environment can maintain. The carrying capacity is a *density dependent* factor. Most organisms have reproductive strategies somewhere between *K*- and *r*-selections.

Speciation is the process by which new species are formed. When gene flow ceases between two sections of a population, speciation begins. Factors which bring about speciation include geographic, seasonal, and behavioral isolation. **Adaptive radiation** occurs when several separate species arise from a single ancestral species, such as the 14 species of Galapagos finches that all evolved from one ancestor.

A species may face a crisis so severe as to cause a shift in the allelic frequencies of the survivors of the crisis. This is called an **evolutionary bottleneck**.

Divergent evolution exists when two or more species evolving from the same group maintain a similar structure from the common ancestor (called a *homologous* structure). However, two species may independently evolve similar structures in **convergent evolution**. Such similar structures are said to be *analogous* or *homoplastic*. An example of homoplasticity is the wings evolved by bats and birds; the two do not share a common ancestor from which they received their wings.

Some phenotypic forms vary gradually within a species, such as height. There are short people, tall people and every height in between short and tall. Other forms are distinct, like flower color, either red or white, or chicken plumage, either barred or non-barred. The occurence of distinct forms is called **polymorphism**.

9.3 Symbiosis

A **symbiosis** is a relationship between two species. The relationship can be beneficial for both, called **mutualism**; beneficial for one and not affect the other, called **commensalism**; beneficial for one and detrimental to the other, called **parasitism**. There is even a symbiosis called *enslavement* where one species enslaves another.

9.4 Hardy-Weinberg Equilibrium

In the early part of the 20th century, there was some question as to how less frequent alleles might be maintained in the population. Hardy and Weinberg came up with the explanation simultaneously. They showed statistically that there should be no change in the gene pool of a sexually reproducing population possessing the five following conditions:

1. **large population**;

2. **mutational equilibrium**;

3. **immigration or emigration must not change the gene pool**;

4. **random mating**; and

5. **no selection for the fittest organism**.

A population with these five characteristics is considered to be in **Hardy-Weinberg equilibrium**. No real population ever possesses these characteristics completely. Small populations are subject to **genetic drift** where one allele may be permanently lost due to the death of all members having that allele. Genetic drift is not caused by selective pressure, so its results are random in evolutionary terms. Mutational equilibrium means that the rate of forward mutations exactly equals the rate of back mutations. This rarely occurs in real populations; however, in the short term, mutations are seldom a major factor in changing allelic frequencies. Any immigration or emigration must not change the gene pool. This condition may occur in some isolated populations and is not typically a major factor in genetic change. The last two conditions probably do not occur in natural populations and are the most influential mechanisms of evolution.
The binomial theorem:

$$p^2 + 2pq + q^2$$

predicts the genotype frequency of a gene with only two alleles in a population in Hardy-Weinberg equilibrium. Imagine that 'A' is the dominant allele and 'a' is the recessive allele, and they are the only alleles for a specific gene. Now imagine that 80% of the alleles are 'A'. This means that 80% of the gametes will be 'A' and 20% will be 'a'. The probability that two 'A's come together is simply $0.8^2 = 0.64$. The probability that two 'a's come together is $0.2^2 = 0.04$. Any remaining zygotes will be heterozygous, leaving 32% heterozygotes. $(2 \times 0.8 \times 0.2 = 0.32)$ Using the formula, we represent 'A' as p and 'a' as q. Since there are only two alleles, **p + q = 1**.

201. If a certain gene possesses only two alleles, and the dominant allele represents 90% of the gene pool, how many individuals display the recessive phenotype.

  A. 0%
  B. 1%
  C. 18%
  D. 10%

202. Which of the following would least likely disrupt the Hardy-Weinberg equilibrium?

  A. emigration of part of a population
  B. a predator that selectively takes the old and sick
  C. a massive flood killing 15% of a large homogeneous population
  D. exposure of the entire population to intense radiation

203. Which of the following is most likely an example of two organisms in the same species?

  A. a cabbage in Georgia and a cabbage in Missouri that mate and produce fertile offspring only in years of unusual weather patterns
  B. two fruit flies on the same Hawaiian island with very different courtship dances
  C. two South American frogs that mate in different seasons
  D. two migratory birds that nest on different islands off the coast of England

204. All of the following factors would most likely favor an *r*-selection reproductive strategy over a *K*-selection strategy EXCEPT:

  A. intense seasonal droughts
  B. a short growing season
  C. limited space
  D. large scale commercial predation by humans

205. If two species are members of the same order, they must also be members of the same:

  A. habitat
  B. family
  C. class
  D. biome

206. The wolf, or *Canis lupus*, is a member of the family Canidae. Which of the following is most likely to be true?

  A. There are more living organisms classified as Canidae than as *Canis*.
  B. There are more living organisms classified as *lupus* than as *Canis*.
  C. An organism may be classified as *Canis* but not as Canidae.
  D. An organism may be classified as *lupus*, but not as *Canis*.

207. If, in a very large population, a certain gene possesses only two alleles and 36% of the population is homozygous dominant, what percentage of the population are heterozygotes?

  A. 16%
  B. 24%
  C. 36%
  D. 48%

208. Although human behavior ensures the success of each new generation of corn, selective breeding by humans has genetically altered corn so that it could not survive in the wild without human intervention. Corn population is controlled, and most of the corn seeds are eaten or become spoiled. The relationship between humans and corn is best described as:

  A. commensalism because humans benefit and corn is neither benefited nor harmed.
  B. commensalism because there is no true benefit to either species.
  C. parasitism because humans benefit and corn is harmed.
  D. mutualism because both species benefit.

9.5 Origin of Life

The universe is 12 to 15 billion years old. According to the *Big Bang Theory*, the universe began as a tiny spec of highly concentrated mass and exploded outward. In the early moments, only hydrogen gas existed. As the universe cooled, helium was able to form. The explosion was irregular, and gravitational forces created clumping of the mass. Heavier elements, and solar systems formed from these clumps of mass.

Our solar system is approximately 4.6 billion years old. The earth itself is about 4.5 billion years old; however, due to the volatile nature of early earth, there are no rocks on earth older than 3.9 billion years old.

Early earth probably had an atmosphere made mainly from nitrogen and hydrogen gas, and very little oxygen gas. One theory holds that the atmosphere contained clouds of H_2S, NH_3, and CH_4 creating a reducing environment. From this environment, the formation of carbon based molecules that we associate with life required little energy to form. Experiments attempting to recreate the atmosphere of early earth have resulted in the autosynthesis of molecules such as urea, amino acids, and even adenine. **The Urey-Miller experiment** was one of the early experiments to make such an attempt.

The first cells are thought to have evolved from **coacervates**, lipid or protein bilayer bubbles. Coacervates spontaneously form and grow from fat molecules suspended in water.

Organisms may have initially assimilated carbon from methane and carbon dioxide in the early atmosphere.

The earliest organisms were probably heterotrophs subsisting on preformed organic compounds in their immediate surroundings. Fossils of these organisms have been dated at **3.6 billion years old**. As preformed compounds became scarce, some of these organisms developed chemosynthetic autotrophy followed by photosynthetic autotrophy.

Around 2.3 billion years ago, the ancestors of cyanobacteria evolved. They were able to use sunlight and water to reduce carbon dioxide. These were the first oxygen producing, **photosynthetic bacteria**. The atmosphere began to fill with oxygen.

Eukaryotes evolved about 1.5 billion years ago, and did not develop into multicellular organisms until several million years later.

9.6 Chordate Features

Chordata is the phylum containing humans. Chordata does not mean backbone. All chordates have bilateral symmetry. They are **deuterostomes**, meaning their anus develops from or near the **blastopore**. (Compare protostomes, where the mouth develops from or near the blastopore.) Chordates have a **coelom** (a body cavity within mesodermal tissue). At some stage of their development they possess a **notochord** (an embryonic axial support, not the back bone), **pharyngeal slits**, **a dorsal**, **hollow nerve cord**, and a **tail**.

Members from the subphylum **Vertebrata** have their notochord replaced by a segmented cartilage or bone structure. They have a distinct brain enclosed in a skull. Most chordates are vertebrates. Vertebrata is composed of two classes of jawless fish (Agnatha), the cartilaginous fish, bony fish, amphibians, reptiles, birds, and mammals. The agnatha arose first and seperately from the rest about 470 million years ago. Amphibians arose from bony fish. Reptiles arose from amphibians about 300 million years ago. Birds and mammals arose from reptiles. Mammals arose from reptiles about 220 million years ago.

This last section contains bits of science trivia. Although it is probable that the MCAT will ask something from this section, it will be only one question or it will be explained in a passage. Keep this in mind, when you decide how much time you want to spend memorizing the details of this section.

You have now reviewed all the science tested by the MCAT. I suggest that you go back and review all of the tests that you have taken to this point. When you are done, you should pick your weakest area and master it. Then go to your next weakest area, and so on.

Next week is Zen Week. This is an important week of mental preparation. Be sure to attend. If you're not in the class, see the website at www.examkrackers.com for information on Zen Week.

See you!

9.7 Equation Summary

Equations
Hardy-Weinburg Equilibrium
$$p + q = 1$$
The Binomial Theorem
$$p^2 + 2pq + q^2$$

9.8 Terms You Need To Know

Terms	
3 to 1 ratio	Hybrid
Adaptive Radiation	*K*-selection
Allele	Kingdom
Barr Body	Law of Independent Assortment
Blastopore	Law of Segregation
Carrier	Locus
Carrying Capacity	Mammalia
Chordata	Mendelian Ratio
Class	Mutational Equilibrium
Coacervates	Mutualism
Coelom	Natural Selection
Complete Dominance	Niche
Commensalism	Notochord
Convergent Evolution	Order
Deuterostomes	Parasitism
Dihybrid Cross	Phenotype
Divergent Evolution	Phylum
Domains	Polymorphism
Dominant	Punnett Square
Evolution	*R*-selection
Evolutionary Bottleneck	Random Mating
F$_1$ Generation	Recessive
Family	Sex Chromosome
Gene Pool	Sex-linked Trait
Genetic Drift	Sigmoidal Growth Curve
Genotype	Speciation
Genus	Species
Hardy-Weinberg Equilibrium	Symbiosis
Heterozygous	Urey-Miller Experiment
Homozygous	Vertebrata

209. Which of the following was the earliest to evolve on Earth?

    A. plants
    B. prokaryotes
    C. protists
    D. fish

210. All of the following are characteristics of members of the phylum Chordata at some point in their life cycle EXCEPT:

    A. a tail
    B. a notochord
    C. a backbone
    D. gills

211. Which of the following are not members of the phylum Chordata?

    A. tunicates
    B. apes
    C. birds
    D. ants

212. Which of the following was probably not necessary for the origin of life on Earth?

    A. H_2O
    B. hydrogen
    C. O_2
    D. carbon

213. Which of the following is the phylum to which Homo sapiens belong?

    A. Mammalia
    B. Chordata
    C. Vertebrata
    D. Homo

214. Humans are members of the order:

    A. Vertebrata
    B. Chordata
    C. Primata
    D. Homididae

215. What do the results of the Urey-Miller experiment demonstrate?

    A. the existence of life on Earth
    B. that small biological molecules cannot be synthesized from inorganic material
    C. that life may have evolved from inorganic precursors
    D. that humans have evolved from photosynthetic cyanobacteria

216. If the first living organisms on Earth were heterotrophs, where did they get their energy?

    A. from eating each other
    B. from eating naturally formed organic molecules
    C. from the sun
    D. from eating dead organisms.

STOP!

DO NOT LOOK AT THESE EXAMS UNTIL CLASS.

30-MINUTE IN-CLASS EXAM FOR LECTURE 1

Passage I (Questions 1-7)

The three dimensional shape of a protein is ultimately determined by its amino acid sequence. The folding pattern itself is a sequential and cooperative process where initial folds assist in aligning the protein properly for subsequent folds. For many smaller proteins, the amino acid sequence alone can direct protein configuration, but for other proteins assistance in the folding process is necessary. Two types of proteins may assist in the folding of a polypeptide chain: enzymes which catalyze steps in the folding process, and proteins which stabilize partially folded intermediates. Proteins in the latter group are called *chaperones*.

An example of an enzyme which catalyzes the folding process is *protein disulfide isomerase*. This enzyme assists in the creation of disulfide bonds. The enzyme is not specific for any particular disulfide bond in a given chain. Instead, it simply increases the rate of formation of all disulfide combinations, and the most stable disulfide formations predominate.

Chaperones also assist in protein folding. As the protein folds, chaperones bind to properly folded sections and stabilize them. Chaperone synthesis can be induced by application of heat or other types of stress, and they are sometimes referred to as *heat shock proteins or stress proteins*.

Although the amino acid sequence determines the configuration of a protein, attempts at predicting protein configuration based upon amino acid sequence have been unsuccessful.

1. Which of the following statements concerning the function of protein disulfide isomerase in the formation of proteins is true?

    A. *Protein disulfide isomerase* increases only the rate at which disulfide bonds are formed.
    B. *Protein disulfide isomerase* increases only the rate at which disulfide bonds are broken.
    C. *Protein disulfide isomerase* increases both the rate at which disulfide bonds are formed and broken.
    D. *Protein disulfide isomerase* increases the rate at which disulfide bonds are formed and decreases the rate at which disulfide bonds are broken.

2. According to the passage, the folding pattern of a protein is determined by the protein's:

    A. primary structure
    B. secondary structure
    C. tertiary structure
    D. quaternary structure

3. Which of the following is the best explanation for why attempts at predicting protein configuration based upon amino acid sequence have been unsuccessful?

    A. It is impossible to know the amino acid sequence of a protein without knowing the DNA nucleotide sequence.
    B. Enzymes and chaperones help to determine the three dimensional shape of a protein.
    C. The three dimensional shape of a protein is based upon hydrogen and disulfide bonding between amino acids, and the number of possible combinations of bonding amino acids makes prediction difficult.
    D. The amino acid sequence of the same protein may vary slightly from one sample to the next.

4. Chaperones assist in the formation of a protein's:

    A. primary structure.
    B. secondary structure.
    C. tertiary structure.
    D. quaternary structure.

5. Natural selection has resulted in increased chaperone synthesis in the presence of elevated temperatures. How might increased *chaperone* production in the presence of heat be advantageous to a cell?

    A. Heat destabilizes intermolecular bonds making protein configuration more difficult to achieve. *Chaperones* counteract this by stabilizing the partially folded intermediates.
    B. Increased temperatures increase reaction rates creating an excess of fully formed proteins. *Chaperones* stabilize the partially folded intermediates and slow the process.
    C. Increased *chaperone* production requires energy. This energy is acquired from the kinetic energy of molecules and thus cools the cell.
    D. Elevated temperatures result in increased cellular activity requiring more proteins. Chaperones increase the rate of polypeptide formation.

6. Which of the following bonds in a protein is likely to be LEAST stable in the presence of heat?

   A. a disulfide bond
   B. a hydrogen bond
   C. a polypeptide bond
   D. the double bond of a carbonyl

7. *Protein disulfide isomerase* most likely:

   A. lowers the activation energy of the formation of cystine.
   B. raises the activation energy of the formation of cystine.
   C. lowers the activation energy of the formation of proline.
   D. raises the activation energy of the formation of proline.

Passage II (Questions 8-14)

In *polyacrylamide gel electrophoresis* (PAGE), electrically charged proteins are dragged by an electric field through the pores of a highly cross-linked gel matrix at different rates depending upon their size and charge.

In a SDS-PAGE, proteins are separated by size only. Since different proteins have different *native charges*, the protein mixture is first dissolved in SDS (sodium dodecyl sulfate) solution. SDS anions disrupt the noncovalent bonds of the proteins and associate with the peptide chains, approximately one molecule of SDS for every two residues of a typical protein. The resulting net negative charge on the protein is normally much greater than the charge on the native protein. *Mercaptoethanol* is usually added in the presence of heat to reduce disulfide bonds and complete the denaturization process. This solution is then applied to a porous gel and an electric field is applied. The rate of movement through the gel is inversely proportional to the logarithm of the molecular weight of the protein.

The proteins are then stained with a dye such as coomassie blue. Lines are formed at different points along the gel corresponding to the molecular weight of the proteins.

A second type of electrophoresis, called *isoelectric focusing*, distinguishes proteins based upon their isoelectric points. A permanent pH gradient is established within a polyacrylamide gel by applying an electric field to polyacrylamide polymers with different p*I*s. When the native proteins are applied to this gel in the absence of SDS, each protein moves until it reaches its p*I*.

8. SDS PAGE would be least effective in distinguishing between the masses of different:

   A. carbohydrate-rich glycoproteins.
   B. acidic proteins.
   C. polar proteins.
   D. enzymes.

9. What is the purpose of coomassie blue?

   A. Coomassie blue increases the separation of the proteins by increasing their mass.
   B. Coomassie blue increases the separation of the proteins by increasing their charge.
   C. Coomassie blue allows the results of electrophoresis to be visualized.
   D. Coomassie blue stops further movement of the proteins by increasing their mass.

GO ON TO THE NEXT PAGE.

10. Which of the following proteins would move the most slowly through the gel in SDS PAGE?

   A. a large protein
   B. a small protein
   C. a protein with a high native charge
   D. a protein with a low native charge

11. Electrophoresis is also used to analyze nucleic acids. In electrophoresis of nucleic acids, SDS is unnecessary because:

   A. nucleic acids already contain negatively charged phosphate groups in proportion to their size.
   B. nucleic acids are already large enough to separate appreciably on their own.
   C. nucleic acids don't have a quaternary or tertiary structure to disrupt.
   D. nucleic acids do not contain hydrogen bonds.

12. SDS is a detergent that does not cleave covalent bonds. Which protein structure cannot be disrupted by SDS?

   A. primary
   B. secondary
   C. tertiary
   D. quaternary

13. Which of the following would most likely occur if a multisubunit protein were subjected to the electrophoresis techniques used in SDS PAGE?

   A. The protein would remain intact and separate from other proteins according to its native charge.
   B. The protein would remain intact and separate from other proteins according to its size.
   C. Each subunit would separate independently, according to its native charge.
   D. Each subunit would separate independently, according to its size.

14. Which of the following is most likely a limitation to SDS PAGE in identifying different proteins within a protein mixture?

   A. SDS PAGE is likely to be an expensive process.
   B. SDS PAGE cannot easily distinguish between proteins of similar molecular weight.
   C. Any proteins used in SDS PAGE are denatured.
   D. The native charge on a protein does not always correspond to its size.

Passage III (Questions 15-21)

Glycolysis is the metabolic breakdown of glucose into the readily useable form of chemical energy, ATP. For some human cells, such as neurons and erythrocytes, glucose is the only source of chemical energy available under typical circumstances.

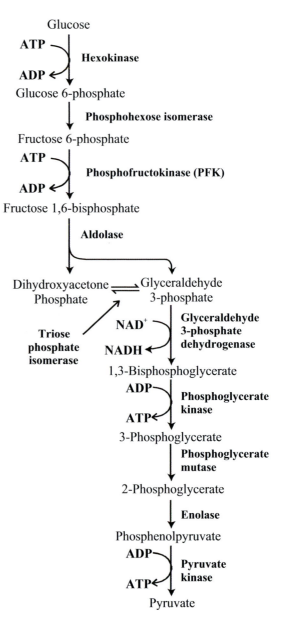

Figure 1 Glycolysis

Each reaction in the glycolytic pathway is governed by an enzyme. Glucose is phosphorylated as it enters the cell in an irreversible reaction with hexokinase. It is not until the reaction governed by phosphofructokinase (PFK), however, that the molecule is committed to the glycolytic pathway. The PFK reaction is called the *committed step*. PFK activity is inhibited when cellular energy is plentiful, and stimulated when energy is low.

Glycolysis can be interrupted by poisons that interfere with glycolytic enzyme activity. Arsenate, a derivative of arsenic, is a deadly poison that acts as a substrate for *glyceraldehyde 3-phosphate dehydrogenase.*

15. The net result of aerobic respiration can be summarized most accurately as:

    A. the oxidation of glucose.
    B. the reduction of glucose.
    C. the elimination of glucose.
    D. the lysis of glucose.

16. The process of the synthesis of ATP in the glycolytic reaction governed by *phosphoglycerate kinase* is called:

    A. oxidative phosphorylation.
    B. substrate-level phosphorylation.
    C. exergonic phosphate transfer.
    D. electron transport.

17. Which of the following gives the net reaction for glycolysis?

    A. Glucose + 4 ADP → pyruvate + 2 ATP
    B. Glucose + 2 ADP + 2 P_i + 2 NAD+ g 2 pyruvate + 2 ATP + 2 NADH
    C. Glucose + O_2 → CO2 + H2O + 2 ATP
    D. Glucose + O_2 → 2 pyruvate + 2 ATP + 2 NADH

18. Which of the following would most likely occur inside a cell in the presence of arsenic?

    A. The concentration of glyceraldehyde 3-phosphate dehydrogenase would decrease.
    B. The concentration of glyceraldehyde 3-phosphate would increase.
    C. The concentration of aldolase would increase.
    D. The concentration of 1,3-bisphosphoglycerate would increase.

19. Why is the PFK reaction, and not the hexokinase reaction, the *committed step*?

    A. The hexokinase reaction is irreversible, but the PFK reaction is not.
    B. The PFK reaction requires the hydrolysis of ATP, but the hexokinase reaction does not.
    C. Glucose 6-phosphate is a higher energy molecule than fructose 1,6-bisphosphate.
    D. Glucose 6-phosphate may be converted into glycogen in some circumstances, but fructose 1,6-bisphosphate has only one possible chemical fate in the cell.

20. From the information in the passage, which of the following might be an allosteric activator of PFK?

    A. citrate
    B. insulin
    C. ADP
    D. ATP

21. The action of arsenate on glyceraldehyde 3-phosphate dehydrogenase is best described as:
    A. competitive inhibition.
    B. noncompetitive inhibition.
    C. allosteric inhibition.
    D. negative feedback.

GO ON TO THE NEXT PAGE.

Questions 22 through 23 are **NOT** based on a descriptive passage.

22. Cofactors are best described as:

   A. nonprotein substances required for all enzyme activity.

   B. small, nonprotein, organic molecules.

   C. metal ions or coenzymes that activate an enzyme by binding tightly to it.

   D. catalysts.

23. The substrate concentration in a reaction which is governed by an enzyme is slowly increased to high levels. As the substrate concentration increases the reaction rate:

   A. continues to increase indefinitely.

   B. continues to decrease indefinitely.

   C. increases at first, then levels off.

   D. does not change.

STOP. IF YOU FINISH BEFORE TIME IS CALLED, CHECK YOUR WORK. YOU MAY GO BACK TO ANY QUESTION IN THIS TEST BOOKLET.

30-MINUTE IN-CLASS EXAM FOR LECTURE 2

Passage I (Questions 24-28)

The giant amoeba-like *Pelomyxa palustris*, the only member of the phylum Caryoblastea, exhibits one of the most primitive forms of cell division in eukaryotes; it does not undergo mitosis. Instead, the nucleus simply splits into two daughter nuclei. Like most other protists, the nucleus of the *Pelomyxa* is bound by a nuclear envelope. During cellular division, the chromosomes of the *Pelomyxa* double in number and assort randomly to the daughter cells. Multiple copies of each chromosome ensure that each daughter cell maintains the necessary amount of genetic material to specify the organism. *Pelomyxa* has no centrioles or mitochondria; however, it does contain two bacterial symbionts which may function similarly to mitochondria. *Pelomyxa* may represent an early stage in the evolution of eukaryotic cells.

Unicellular protists from the phylum Pyrrhophyta, commonly called *dinoflagellates*, undergo a form of mitosis where the nuclear membrane remains intact. Microtubules extending through the nuclear membrane attach to chromosomes. The nuclear membrane grows between attached chromosomes separating them and creating two daughter nuclei. The dinoflagellate then divides with each daughter cell accepting a nucleus. Most dinoflagellates are photosynthetic, and most are protected by cellulose plates.

Mitosis in diatoms from the phylum *chrysophyta* is similar to that in dinoflagellates, though slightly more advanced.

24. Which of the following is true according to the passage?

   A. Eukaryotes that lack centrioles cannot undergo mitosis.
   B. Some prokaryotes undergo mitosis, while others do not.
   C. Not all eukaryotes undergo mitosis.
   D. All protists lack mitochondria.

25. During mitosis, the nuclear envelope of a mammalian cell:

   A. disintegrates during replication of the chromosomes.
   B. disintegrates while crossing over is taking place.
   C. disintegrates while the chromosomes condense.
   D. remains intact.

26. As presented in the passage, the theory that *Pelomyxa* represents an early stage in the evolution of eukaryotic cells would be most *weakened* by which of the following?

   A. A protist containing mitochondria and undergoing mitosis, proved to be an ancestor to *Pelomyxa*.
   B. Diatoms were found to have evolved from *Pelomyxa*.
   C. New evidence placed dinoflagellates in the kingdom plantae.
   D. A second species belonging to the phylum Caryoblastea is discovered and found to undergo mitosis.

27. The separation of duplicate chromosomes of dinoflagellates most closely resembles which two phases in mitosis?

   A. prophase and metaphase
   B. metaphase and anaphase
   C. anaphase and telophase
   D. anaphase and cytokinesis

28. The bacterial symbionts in *Pelomyxa* most likely:

   A. parasitically infect the *Pelomyxa*.
   B. provide energy for the *Pelomyxa* from the metabolism of absorbed nutrients.
   C. reproduce independently from the *Pelomyxa* through a primitive mitosis.
   D. function in lipid synthesis.

Passage II (Questions 29-34)

For many genes in eukaryotes, the mRNA initially transcribed is not translated in its entirety. Instead, it is processed in a series of steps occurring within the nucleus. The initial unprocessed mRNA is called pre-mRNA. Just after transcription begins, the 5′ end of the pre-mRNA is *capped* with GTP forming a 5′–5′ triphosphate linkage. The *cap* protects the mRNA from exonucleases acting on its 5′ end, and acts as an attachment site for ribosomes. At the 3′ end, the *polyadenylation signal*, AAUAA, signals the cleavage of the pre-mRNA 10 to 20 nucleotides downstream. Several adenosine residues are added to the 3′ end to form a *poly-A tail*. The poly-A tail wraps tightly around an RNA-protein complex protecting the 3′ end from degradation.

As much as 90% of the pre-mRNA may be removed from the nucleotide sequence as *introns*. The remaining *exons* are spliced together in a single mRNA strand, and leave the nucleus to be translated in the cytosol or on the rough endoplasmic reticulum.

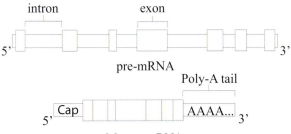

Figure 1 Post-transcriptional processing of mRNA

In *R looping*, a technique used to identify introns, a fully processed mRNA strand is hybridized with its double stranded DNA counterpart under conditions which favor formation of hybrids between DNA and RNA strands. In this process, the RNA strand displaces one DNA strand and binds to the other DNA strand along complementary sequences. The results can be visualized through electron microscopy and are shown in Figure 2.

In prokaryotes, mRNA is normally translated without modification.

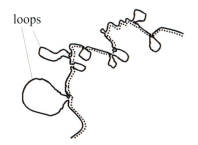

Figure 2 R looping

29. Which of the following would most strongly indicate that the *poly-A tail* is added after transcription and not during transcription?

- **A.** evidence of an enzyme in the nucleus that catalyzes the synthesis of a sequence of multiple adenosine phosphate molecules
- **B.** a sequence of multiple thymine nucleotides following each gene in eukaryote DNA
- **C.** a sequence of multiple uracil nucleotides following each gene in eukaryote mRNA
- **D.** a sequence of multiple adenosine nucleotides following each gene in eukaryote DNA

30. Small nuclear ribonucleoproteins (snRNPs) catalyze the splicing reaction in the post-transcriptional processing of RNA. Based upon the information in the passage, in which of the following locations would snRNPs most likely be found?

- **A.** the cytosol of a prokaryotic cell
- **B.** the cytosol of a eukaryotic cell
- **C.** the lumen of the endoplasmic reticulum
- **D.** the nucleus of the eukaryotic cell

31. The loops in Figure 2 represent:

- **A.** DNA introns.
- **B.** DNA exons.
- **C.** RNA introns.
- **D.** RNA exons.

32. In *R looping* the displaced DNA strand contains the complementary nucleotide sequence for:

- **A.** the entire DNA strand shown in Figure 2.
- **B.** the entire RNA strand shown in Figure 2.
- **C.** the DNA introns only.
- **D.** the RNA introns only.

33. The intron sequences of identical genes in closely related species are often very different. Which of the following is most strongly suggested by this evidence?

- **A.** Identical genes in closely related species may code for different proteins.
- **B.** Changes in the amino acid sequence of a protein do not necessarily change protein function.
- **C.** Intron sequences are heavily characterized by selective pressure.
- **D.** Selective pressure has little or no role in the development of intron sequences.

GO ON TO THE NEXT PAGE.

34. Which of the following represents the 5′ end of eukaryotic mRNA?

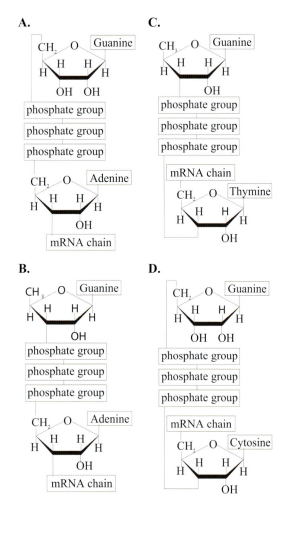

A.

C.

B.

D.

An alteration in cellular DNA other than genetic recombination is called a *mutation*. Mutations can occur in both germ and somatic cells. Germ line mutations are transmitted to offspring and can result in genetic diseases. Somatic mutations lead to *neoplastic diseases*, which are responsible for 20% of all deaths in industrialized countries.

The *basal mutation rate* is the natural rate of change in the nucleotide sequence in the absence of environmental mutagens. Such mutations result mainly from errors in DNA replication. A common error during replication is a *tautomeric shift*. Thymine, in its keto form, pairs with adenine; however, in its enol form, it prefers guanine. Adenine usually exists in an amino form that prefers thymine, and, rarely, in an imino form that pairs with cytosine. Tautomeric shifts result in mismatched base pairs.

Environmental mutagens such as chemicals and radiation can increase the mutation rate. Ionizing radiation, such as gamma rays and x-rays, excites electrons creating radicals inside the cell that react with DNA. Ultraviolet radiation from normal sunlight cannot penetrate the skin nor make radicals; however, it can form pyrimidine dimers from adjacent pyrimidines. Chemicals, such as deaminating agents, can convert adenine, guanine, and cytosine into hypoxanthine, xanthine, and uracil, respectively, which lead to mismatched base pairs and thus errors during replication. Though deamination of adenine and guanine are rare, deamination of cytosine is fairly common.

Because many mutations occur initially to only one strand of a DNA double helix, they can be repaired. In a healthy cell there are specialized enzyme systems which monitor and repair DNA. However, once the DNA is replicated, these enzymes can no longer recognize the mutation and repair becomes very unlikely. For errors occurring during replication there are enzymes called *glycosylases* that can remove mismatches and small insertions and deletions. Glycosylases, which also work on the deaminated bases, must be capable of identifying mismatched base pairs and distinguishing between the old and new DNA strand.

35. It has been hypothesized that in ancestral organisms DNA contained uracil and not thymine. Why might DNA that contains thymine have a natural selective advantage over DNA that contains uracil?

  A. DNA that contains uracil would be unable to utilize glycosylases that repair mutations caused by the deamination of cytosine.
  B. Uracil does not undergo a tautomeric shift and thymine does.
  C. DNA that contains uracil cannot form a double helix.
  D. Uracil is a purine, and thymine is a pyrimidine.

36. Radiation therapy is used to treat some forms of cancer by damaging DNA and thus killing the rapidly reproducing cancerous cells. Why might radiation treatment have a greater effect on cancer cells than normal cells?

  A. Normal cells have more time between S phases to repair damaged DNA.
  B. Damaged DNA is more reactive to radiation.
  C. Cancer cells have lost the ability to repair damaged DNA.
  D. The effect of radiation is the same, but there are more cancer cells than normal cells.

37. From the information in the passage, which of the following is LEAST likely to be true concerning neoplastic diseases?

  A. Tumors may produce neoplastic diseases.
  B. Neoplastic diseases are hereditary.
  C. Neoplastic diseases begin with the mutation of a normal cell.
  D. Exposure to chemical mutagens may lead to a neoplastic disease.

38. Ionizing radiation can create double stranded breaks in DNA. Eukaryotes are able to repair some of these breaks, but prokaryotes are not. Which of the following gives the most likely explanation for this difference?

  A. Prokaryotes do not possess a ligase enzyme to join separate DNA molecules.
  B. Prokaryotic DNA is single stranded.
  C. Eukaryotes have matching pairs of chromosomes to act as a template for repair.
  D. Eukaryotes contain more DNA making the consequences of a break less severe.

39. Which of the following pairs of nitrogenous bases might form a dimer in DNA when exposed to UV radiation?

  A. thymine - thymine
  B. thymine - adenine
  C. thymine - guanine
  D. uracil – uracil

40. 5-bromouracil resembles thymine enough to become incorporated into DNA during replication. Once incorporated, however, it rearranges to resemble cytosine. If 5-bromouracil were present during the replication of the sense strand shown below, which of the following might be the sense strand formed in the following replication?

$$5'\text{-GGCGTACG-}3'$$

  A. 5'-GGCGCACG-3'
  B. 5'-GGCGATCG-3'
  C. 3'-CCGCAGGC-5'
  D. 5'-GGCGTGCG-3'

41. According to the passage, in the absence of *glycosylases*, a *tautomeric shift* would most likely result in which of the following mutations?

  A. a frameshift mutation
  B. a base pair insertion
  C. a base pair substitution
  D. a chromosomal aberration

42. According to the information in the passage, which of the following is true concerning mutations?

  A. Mutations do not occur in the absence of radiation or chemical mutagens.
  B. Industrial countries have a higher basal mutation rate than non-industrial countries.
  C. Hypoxanthine is an example of a chemical mutagen.
  D. Mutations occurring in rapidly reproducing cells are more likely to become a permanent part of the cell genome.

GO ON TO THE NEXT PAGE.

43. DNA replication occurs during:

A. prophase.
B. metaphase.
C. telophase.
D. interphase.

44. Turner's syndrome occurs due to nondisjunction at the sex chromosome resulting in an individual with one X and no Y chromosome. Color-blindness is a sex-linked recessive trait. A color-blind man marries a healthy woman. They have two children both with Turner's syndrome. One of the children is color-blind. Which of the following is true?

A. Nondisjunction occurred in the father for both children.
B. Nondisjunction occurred in the mother for both children.
C. Nondisjunction occurred in the mother for the color-blind child and in the father for the child with normal vision.
D. Nondisjunction occurred in the father for the color-blind child and in the mother for the child with normal vision.

45. A primary spermatocyte is:

A. haploid and contains 23 chromosomes.
B. haploid and contains 46 chromosomes.
C. diploid and contains 23 chromosomes.
D. diploid and contains 46 chromosomes.

46. Crossing over occurs in:

A. mitosis, prophase.
B. meiosis, prophase I.
C. meiosis, prophase II.
D. interphase.

STOP. IF YOU FINISH BEFORE TIME IS CALLED, CHECK YOUR WORK. YOU MAY GO BACK TO ANY QUESTION IN THIS TEST BOOKLET.

30-MINUTE IN-CLASS EXAM FOR LECTURE 3

Passage I (Questions 47-53)

Disease-causing microbial agents are called pathogens. Microbial growth is affected by temperature, O_2, pH, osmotic activity, and radiation. Physical methods of pathogen control include chemical, heat, filtration, ultraviolet radiation, and ionizing radiation. Moist heat is generally more effective than dry heat.

Sterilization is the removal or destruction of all living cells, viable spores, viruses, and viroids. Sometimes it is only deemed necessary to kill or inhibit pathogens. This is called disinfection. *Sanitation* reduces the number of microbes to levels considered safe by public health standards.

The *decimal reduction time (D)* or *D value* is the time required to kill 90% of microorganisms or spores in a sample at a specified temperature. Environmental factors may affect D values. The subscript on the D value indicates the temperature at which it applies. Increasing the temperature decreases the D value. The z value is the increase in temperature necessary to reduce a D value by a factor of 10. Table 1 shows D and z values for some food-borne pathogens.

Bacteria	Substrate	D value (°C) in minutes	z value (°C)
C. botulinum	Phosphate buffer	$D_{121}=0.20$	10
C. perfringens	Culture media	$D_{90}=3\text{-}5$	6-8
Salmonella	Chicken a la king	$D_{60}=0.40$	5.0
S. aureus	Chicken a la king	$D_{60}=5.4$	5.5
S. aureus	Turkey stuffing	$D_{60}=15$	6.8
S. aureus	0.5% NaCl	$D_{60}=2.0\text{-}2.5$	5.6

Table 1 D and z values for some food-borne pathogens

Like population growth, population death is exponential. However, as the population reduces to very low levels, the proportion of resistant strains increases slowing the overall rate of death.

The food processing industry relies on D and z values for guidelines in controlling contamination. For instance, after being canned, food must be heated sufficiently to destroy any endospores of *Clostridium botulinum*, the bacteria responsible for botulism. *C. botulinum* is an obligatory anaerobic, gram positive bacterium found in soil and aquatic sediments. When food containing endospores of *C. botulinum* is stored, the endospores may germinate and produce a deadly neurotoxin. Although the disease is fatal to 1/3 of untreated patients, it can be effectively treated with an injection of antibodies produced by horses.

47. Certain eating utensils are treated with a sanitizer. After 3 minutes the number of microbes is reduced from 6.25×10^{12} to 2.5×10^{11}. According to the passage, 3 minutes more exposure to the sanitizer would reduce the number of microbes to:

   A. 1.0×10^{10}.
   B. 5.0×10^{10}.
   C. 6.0×10^{10}.
   D. 1.9×10^{11}.

48. From Table 1, the D value for *Salmonella* in Chicken a la king at 70°C is:

   A. 0.24 seconds.
   B. 2.4 seconds.
   C. 24 seconds.
   D. 40 minutes.

49. If federal regulations require that canned food be heated at 121°C long enough to reduce a colony of *C. botulinum* in a phosphate buffer from 10^{12} bacteria to 1 bacterium, how long must canned food be heated?

   A. 12 seconds
   B. 2.4 minutes
   C. 12 minutes
   D. 2.0×10^{11} minutes

50. A nutrient rich agar is seeded with a few bacteria, which quickly become a thriving colony. Which of the following most accurately depicts the exponential portion of the population growth on a logarithmic scale?

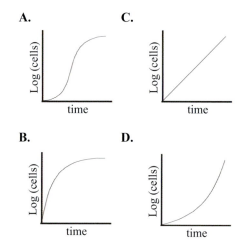

51. According to the passage, which of the following is true concerning a surface that has been disinfected?

    A. No living microorganisms exist on the surface.

    B. Some living microbes remain on the surface, but all or most microbes capable of producing disease have been destroyed or reduced.

    C. The surface has been cleansed of bacteria, but not necessarily viruses or viroids.

    D. All pathogens on the surface are destroyed or removed, while all nonpathogens on the surface remain alive.

52. Which of the following statements is best supported by the data in Table 1?

    A. A bacterium's resistance to heat is directly related to its z value.

    B. *C. botulinum* is more likely to contaminate commercially processed food than *S. aureus*.

    C. A bacterium's resistance to heat may vary depending upon its environment.

    D. In chicken a la king, *Salmonella* is more resistant to heat than *S. aureus*.

53. Which of the following statements does NOT contradict the information in the passage concerning *C. botulinum*?

    A. *C. botulinum* thrives in the presence or absence of oxygen.

    B. *C. botulinum* has a lipid bilayer outside its peptidoglycan cell wall.

    C. A glass containing *C. botulinum* may be disinfected by immersion in boiling water.

    D. Animals have no natural defense against the neurotoxin produced by *C. botulinum*.

Passage II (Questions 54-58)

Even within mature cells, reshuffling of genetic material can take place. The vehicles by which nucleotides move from one position on a chromosome to another, or from one chromosome to another, are collectively termed *transposable genetic elements (TGE)*. TGEs in prokaryotes include *insertion sequence (IS)* elements, and *transposons*.

IS elements are segments of double stranded DNA. Each single strand of DNA in an IS element possesses a nucleotide sequence at one end that is complementary to the reverse sequence of nucleotides at its other end. These end sequences are termed *inverted repeat (IR) sequences*. Between the IR sequences is a *transposase* gene which codes for an enzyme that allows the IS element to integrate into the chromosome. IS elements do not contain entire genes other than the transposase gene. Insertion of an IS element into a gene activates or deactivates that gene, depending upon the location of the insertion and the orientation of the IS element.

The TGEs known as transposons exist mainly as a series of complete genes sandwiched between two opposite oriented IS elements. As well as existing in prokaryotes, transposons are the main type of TGEs found in eukaryotes. They are similar to retroviruses. Both IS elements and transposons are found in bacterial plasmids.

A TGE may move as a complete entity, called *conservative transposition*, or it may move a duplicate copy of itself, called *replicative transposition*.

54. One strand of an IS element begins with the nucleotide sequence 5´-ACTGTTAAG-3´. The same strand must end with the nucleotide sequence:

    A. 5´-GAATTGTCA-3´

    B. 5´-TGACAATTC-3´

    C. 5´-CTTAACAGT-3´

    D. 5´-ACTGTTAAG-3´

55. When the passage states that *transposons* are similar to retroviruses, to what aspect of the life cycle of a retrovirus is the passage most likely referring?

    A. reverse transcription of viral RNA

    B. proliferation of multiple copies of viral genome leading to the lyses of the host cell

    C. capsid formation

    D. the procedure by which the viral genome integrates into the host genome

GO ON TO THE NEXT PAGE.

56. Which of the following is not typically associated with genetic recombination in prokaryotes?

  A. transduction
  B. TGEs
  C. binary fission
  D. transformation

57. Which of the following mechanisms of genetic recombination between prokaryotes involves plasmids?

  A. transduction
  B. conjugation
  C. meiosis
  D. binary fission

58. Bacterium A is able to live on a histidine deficient medium. Bacterium B is not. After initiating conjugation with bacterium B, Bacterium A is unable to live on a histidine deficient medium. Which of the following statements is most likely true concerning the conjugation?

  A. *Conservative transposition* occurred during conjugation where bacterium A transferred the his+ gene to bacterium B.
  B. *Replicative transposition* occurred during conjugation where bacterium A transferred the his+ gene to bacterium B.
  C. *Conservative transposition* occurred during conjugation where bacterium B transferred the his+ gene to bacterium A.
  D. *Replicative transposition* occurred during conjugation where bacterium B transferred the his+ gene to bacterium A.

Passage III (Questions 59-65)

λ phage (pronounced lambda phage) is a double stranded DNA bacteriophage exhibiting both a lytic and a lysogenic lifecycle. The virion injects a single molecule of double stranded DNA containing 40 genes into the host cell. The two ends of the DNA molecule are single strands of DNA that are reverse complements of each other. Once inside the cell, the host cell DNA ligase links the single stranded ends together forming a small circle of DNA. The virus may now enter a lytic phase, or it may insert itself between the galactose and biotin operons. Insertion into the host cell genome requires the viral enzyme *integrase*. Integrase is translated soon after infection of the host cell.

At the same time that integrase is translated, a protein called λ *repressor* is also translated. λ repressor prevents the transcription of all λ phage genes except its own. For as long as the concentration of λ repressor is maintained above a critical limit, the virus remains lysogenic. This can last hundreds of thousands of cell generations.

Damaged DNA activates a cellular protease that degrades λ repressor. When the concentration of λ repressor falls below the critical limit, the gene for *excisionase* is transcribed. Excisionase cuts the viral DNA from the host cell chromosome. The virus then switches from the lysogenic to the lytic pathway which ultimately results in the lysis of the host cell.

59. While integrated into the host cell DNA a λ phage is called a:

  A. virion.
  B. prophage.
  C. chromosome.
  D. plasmid.

60. λ phage most likely integrates into the host cell DNA in the:

  A. host cell nucleus.
  B. host cell mitochondria.
  C. lumen of the host cell endoplasmic reticulum.
  D. cytoplasm of the host cell.

61. Which of the following would most likely lead to lysis of a host cell infected with λ phage in the lysogenic stage?

   A. infection by a second λ phage
   B. binary fission
   C. exposure to ultraviolet radiation
   D. mitosis

62. Which of the following must be true in order for infection with λ phage to take place?

   A. The host cell must be diseased or weakened.
   B. The host cell membrane or wall must contain a specific receptor protein.
   C. Some type of mutagen must be present.
   D. Viral DNA must penetrate the host cell nuclear membrane.

63. Which of the following describes in the correct chronological order the events of an infection with λ phage?

   A. viral DNA is injected into the host cell—viral DNA is translated and replicated—the capsid is formed—the host cell lyses releasing hundreds of viral progeny.
   B. viral DNA is injected into the host cell—viral DNA is transcribed—viral RNA is translated and reverse transcribed—the capsid is formed—the host cell lyses releasing hundreds of viral progeny.
   C. viral DNA is injected into the host cell—viral DNA is transcribed and replicated—the capsid is formed—the host cell lyses releasing hundreds of viral progeny.
   D. viral DNA is injected into the nucleus—viral DNA is transcribed and replicated—the capsid is formed—the host cell lyses releasing hundreds of viral progeny.

64. Which of the following is a likely host cell for a λ phage?
   A. an E. coli bacterium
   B. T-lymphocyte
   C. a paramecium
   D. a neuron

65. Which of the following is never found inside the capsid of a virus?

   A. single stranded RNA only
   B. double stranded RNA only
   C. single stranded DNA only
   D. RNA and DNA

GO ON TO THE NEXT PAGE.

66. Bacteria contain all of the following EXCEPT:

    A. membrane bound organelles
    B. double stranded DNA
    C. ribosomes
    D. circular DNA

67. Which of the following statements are true concerning yeast?

    I. Yeasts are eukaryotic.
    II. Yeasts are unicellular.
    III. Yeasts are facultative anaerobes.

    A. I only
    B. I and II only
    C. I and III only
    D. I, II, and III

68. The proteins and glycoproteins which make up the capsid, envelope and spikes of a virus determine the infective properties of that virus. All of the following are true concerning viruses EXCEPT:

    A. Some complex viruses replicate without the synthetic machinery of the host cell.
    B. Animal viruses enter their host via endocytosis.
    C. A bacteriophage sheds its protein coat outside the host cell and injects its nucleic acids through the host cell wall.
    D. A latent bacteriophage consisting only of a DNA fragment is called a prophage.

69. Which of the following is found either in prokaryotes or eukaryotes, but not in both?

    A. a cell wall
    B. ribosomes
    C. RNA
    D. centrioles

STOP. IF YOU FINISH BEFORE TIME IS CALLED, CHECK YOUR WORK. YOU MAY GO BACK TO ANY QUESTION IN THIS TEST BOOKLET.

30-MINUTE IN-CLASS EXAM FOR LECTURE 4

Passage I (Questions 70-76)

Smooth muscle and visceral organs of the body are innervated by the autonomic nervous system, which is controlled mainly by centers within the spinal cord, brain stem, and hypothalamus. The two branches of the autonomic nervous system are the sympathetic and parasympathetic. Most visceral organs are innervated by both branches.

The sympathetic nervous system is composed of motor pathways consisting of two neurons, the preganglionic and the postganglionic. Preganglionic nerve fibers exit the spinal cord between segments T1 and L2. The neuronal cell bodies of the postganglionic neurons are mainly contained in the sympathetic paravertebral chain ganglion located on either side of the spinal cord. One exception to this rule is the nerves innervating the adrenal medulla, which synapse directly onto the *chromaffin cells*. Chromaffin cells are themselves modified postganglionic neurons.

The parasympathetic nervous system is also composed of motor pathways consisting of two neurons; however, the bodies of the postganglionic cells of the parasympathetic system are located on or near the effector organs. Most parasympathetic nerves exit the central nervous system through cranial nerves, the vagus nerve containing 75 percent of all parasympathetic nerve fibers.

The delivery of neurotransmitter to effector organs by the autonomic nervous system is less precise than in neuromuscular junctions of the somatic nervous system.

70. From the information in the passage, it can be presumed that *chromaffin cells* secrete:

   A. cortisol.
   B. epinephrine.
   C. ACTH.
   D. renin.

71. Which of the following is not innervated by the autonomic nervous system?

   A. the arteries of the heart
   B. the iris musculature of the eye
   C. the diaphragm
   D. sweat glands

72. Which of the following cell types most likely contain adrenergic receptors (receptors that respond to epinephrine)?

   A. chromaffin cells
   B. sympathetic postganglionic neurons
   C. parasympathetic postganglionic neurons
   D. cardiac muscle cells

73. Which of the following is not an autonomic response to temperature change?

   A. piloerection
   B. shivering
   C. constriction of cutaneous vessels
   D. sweating

74. Which of the following nervous systems is responsible for the simple reflex arc?

   A. sympathetic
   B. parasympathetic
   C. both sympathetic and parasympathetic
   D. somatic

75. Upon arrival to a high altitude environment, individuals may experience symptoms of nausea and vertigo. These symptoms generally subside after a few days exposure to the environment. The system most likely responsible for acclimatizing the body in this case is:

   A. the somatic nervous system.
   B. the sympathetic nervous system.
   C. the parasympathetic nervous system.
   D. the endocrine system.

76. Amphetamines cause epinephrine to be released from the ends of associated neurons. Which of the following is most likely NOT a symptom of amphetamine usage?

   A. increased heart rate
   B. elevated blood glucose levels
   C. constricted pupils
   D. increased basal metabolism

Passage II (Questions 77-83)

In a healthy cell, smooth endoplasmic reticulum (ER) performs several functions including carbohydrate metabolism, lipid synthesis, and oxidation of foreign substances such as drugs, pesticides, toxins and pollutants.

The liver maintains a relatively stable level of glucose in the blood via its glycogen stores. An increase of cyclicAMP activates protein kinase A, which leads to the formation of glucose 1-phosphate from glycogen. Glucose 1-phosphate is converted to glucose 6-phosphate, which still cannot diffuse through the cell membrane. Glucose 6-phosphatase, associated with smooth ER, hydrolyzes glucose 6-phosphate to glucose, which is then transported from the cell into the blood stream.

The smooth ER synthesizes several classes of lipids, including triacylglycerols which are stored in the ER lumen, cholesterol and its steroid hormone derivatives, and phospholipids for incorporation into the various membranous cell structures. Phospholipids are synthesized only on the cytosol side of the ER. They are then selectively flipped to the other side by phospholipid translocators.

Most detoxification reactions in the smooth ER involve oxidation. Such reactions usually involve the conversion of hydrophobic compounds into hydrophilic compounds, and are governed by a system of enzymes called *mixed-function oxidases*. Cytochrome P-450, a group of iron-containing integral membrane proteins, is a central component of one mixed-function oxidase system. Mixed-function oxidases also govern the oxidation of steroids and fatty acids. Ingestion of the depressant *phenobarbital* triggers an increase in smooth ER and mixed-function oxidases but not in other ER enzymes.

77. Which of the following tissues would be expected to have especially well developed smooth ER?

A. skeletal muscle
B. adrenal cortex
C. intestinal epithelium
D. cardiac muscle

78. Where are many of the enzymes necessary for phospholipid synthesis likely to be located?

A. ER lumen
B. cytosol
C. lysosome
D. Golgi complex

79. Compared to a healthy individual, an individual who ingests large amounts of *phenobarbital* over an extended period of time will most likely:

A. detoxify dangerous drugs more slowly.
B. be more able to maintain steady blood glucose levels.
C. degrade phenobarbital more quickly.
D. be more responsive to therapeutically useful drugs such as antibiotics.

80. A given type of phospholipid may exist in different concentrations on either side of the same membranous structure. Which of the following enzymes most likely contributes to such an asymmetric arrangement?

A. glucose 6-phosphatases
B. phospholipid translocators
C. cytochrome P-450
D. protein kinase A

81. The primary structure of *mixed function* oxidases is most likely synthesized at the:

A. smooth ER.
B. rough ER.
C. Golgi apparatus.
D. cellular membrane.

82. In the final step of the reactions governed by mixed-function oxidases, oxygen is converted to water. In this step, the iron atom in P-450 is most likely:

A. reduced
B. oxidized
C. removed
D. inverted

83. Some cells, called adipocytes, specialize in the storage of triacylglycerols synthesized by the smooth ER. The primary function of adipocytes is to:

A. maintain chemical homeostasis of the body.
B. filter and remove toxins.
C. provide for cholesterol synthesis.
D. serve as a reservoir of stored energy.

GO ON TO THE NEXT PAGE.

Passage III (Questions 84-90)

The action potential of cardiac muscle differs from the action potential in skeletal muscle in two important ways. First, depolarization in skeletal muscle is created by the opening of fast Na^+ voltage-gated channels; depolarization of cardiac muscle is effected by both fast Na^+ voltage-gated channels and slow Ca^{2+}–Na^+ voltage-gated channels. Fast Na^+ voltage-channels exhibit three stages: closed; open; and inactivated. Upon an increase in membrane potential they open for a fraction of a second allowing Na^+ ions to rush into the cell, and then immediately become inactivated. Slow Ca^{2+}–Na^+ voltage-gated channels open more slowly and close more slowly allowing both Na^+ and Ca^{2+} to enter the cell. The second major difference between skeletal and cardiac action potentials is that, upon depolarization, the cardiac muscle membrane becomes highly impermeable to K^+. As soon as Ca^{2+}–Na^+ voltage-gated channels are closed, the membrane suddenly becomes very permeable to K^+.

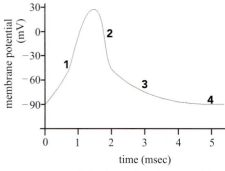

Figure 1 a skeletal muscle action potential

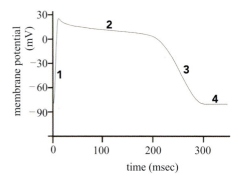

Figure 2 a cardiac muscle action potential

84. Influx of Ca^{2+} ions contribute most to which of the numbered sections from the cardiac action potential in Figure 2?

   A. 1
   B. 2
   C. 3
   D. 4

85. In Figure 1 the numbered section of the skeletal muscle action potential which best represents depolarization is:

   A. 1
   B. 2
   C. 3
   D. 4

86. In a lab experiment a student placed a beating frog heart into saline solution. Which of the following is true if the student adds acetylcholine to the same solution?

   A. Section number 2 in Figure 2 will lengthen.
   B. Section number 4 in Figure 2 will lengthen.
   C. Section number 4 in Figure 1 will shorten.
   D. Section number 4 in Figure 2 will shorten.

87. Which of the following contributes most to section number 3 of the action potential shown in Figure 1?

   A. Na^+ ions are diffusing into the cell.
   B. K^+ ions are diffusing into the cell.
   C. K^+ ions are diffusing out of the cell.
   D. Ca^{2+} ions are diffusing out of the cell.

88. Between action potentials, a potential difference called the resting potential exists across the neuron cell membrane. All of the following help to establish the resting potential of a neuronal membrane EXCEPT:

   A. selective permeability of the cell membrane.
   B. the Na^+/K^+ pump.
   C. the Na^+ voltage gated channels.
   D. the electrochemical gradient of multiple ions.

89. Which of the following cell types most likely contain Na^+ voltage gated channels?

    A. an epithelial cell from the proximal tubule of a nephron
    B. a parietal cell from the lining of the stomach
    C. an α-cell from the islet of Langerhans in the pancreas
    D. a muscle fiber from the gastrocnemius

90. According to Figure 1, at 2 msec after the action potential begins, Na^+ voltage gated channels are most likely:

    A. open.
    B. closed.
    C. inactivated.
    D. activated.

Questions 91 through 92 are **NOT** based on a descriptive passage.

91. All of the following are true concerning a typical motor neuron EXCEPT:

    A. the K^+ concentration is greater inside the cell than outside the cell.
    B. K^+ voltage-gated channels are more sensitive than Na^+ voltage-gated channels to a change in membrane potential.
    C. Cl^- concentrations contribute to the membrane resting potential.
    D. the action potential begins at the axon hillock.

92. In saltatory conduction:

    A. an action potential jumps along a myelinated axon from one node of Ranvier to the next.
    B. an action potential moves rapidly along the membrane of a Schwann cell, which is wrapped tightly around an axon.
    C. an action potential jumps from the synapse of one neuron to the next.
    D. ions jump from one node of Ranvier to the next along the axon.

STOP. IF YOU FINISH BEFORE TIME IS CALLED, CHECK YOUR WORK. YOU MAY GO BACK TO ANY QUESTION IN THIS TEST BOOKLET.

STOP.

30-MINUTE IN-CLASS EXAM FOR LECTURE 5

Passage I (Questions 93-100)

General hormones which circulate in the blood can be divided into 3 categories. 1) Steroid hormones, synthesized in the smooth endoplasmic reticulum of endocrine glands and secreted from the cell via exocytosis, bind to proteins for transport to their target tissue. They then diffuse through the target cell membrane and normally bind to a large receptor protein in the cytosol, which in turn carries them into the nucleus where they exert their effect directly at the transcriptional level. 2) Peptide hormones bind to membrane bound receptors and act via a second messenger system. 3) Tyrosine derivatives are further divided into thyroid hormones and the *catecholamines*, epinephrine and norepinephrine. Thyroid hormones diffuse through the cell membrane and bind to receptors in the nucleus. They also act directly at the transcription level. The catecholamines act at the cell membrane through a second messenger system.

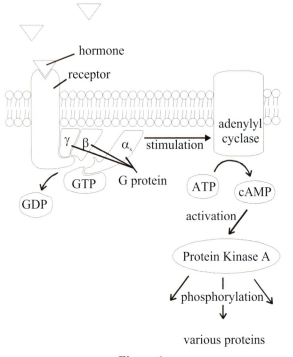

Figure 1

One second messenger system, shown in Figure 1, works as follows: the activated membrane bound hormone receptor activates a protein inside the cell, called a *G-protein*. Once activated, the G-protein exchanges GDP for GTP, which causes a portion of the G-protein to dissociate. Depending upon the type of G-protein, the dissociated portion may stimulate or inhibit *adenylyl cyclase*, another membrane bound protein. Inhibitory G-proteins are called G_i-proteins; stimulating G-proteins are called G_s-proteins. After activating adenylyl cyclase, the dissociated portion of the G-protein must hydrolyze GTP to GDP in order to become inactive and recombine with its other portion. Adenylyl cyclase converts ATP to cyclic AMP. Cyclic AMP activates *protein kinase A*. Kinases are a family of enzymes that catalyze the transfer of γ-phosphate from ATP to select protein amino acid residues. The biological effects of cyclic AMP are mediated by changes in protein phosphorylation.

In 1957, Earl Sutherland found that liver homogenates incubated with either epinephrine or glucagon would stimulate the activity of glycogen phosphorylase, an enzyme which governs the conversion of glycogen to glucose-1-phosphate. However, if the membranes present in the homogenate were removed by centrifugation, glycogen phosphorylase could no longer be activated by epinephrine or glucagon, but could still be activated by the addition of cyclic AMP.

93. Cortisol most likely binds to a receptor protein:

　　A. bound to the cell membrane.
　　B. in the cytosol.
　　C. on the nuclear membrane.
　　D. just outside the cell.

94. Phosphodiesterase breaks down cyclic AMP to AMP. Caffeine, a drug abundant in coffee, suppresses the activity of phosphodiesterase. According to the information in the passage, which of the following would most likely be found in a blood sample of someone who has recently drunk large amounts of coffee?

　　A. high ADH levels
　　B. low insulin levels
　　C. high glucose levels
　　D. low glucose levels

95. A certain mutant tumor cell line has normal epinephrine receptors and normal adenylyl cyclase; however, it fails to increase cAMP in the presence of epinephrine. The most likely explanation for this is:

　　A. the G_s-protein in the cell is either missing or malfunctioning.
　　B. the G_i-protein in the cell is either missing or malfunctioning.
　　C. epinephrine diffuses directly into the cell to act on protein kinase A.
　　D. the mutation results in a change in the structure of protein kinase A.

96. Glucagon works via a G-protein system. Which of the following best explains why glucagon stimulates glycogen breakdown in liver cells, but stimulates lipid break down in fat cells?

   A. Fat cells contain a G_s-protein while liver cells contain a G_i-protein.
   B. Liver cells don't contain *adenylyl cyclase*.
   C. The two cell types contain a different set of proteins phosphorylated by protein kinase A.
   D. Glycogen and lipid breakdown are not governed by cyclic AMP levels.

97. Which of the following was most likely not part of the membrane presence removed by Sutherland during his experiment?

   A. glucagon and epinephrine receptor proteins
   B. adenylyl cyclase
   C. cyclic AMP
   D. G-protein

98. Epinephrine binds to several types of receptors called *adrenergic* receptors. Heart muscle cells contain β_1-adrenergic receptors, where as the smooth muscle cells of the gut contain many α_2-adrenergic receptors. Which of the following is most likely true concerning these two types of receptors?

   A. β_1-adrenergic receptors activate G_s-proteins while α_2-adrenergic receptors activate G_i-proteins.
   B. β_1-adrenergic receptors activate G_i-proteins while α_2-adrenergic receptors activate G_s-proteins.
   C. Both β_1-adrenergic receptors and α_2-adrenergic receptors activate G_i-proteins.
   D. Both β_1-adrenergic receptors and α_2-adrenergic receptors activate G_s-proteins.

99. *Vibrio cholerae*, the bacterium responsible for Cholera, releases an enterotoxin which acts on G_s-proteins of the intestinal mucosa inhibiting their GTPase activity. Which of the following is most likely to occur inside an intestinal mucosal cell of an individual infected with cholera?

   A. decreased activation of protein kinase A
   B. increase electrolyte concentration
   C. decreased rate of hormones binding to membrane bound receptors
   D. increased concentration of cyclic AMP

100. Which of the following is the most likely effect of aldosterone?

   A. activation of ion channels via binding to membrane-bound protein channels
   B. increase of cyclic AMP
   C. activation of protein kinase A
   D. increased production of membrane bound protein.

GO ON TO THE NEXT PAGE.

Passage II (Questions 101-106)

Human chorionic gonadotropin (HCG) is secreted by the placenta. It can be detected in maternal plasma or urine within 9 days of conception, which is shortly after the *blastocyst* implants in the uterine wall. Maternal blood levels of HCG increase exponentially for the first 10 to 12 weeks of pregnancy and decline to a stable plateau for the remainder of the pregnancy.

HCG stimulates the *corpus luteum* to secrete estrogens and progesterone until the placenta assumes the synthesis of these steroids. During this time the corpus luteum grows to approximately twice its initial size. After 13 to 17 weeks, the corpus luteum involutes.

HCG acts to stimulate the testes of the male fetus to secrete testosterone. It is this testosterone that accounts for the male sex organs.

HCG acts on a G-protein-coupled receptor on the target cell membrane. LH acts on the same receptor and FSH acts on a very similar receptor. Once activated, the G protein stimulates an increase in cyclicAMP, which activates protein kinase A.

101. HCG acts most like which of the following hormones?

- A. LH
- B. FSH
- C. estrogen
- D. progesterone

102. Which of the following hormones is most responsible for preventing the sloughing off of the uterine wall during pregnancy?

- A. FSH
- B. HCG
- C. estrogen
- D. progesterone

103. Syncytiotrophoblast cells in the placenta are responsible for manufacture and release of HCG. Which of the following statements most accurately describes this process?

- A. HCG is manufactured in the rough ER and modified in the Golgi apparatus before secretion.
- B. HCG is manufactured by the smooth ER and diffused into the blood stream.
- C. HCG is manufactured in the nucleus and transported via a protein carrier into the blood.
- D. HCG is manufactured at the membrane of the syncytiotrophobast and released into the plasma.

104. The corpus *luteum* mentioned in the passage is:

- A. a permanent functional part of any healthy ovary.
- B. the remainder of the follicle which produced the ovum.
- C. a gland developed in the fetal ovary during the third trimester of pregnancy.
- D. a group of neuronal cells in the hypothalamus.

105. Early pregnancy tests use an antibody to bind to a hormone in the urine. Pregnancy is indicated when binding occurs. To which hormone does the antibody most likely bind?

- A. progesterone
- B. estrogen
- C. HCG
- D. LH

106. Which of the following hormones most likely acts as the substrate molecule for synthesis of cortisol and aldosterone in the fetal adrenal gland?

- A. HCG
- B. estrogen
- C. LH
- D. FSH

Passage III (Questions 107-114)

Neurons of the hypothalamus secrete gonadotropin-releasing hormone (GnRH), a 10-amino acid peptide which stimulates the anterior pituitary to release the gonadotropins LH and FSH. LH secretion keeps close pace with the pulsatile release of GnRH, whereas FSH changes more gradually in response to long term changes in GnRH levels. Both FSH and LH act by changing intracellular cyclicAMP levels. In women, GnRH secretion occurs monthly guiding the menstrual cycle. In men, GnRH secretion occurs in bursts throughout each day to maintain a relatively steady blood level of the hormone.

Production of spermatozoa occurs in the seminiferous tubules of the testes where a single *Sertoli cell* envelopes and nurtures the developing spermatozoa. FSH stimulates Sertoli cells, which, in turn, secret *inhibin* that acts at the pituitary level to inhibit production of FSH independently of LH. LH stimulates the *Leydig cells* located between the seminiferous tubules to secrete testosterone, which acts on the hypothalamus to inhibit GnRH. Testosterone stimulates spermatogenesis. Both FSH and LH are required for spermatogenesis.

Oogenesis occurs in the ovaries of the female. The oocyte develops surrounded by *theca* and *granulosa cells*. The entire structure is called a follicle. As a follicle develops, granulosa cells secrete a viscous glycoprotein layer, called the *zona pellucida*, which surrounds the oocyte. The granulosa cells remain in contact with the oocyte via thin strands of cytoplasm. Theca cells differentiate from interstitial cells and form a thin layer surrounding the granulosa cells. The follicle does not require FSH to reach this stage. When stimulated by LH, theca cells supply granulosa cells with androgen, which is then converted to estradiol and secreted into the blood along with inhibin. Like testosterone, estradiol inhibits GnRH secretion from the hypothalamus.

107. The follicle in its earliest stages is called a primordial follicle. The primordial follicle most likely contains a:

    A. haploid, primary oocyte.
    B. diploid, primary oocyte.
    C. haploid, secondary oocyte.
    D. diploid, secondary oocyte.

108. Which of the following two cell types possess the most similar phylogeny (developmental history)?

    A. theca cells and Sertoli cells
    B. Leydig cells and granulosa cells
    C. Leydig cells and theca cells
    D. granulosa cells and oocytes

109. Androgens are sometimes taken by athletes to improve performance. Which of the following may be a side effect of taking large quantities of androgens?

    A. infertility due to decreased endogenous testosterone production
    B. infertility due to decreased secretion of GnRH
    C. increased fertility due to decreased endogenous testosterone production
    D. increased fertility due to increased Sertoli cell activity

110. Which of the following hormones would most likely be found in the nucleus of a somatic cell of a pregnant woman?

    A. GnRH
    B. FSH
    C. LH
    D. testosterone

111. Some post-menopausal women suffer from osteoporosis, a lowering in density of the bones. Administration of estrogens is an effective treatment for this disease. This demonstrates that estrogen most likely inhibits the activity of:

    A. osteoblasts
    B. osteoclasts
    C. osteocytes
    D. hemopoietic stem cells

112. Which of the following are released into the fallopian tube during ovulation?

    I. the follicle
    II. the secondary oocyte
    III. some granulosa cells

    A. I only
    B. II only
    C. II and III only
    D. I, II, and III

GO ON TO THE NEXT PAGE.

113. Which of the following hormones is most closely associated with ovulation?

   A. estrogen
   B. FSH
   C. LH
   D. GnRH

114. A vaccine that stimulates the body to produce antibodies against a hormone has been suggested as a long term male contraceptive. In order to insure that the vaccine has no adverse effects on androgen production, which hormone should be targeted?

   A. FSH
   B. LH
   C. GnRH
   D. testosterone

Question 115 is **NOT** based on a descriptive passage.

115. A competitive inhibitor of TSH binding to TSH receptors on the thyroid would lead to a rise in the blood levels of which of the following:

   A. TSH
   B. Thyroxine
   C. PTH
   D. epinephrine

STOP. IF YOU FINISH BEFORE TIME IS CALLED, CHECK YOUR WORK. YOU MAY GO BACK TO ANY QUESTION IN THIS TEST BOOKLET.

30-MINUTE IN-CLASS EXAM FOR LECTURE 6

Passage I (Questions 116-122)

Dietary fat is mainly composed of triglycerides and smaller amounts of cholesterol and phospholipids. Lingual lipase secreted in the mouth digests a very small portion of fat while in the mouth and small intestine, but mainly in the stomach. Enterocytes in the small intestine also release tiny amounts of enteric lipase. However, the most important enzyme for the digestion of fats is pancreatic lipase.

Since fats are not soluble in the aqueous solution in the small intestine, fat digestion would be very inefficient were it not for bile salts and lecithin, which increase the surface area upon which lipase can act. In addition, bile forms micelles with the fatty acid and monoglyceride products of enzymatic hydrolysis of triglycerides, and carries these micelles to the brush border of the intestine where they are absorbed by an enterocyte.

Once inside the enterocyte, the fatty acids are taken up by the smooth endoplasmic reticulum and new triglycerides are formed. These triglycerides combine with cholesterol and phospholipids to form new globules called chylomicrons that are secreted through exocytosis to the basolateral side of the enterocyte. From there, the chylomicrons move to the lacteal in the intestinal villus.

Chylomicrons are just one member of the lipoprotein families which transport lipids through the blood. The other members are very low density lipoproteins (VLDL), low density lipoproteins (LDL), and high density lipoproteins (HDL). VLDLs are degraded by lipases to LDL. LDLs account for approximately 60-75% of plasma cholesterol and their levels are directly related to cardiovascular risk. HDLs, on the other hand, account for only 20-25% of plasma cholesterol and HDL plasma levels are inversely related to cardiovascular risk. HDL levels are positively correlated with exercise and moderate alcohol intake and inversely related to smoking, obesity and use of progestin-containing contraceptives.

116. Bile allows lipases to work more efficiently by increasing the surface area of fat. If fat globules are assumed to be spherical, then each time bile decreases the diameter of all the fat globules in the small intestine by a factor of two, the surface area of the fat is increased by a factor of:

    A. 2
    B. 4
    C. 8
    D. 16

117. Most dietary fat first enters the blood stream:

    A. from the right lymphatic duct into arterial circulation.
    B. from the thoracic duct into venous circulation.
    C. from the small intestine into capillary circulation.
    D. from the intestinal enterocyte into the lacteal.

118. Most tests for serum cholesterol levels do not distinguish between HDLs and LDLs. Life insurance rates generally increase with increasing serum cholesterol levels. Which of the following supports the claim that it is important to determine whether HDLs or LDLs are responsible for high serum cholesterol when evaluating a patients risk for coronary heart disease?

    A. The risk from coronary heart disease doubles from an HDL level of 60mg/100ml to 30mg/100ml.
    B. The incidence of coronary heart disease rises in linear fashion with the level of serum cholesterol.
    C. The optimal serum cholesterol for a middle aged man is probably 200mg/100ml or less.
    D. VLDL is the main source of plasma LDL.

119. Lingual lipase most likely functions best at a pH of:

    A. 2
    B. 5
    C. 7
    D. 9

120. Pancreatic enzymes are released from the pancreas in an inactive form called a zymogen. Which of the following activates the zymogen form of pancreatic lipase?

    A. bile salts
    B. trypsin
    C. low pH
    D. high pH

121. The process by which bile increases the surface area of dietary fat is called:

A. lipolysis.
B. adipolysis.
C. malabsorption.
D. emulsification.

122. A patient that has had his pancreas surgically removed would most likely need to supplement his diet with the following enzymes EXCEPT:

A. lactase.
B. lipase.
C. amylase.
D. proteases.

Passage II (Questions 123-130)

The rate at which a solute is excreted in the urine is given by the product of the concentration (U) of the solute in the urine and the urine flow rate (V). Dividing this number by the blood plasma concentration (P) of the solute, gives the minimum volume of blood plasma necessary to supply the solute. This number is called the *renal clearance* (C).

$$C = \frac{U \times V}{P}$$

Because almost no solute is completely filtered from the blood in a single pass through the renal corpuscles, the renal clearance does not represent an actual volume of plasma. However, the organic dye, *p*-aminohippuric acid (PAH), is not only filtered but also secreted by the kidney. As a result, 90% of PAH that enters the kidney is excreted. Thus, the clearance of PAH is an approximation of the *renal plasma flow* (RPF) to within 10%.

Due to secretion, resorption, and metabolism in the nephron, the clearance of a solute may not be equal to its rate of filtration by the glomerulus. *Inulin*, a nonmetabolizable polysaccharide, is neither secreted nor resorbed. Thus the clearance of inulin is equal to the glomerular filtration rate (GFR). In a healthy adult, the GFR for both kidneys is approximately 125 ml/min.

The filtration fraction (FF) is the fraction of the plasma that filters through the glomerular membrane. The filtration fraction is given by the equation:

$$FF = \frac{GFR}{RPF}$$

123. Which of the following must be true of a solute that has a *renal clearance* greater than the GFR?

A. The solute is being resorbed by the nephron.
B. The solute is being secreted by the nephron.
C. The plasma concentration of the solute must be greater than the plasma concentration of inulin.
D. The solute has exceeded its transport maximum.

GO ON TO THE NEXT PAGE.

124. Since *inulin* is neither secreted nor resorbed, what must also be true about *inulin*, if its clearance is to accurately represent the GFR?

    A. Inulin lowers the renal blood flow.
    B. Inulin raises the filtration rate.
    C. The filtered concentration of inulin is exactly equal to its concentration in the plasma.
    D. 100% of inulin is filtered in a single pass through the kidney.

125. *Inulin* is most likely:

    A. smaller than an amino acid.
    B. the same size as albumin.
    C. the size of a red blood cell.
    D. larger than glucose.

126. Which of the following would LEAST affect the *renal clearance* of a solute?

    A. size and charge
    B. plasma concentration
    C. glomerular hydrostatic pressure
    D. the concentration of PAH

127. Based upon the information in the passage, the renal clearance of glucose in a healthy adult is most likely:

    A. 0 ml/min
    B. 60 ml/min
    C. 125 ml/min
    D. 145 ml/min

128. If a patient has a PAH clearance of 625 ml/min, and an inulin clearance of 125 ml/min, then approximately what percent of plasma is filtered by the patient's kidneys each minute?

    A. 0 %
    B. 0.2 %
    C. 20%
    D. 50 %

129. The blood concentration of creatinine, a naturally occurring metabolite of skeletal muscle, tends to remain constant. Creatinine flows freely into Bowman's capsule, and only negligible amounts of creatinine are secreted or absorbed. A patient with a normal GFR has a urine output of 1 ml/min with a creatinine concentration of 2.5 mg/ml. What is the patient's plasma concentration of creatinine?

    A. 2 mg/100ml
    B. 5 mg/100ml
    C. 2 mg/ml
    D. 5 mg/ml

130. If an individual has a hematocrit of 50%, and a PAH clearance of 650 ml/min, what is his *renal blood flow* (RBF)?

    A. 325 ml/min
    B. 650 ml/min
    C. 1300 ml/min
    D. 2600 ml/min

Passage III (Questions 131-136)

Like other circulatory systems, renal perfusion is autoregulated. The *juxtaglomerular apparatus* (JGA) assists in regulating the volume and pressure of the renal tubule and the glomerular arterioles. The JGA connects the arterioles of the glomerulus with the distal convoluted tubule of the nephron via specialized smooth muscle cells, called juxtaglomerular cells. Renin, a proteolytic enzyme, is synthesized and stored in these cells. Renin release is stimulated by low plasma sodium, low blood pressure, and sympathetic stimulation via β-adrenergic receptors on the juxtaglomerular cells.

Once released into the blood, renin acts on angiotensinogen, a plasma protein, to form angiotensin I. Angiotensin I, an inactive decapeptide, is cleaved in the lungs to form angiotensin II. Angiotensin II stimulates the release of aldosterone, causes systemic vasoconstriction, enhances neurotransmitter release from sympathetic nerve endings, stimulates ADH release, and acts in the brain to cause thirst. Angiotensin II preferentially constricts the efferent arterioles of the glomerulus.

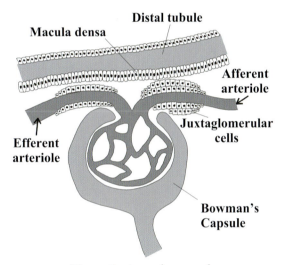

Figure 1 A renal corpuscle

Specialized epithelial cells of the distal tubule, called *macula densa*, are contiguous with the juxtaglomerular cells. The presence of low sodium chloride concentration causes the macula densa to signal the juxtaglomerular cells to release renin into the blood, and to lower the resistance of the afferent arterioles. A low resistance in the afferent arterioles increases the glomerular filtration rate.

When functioning properly, the JGA operates to maintain the glomerular filtration rate in the face of large fluctuations in arterial pressure.

131. Amino acids are reabsorbed in the proximal tubule by a secondary active transport mechanism down the concentration gradient of sodium. A high protein diet would most likely lead to:

   A. a decrease in glomerular filtration rate and renal blood flow.
   B. an increase in glomerular filtration rate and renal blood flow.
   C. a decrease in glomerular filtration rate and an increase in renal blood flow.
   D. an increase in glomerular filtration rate and a decrease in renal blood flow.

132. Renal artery stenosis (partial blockage of the renal artery) leads to activation of the renin-angiotensin cascade resulting in:

   A. high renal blood pressure only.
   B. low renal blood pressure only.
   C. high systemic blood pressure.
   D. low systemic blood pressure.

133. Which of the following is *least* likely to be affected by angiotensin II?

   A. the adrenal cortex
   B. the thyroid
   C. the posterior pituitary
   D. the autonomic nervous system

134. The role of renin in the conversion of angiotensinogen to angiotensin I is most likely to:

   A. act as a cofactor.
   B. bind to angiotensinogen making it soluble in the aqueous solution of the blood.
   C. lower the energy of activation of the reaction.
   D. add to angiotensinogen in a hydrolytic reaction.

GO ON TO THE NEXT PAGE.

135. The renin-angiotensin cascade increases all of the following EXCEPT:

A. urine volume.
B. blood pressure.
C. Na^+ reabsorption.
D. K^+ excretion.

136. Angiotensin II most likely:

A. diffuses into the nucleus of the target cell.
B. attaches to a receptor on the membrane of the target cell and acts through a second messenger system.
C. acts on the cells of distal tubule increasing transcription of sodium transport proteins.
D. creates a chemical reaction in the plasma that increases the permeability of the collecting duct.

Questions 137 through 138 are **NOT** based on a descriptive passage.

137. The ability to produce a concentrated urine is primarily based on the presence of functional kidney nephrons. The most important structure involved in concentrating urine within the nephron is the:

A. glomerulus.
B. proximal convoluted tubule.
C. loop of Henle.
D. Bowman's capsule.

138. Which of the following statements about digestion is NOT true?

A. Carbohydrate metabolism begins in the mouth.
B. Most dietary protein is absorbed into the body in the stomach.
C. The large intestine is a major source of water reabsorption.
D. The liver produces bile which is stored in the gallbladder.

STOP. IF YOU FINISH BEFORE TIME IS CALLED, CHECK YOUR WORK. YOU MAY GO BACK TO ANY QUESTION IN THIS TEST BOOKLET.

30-MINUTE IN-CLASS EXAM FOR LECTURE 7

Passage I (Questions 139-145)

The immune system has several methods of selecting and destroying particles that are foreign to the body. One such mechanism works as follows: Large phagocytotic cells called macrophages engulf some antigens and process them internally, ultimately producing antigen fragments that protrude from the outer surface of the membrane of the macrophage. The antigen fragments are held to the membrane by MHC class II molecules. A special type of T-cell, called a *helper T-cell*, recognizes and binds to the MHC class II-antigen complex. The helper T-cell produces protein local mediators called *interleukins* which stimulate the T-cell to divide. A newly formed helper T-cell activates a B-cell capable of producing antibodies that specifically bind to the antigen. The B-cell begins cell division into *plasma cells* and *memory B cells*. Memory B cells resemble unstimulated B cells and do not secrete antibodies. Plasma cells produce antibodies that bind to the antigen. Once bound to the antigen, the antibody may initiate a chemical chain reaction which results in lysis of the antigen carrying cell or, if the antigen is a chemical such as a poison, the antibody may simply inactivate it. In addition, cells with antibodies bound to their surfaces may be engulfed by macrophages or punctured by natural killer cells.

In an effort to discover which cell type creates antibodies, a scientist performed the following experiment. A *nude mouse*, which lacks a thymus and cannot form antibodies, was injected with healthy lymphocytes from the thymus of a donor mouse. The host mouse was then injected with an antigen. Antibodies were produced in the host mouse. Lymphocyte samples were then removed from the spleen of the host mouse and the host cells or the donor cells were selectively destroyed from separate samples. The scientist found that the host cell samples were still able to produce antibodies against the antigen while the donor cell samples were not.

139. Which of the following is most likely to create the immune system response mediated by MHC class II molecules?

   A. a bacterial infection
   B. a cell infected by a virus
   C. a tumor
   D. a foreign tissue graft

140. Which of the following is a cell that a nude mouse would be unable to produce?

   A. an antibody
   B. a T-cell
   C. a B-cell
   D. a macrophage

141. The experiment demonstrated that:

   A. antibodies are produced by T-cells.
   B. antibodies are produced by B-cells only after exposure to T-cells.
   C. nude mice are unable to produce antibodies.
   D. nude mice are unable to produce T-cells.

142. *Interleukins* most likely act:

   A. via a second messenger by binding to a membrane bound protein receptor.
   B. by diffusing through the membrane of the helper T-cell and binding to a receptor in the cytosol.
   C. at the transcriptional level by binding directly to nuclear DNA.
   D. via the nervous system.

143. All of the following arise from the same stem cells in the bone marrow EXCEPT:

   A. helper T-cells.
   B. B lymphocytes.
   C. erythrocytes.
   D. osteoblasts.

144. Which of the following is a foreign particle capable of provoking an immune response?

   A. antigen
   B. antibody
   C. interleukin
   D. histamine

145. The function of the *memory B-cell* is most likely to:

   A. remain as an immune system reserve against different antigens.
   B. attract killer T-cells to the infected area.
   C. magnify the immune response by releasing antibodies into the blood stream.
   D. allow for rapid production of antibodies in the case of reinfection with the same antigen.

Passage II (Questions 146-152)

The function of plasma cells in the immune system is to secrete specific proteins called antibodies, also called immunoglobulins, which bind to antigens and mark them for destruction. One plasma cell can only make antibodies that are specific for a single *epitope* for a single antigen.

Antibodies exist mainly in blood plasma but are also present in tears, milk, saliva, and respiratory and intestinal tract secretions.

Treatment of antibodies with the protease "papain" yields three fragments: two F_{ab} fragments and one F_c fragment. The F_c fragment is nearly identical in all antibodies. *Mercaptoethanol* cleaves disulfide bonds. Treatment of antibodies with mercaptoethanol results in two *light chain* polypeptides (25,000 daltons) and two *heavy chain* polypeptides (50,000 daltons). Both the light and heavy chains possess constant, variable, and hypervariable regions in their amino acid structures. The light chain contains no part of the F_c region.

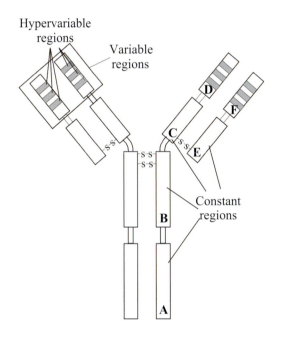

Figure 1 An antibody

146. What level of protein structure is disrupted by mercaptoethanol?

   A. primary
   B. secondary
   C. tertiary only
   D. tertiary and quaternary

147. Permanent immunity may be imparted to an individual by a vaccine containing:

   A. antibodies.
   B. antigens.
   C. plasma cells.
   D. white blood cells.

148. The light chain is represented by which of the following labeled segments in Figure 1?

   A. A, B, C, and D
   B. C, D, E, and F
   C. C and D
   D. E and F

149. Which of the following labeled segments from the antibody in Figure 1 most likely attaches to an antigen?

   A. A
   B. D and F
   C. D but not F
   D. C and D

150. Which of the following structures in the plasma cell most likely produces antibodies?

   A. ribosomes in the cytosol
   B. rough endoplasmic reticulum
   C. smooth endoplasmic reticulum
   D. cellular membrane

151. Which of the following statements concerning the immune system is NOT true?

   A. Plasma cells which provide immunity against one disease may also provide immunity against a closely related disease.
   B. Two antibodies from the same plasma cell must bind to the same antigen type.
   C. Plasma cells arise from T lymphocytes.
   D. Memory B cells help the immune system to respond to the same antigen more quickly during a secondary immune response.

152. Which of the following labeled segments from Figure 1 is part of the F_c region of an antibody?

   A. A and B
   B. A, B, C, and D
   C. A, B, C, and E
   D. C, D, E, and F

Oxygen in erythrocytes is stored by hemoglobin. In adults, hemoglobin consists of four polypeptide chains, two alpha (α) and two beta (β), each held together by noncovalent interactions. The interior of each folded chain consists almost entirely of nonpolar residues. Each polypeptide chain contains a heme group with a single oxygen binding site. The heme group is a nonpolypeptide unit with an iron atom at its center. The organic portion of the heme, *protoporphyrin*, binds the iron atom at its center with four nitrogen atoms, leaving the iron atom with a +2 or +3 oxidation state and capable of making two more bonds. Carbon monoxide is a byproduct of the break down of the heme group. An isolated heme group binds to CO 25,000 times as strongly as it binds to O_2. However, the binding affinity of hemoglobin and myoglobin for CO is only about 200 times as great as for O_2.

The oxygen carrier in skeletal muscle tissue is myoglobin. Myoglobin is very similar to hemoglobin, except that it consists of only a single polypeptide chain. Myoglobin does not show a decreased affinity for O_2 over a broad range of pH nor in the presence of CO_2. Both decreased pH and increased CO_2 enhance the release of O_2 in hemoglobin. Hemoglobin has a lower affinity for oxygen than does myoglobin, partially due to BPG, a chemical in red blood cells. BPG affects the characteristic sigmoidal oxygen dissociation curve of hemoglobin.

Although the polypeptide chain of myoglobin is identical to the α chain of hemoglobin only at 24 of 141 amino acid positions, the three dimensional shapes of myoglobin and hemoglobin α chains are very similar. Inter-species comparison of the three dimensional shape of hemoglobin reveals this same similarity. A further comparison of the amino acid sequences in the hemoglobin of different species reveals that nine positions are the same in nearly all known species. Several of these invariant residues affect the oxygen binding site.

153. Which of the following most likely acted as the evolutionary selective pressure which led to the decreased affinity of hemoglobin and myoglobin for CO as compared to the isolated heme group?

   A. The emergence of industrial societies which significantly increased the level of environmental CO.
   B. The products of aerobic respiration.
   C. The endogenous production of CO due to the breakdown of heme.
   D. Carbonic acid in the blood.

154. Hemoglobin and myoglobin are:

   A. similar in their primary and tertiary structure.
   B. similar in their quaternary structure but differ in their primary structure.
   C. similar in their tertiary structure but differ in their primary structure.
   D. similar in their primary structure but differ in their quaternary structure.

155. Which of the following gives the oxygen dissociation curves for hemoglobin H and myoglobin M?

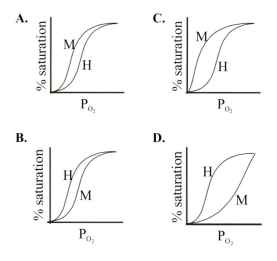

156. Increased concentration of lactic acid due to active skeletal muscle most likely results in which of the following?

  A. increased off-loading of O_2 by both hemoglobin and myoglobin
  B. increased off-loading of O_2 by hemoglobin but not myoglobin
  C. decreased off-loading of O_2 by both hemoglobin and myoglobin
  D. decreased off-loading of O_2 by hemoglobin but not myoglobin

157. In a healthy human body, where is BPG likely to have its greatest effect?

  A. in the alveoli
  B. in the capillaries of the lungs
  C. in the muscle cells of the heart
  D. in the capillaries of skeletal muscle

158. The tertiary structure of human myoglobin most likely:

  A. represents a fundamental design for oxygen carriers in nature.
  B. arose relatively late in human evolution.
  C. varies among individuals of the same species.
  D. allows for more than one heme group.

159. Which of the following is LEAST likely to increase the breathing rate of an individual?

  A. increased pH in the blood
  B. low oxygen levels in the blood
  C. increased muscle activity
  D. increased CO_2 levels

Questions 160 through 161 are **NOT** based on a descriptive passage.

160. Which of the following is NOT true concerning the lymphatic system?

  A. The lymphatic system removes large particles and excess fluid from the interstitial spaces.
  B. The lymphatic system is a closed circulatory system.
  C. The lymphatic system contains lymphocytes that function in the bodies immune system.
  D. Most fatty acids in the diet are absorbed by the lymphatic system before entering the blood stream.

161. Which of the following is a cell that does NOT contain a nucleus?

  A. erythrocyte
  B. platelet
  C. macrophage
  D. B lymphocyte

STOP. IF YOU FINISH BEFORE TIME IS CALLED, CHECK YOUR WORK. YOU MAY GO BACK TO ANY QUESTION IN THIS TEST BOOKLET.

STOP.

30-MINUTE IN-CLASS EXAM FOR LECTURE 8

Passage I (Questions 162-169)

The functional unit of a skeletal muscle cell is the *sarcomere*. The protein polymers actin and myosin lie lengthwise along a sarcomere creating the various regions shown in Figure 1.

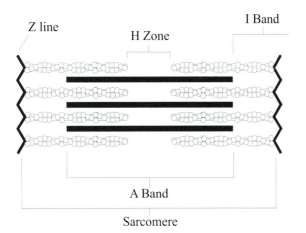

Sarcomere

A group of muscle cells within a muscle may be innervated by a single neuron making up a *motor unit*. The neuron carries an action potential to each muscle cell in the motor unit. The action potential is delivered deep into each muscle cell via tubular invaginations in the sarcolemma or cell membrane called T-tubules. The change in membrane potential is transferred to the sarcoplasmic reticulum, which causes it to become permeable to Ca^{2+} ions and to release its large stores of calcium into the cytosol. Once in the cytosol, the Ca^{2+} ions cause a conformational change in the protein troponin, which in turn acts upon a second protein, tropomyosin, exposing an actin-myosin binding site. Upon binding, the actin and myosin slide past each other shortening the sarcomere and creating a muscular contraction. ATP then binds to myosin releasing it from actin. Myosin immediately hydrolyzes ATP using the energy to return to its ready position. The Ca^{2+} ions are removed from the cytosol by extremely efficient calcium pumps via active transport. Once the Ca^{2+} has been sequestered back into the lumen of the sarcoplasmic reticulum, the Ca^{2+} ions are bound by *calsequestrin*.

Although the direct source of energy for muscle contraction comes from ATP, the ATP concentration in actively contracting muscle remains virtually constant. In addition, it has been shown that inhibitors of glycolysis and cellular respiration have no effect on ATP levels in actively contracting muscle over the short term. Instead, *phosphocreatine* donates its phosphate group to ADP in a reaction catalyzed by *creatine kinase*. Phosphocreatine levels are replenished via ATP from glycolysis and cellular respiration.

For active muscle cells the delivery of oxygen to the cell is far too slow to maintain sufficient energy levels, thus many muscle cells contain large amounts of myoglobin, which along with the cytochromes in the mitochondrial membrane impart a red hue to the muscle.

162. Which of the following proteins is an ATPase?

A. actin
B. myosin
C. phosphocreatine
D. calsequestrin

163. What is the function of *calsequestrin*?

A. to lower the free Ca^{2+} ion concentration inside the lumen of the sarcoplasmic reticulum
B. to raise the free Ca^{2+} ion concentration inside the lumen of the sarcoplasmic reticulum
C. to make Ca^{2+} for release into the cytosol
D. to pump Ca^{2+} into the sarcoplasmic reticulum

164. Two types of skeletal muscle are named after their characteristic color, red muscle and white muscle. Which of the following statements is most likely true concerning these muscle types?

A. Both muscle types contain large amounts of myoglobin and mitochondria.
B. White muscle contains more mitochondria than red muscle.
C. White muscle is capable of longer periods of contraction than red muscle.
D. Red muscle is capable of longer periods of contraction than white muscle.

165. Which of the following concentrations changes the most within the cytosol during the contraction of a muscle cell?

A. ATP
B. Myosin
C. Actin
D. Ca^{2+}

166. If a *creatine kinase* inhibitor is administered to an active muscle cell, which of the following would most likely occur?

    A. ATP concentrations would diminish while muscle contractions continued.

    B. Phosphocreatine concentrations would diminish while muscle contractions continued.

    C. ATP concentrations would remain constant while the percent saturation of myoglobin with oxygen would diminish.

    D. Cellular respiration and glycolysis would increase to maintain a constant ATP concentration.

167. Muscle cell T-tubules function to:

    A. create an action potential within a muscle cell.

    B. receive the action potential from the presynaptic neuron.

    C. deliver the action potential directly to the sarcomere.

    D. supply Ca^{2+} to the cytosol during an action potential.

168. All of the following are true concerning skeletal muscle cells EXCEPT:

    A. skeletal muscle cells contain more than one nucleus.

    B. human skeletal muscle cells continue normal cell division via mitosis throughout adult life.

    C. skeletal muscle cells contain more than one sarcomere.

    D. during muscle contraction, only the H band and the I band change length.

169. Which of the following is most likely NOT true concerning the uptake of Ca^{2+} ions from the cytosol during muscle contraction?

    A. It requires ATP.

    B. The mechanism involves an integral protein of the sarcolemma.

    C. It occurs against the concentration gradient of Ca^{2+}.

    D. It is rapid and efficient.

Passage II (Questions 170-175)

Connective tissues secrete large molecules that make up their *extracellular matrix*. Specialized cells called fibroblasts play the major role in the formation of the matrix. Three types of molecules characterize a matrix: *proteoglycans*; *structural proteins*; and *adhesive proteins*. Proteoglycans contain hydrated protein and carbohydrate chains, and can be very large. They typically create a gelatinous structure between the cells. Structural proteins add strength and flexibility to the matrix, and the adhesive proteins hold the cells together.

The main component of both bone and cartilage is the matrix. The cellular element of cartilage is called a *chondrocyte*. In their immature form, they secrete the matrix which includes collagen, the most abundant protein in the body. Three polypeptide chains wrap around each other to form the triple helix which gives collagen much of its strength and flexibility. Cartilage is not innervated and has no blood supply.

The human fetus has a cartilaginous endoskeleton which is gradually replaced with bone before and after birth until adulthood. Bone forms within and around the periphery of small cartilaginous replicas of adult bones. Bone forming *osteoblasts* differentiate from *fibroblasts* of the perichondrium. In long bone formation, *periosteum* ossifies around the shaft or *diaphysis* of the bone. Chondrocytes enlarge within the diaphysis and then break down leaving a honeycombed cartilage. Calcium deposits form and vascular connective tissue invades the area. Ossification of the inner portion of the bone proceeds from the center toward each end. *Osteoclasts* differentiate from certain phagocytotic blood cells and begin to burrow through older bone. The osteoblasts line up around the periphery of the newly formed tunnels and deposit concentric layers of bone. Vascular connective tissue and nerves move into these tunnels which are called osteons. In long bones, most *osteons* form along the length of the bone, and the columnar shape of the osteons gives these bones tremendous strength.

170. Each concentric layer of bone formed by osteoblasts in an osteon is called a:

    A. lacuna.

    B. lamella.

    C. collagen fibril.

    D. diaphysis.

171. Osteoblasts which become trapped in small spaces within the bone and mature are called:

    A. fibroblasts.

    B. chondrocytes.

    C. osteoclasts.

    D. osteocytes.

GO ON TO THE NEXT PAGE.

172. Which of the following matrix components best describes collagen?

    A. structural protein
    B. proteoglycan
    C. adhesive protein
    D. glycosaminoglycan

173. Which of the following hormones most likely stimulates osteoclasts?

    A. parathyroid hormone
    B. calcitonin
    C. epinephrine
    D. prostaglandin

174. All of the following statements are true concerning bone EXCEPT:

    A. Bone is connective tissue.
    B. Bone is innervated and has a blood supply.
    C. Yellow bone marrow stores triglycerides as a source of energy for the body.
    D. Bone is the only nonliving tissue in the body.

175. Which of the following cells most likely arises from the same stem cell in the bone marrow as an erythrocyte?

    A. osteoblast
    B. fibroblast
    C. osteoclast
    D. chondrocyte

Passage III (Questions 176-182)

The strength of muscle contraction is directly related to the cross-sectional area of the muscle, with a maximal force of 3 to 4 kg/cm³. The rate at which a muscle can perform work, or the muscle power, varies over time and is given in Table 1.

Time	Power (kg m/min)
First 10 seconds	7000
Next 1.5 minutes	4000
Next 30 minutes	1700

Table 1 Muscle power variance over time

Energy for muscle contraction is derived directly from ATP and ADP. However, a muscle cell's original store of ATP is used up in less than 4 seconds by maximum muscle activity. Three systems work to maintain a nearly constant level of ATP in a muscle cell during muscle activity: *the phosphagen system; the glycogen-lactic acid system*; and the *aerobic system*. *Phosphocreatine* contains a higher energy phosphate bond than even ATP and is used to replenish the ATP stores from ADP and AMP. This is the phosphagen system and it can sustain peak muscular activity for about 10 seconds.

The glycogen-lactic acid system is relied upon for muscular activity lasting beyond 10 seconds but not more than 1.6 minutes. This system produces ATP from glycolysis.

Aerobic metabolism of glucose, fatty acids and amino acids can sustain muscular activity for as long as the supply of nutrients lasts.

The recovery of muscle after exercise involves replacement of oxygen and glycogen. Before exercise, the body contains approximately 2.5 liters of oxygen in the lungs, hemoglobin, myoglobin, and the body fluids. This oxygen is used up in approximately 1 minute of heavy exercise. In a resting adult having just finished 4 minutes of heavy exercise, the oxygen uptake is increased dramatically at first and then levels back down to normal over a 1 hour period. The extra oxygen taken in is called the *oxygen debt*, and is about 11.5 liters. Glycogen stores are replenished in 2 days for individuals on a high carbohydrate diet, but can take several more days for those on a high fat or high protein diet.

176. Based upon the information in Table 1, which of the following are the most likely rates of molar production of ATP for the phosphagen system, the glycogen-lactic acid system, and the aerobic system respectively?

    A. 1.7 M/min, 4 M/min, 7 M/min
    B. 2 M/min, 2 M/min, 2 M/min
    C. 4 M/min, 3 M/min, 2.5 M/min
    D. 4 M/min, 2.5 M/min, 1 M/min

177. Why is the *oxygen debt* greater than the amount of oxygen stored in the body?

    A. Exercise increases the hemoglobin content of the blood so that it can store more oxygen.
    B. Heavy breathing after exercise takes in more oxygen.
    C. In addition to replenishing the stored oxygen, oxygen is used to reconstitute the phosphagen and lactic acid systems.
    D. In addition to replenishing the stored oxygen, oxygen is used to reconstitute the phosphagen, lactic acid system, and aerobic systems.

178. According to the information in the passage, in order for an individual on a high carbohydrate diet to perform at his peak in an athletic event, he should not engage in heavy exercise for at least:

    A. 1.6 minutes before the event.
    B. 30 minutes before the event.
    C. 48 hours before the event.
    D. 10 days before the event.

179. Most of the lactic acid produced by muscle activity is reconverted into glucose via the Cori cycle. Which organ plays the major role in the reconstitution of lactic acid to glucose?

    A. kidney
    B. liver
    C. spleen
    D. muscle tissue

180. Fast twitch muscle fibers contract with more force than slow twitch muscle fibers. However, slow twitch muscle fibers account for most of the workload in an endurance event. What organelle is most likely more abundant in slow twitch muscle fibers than fast twitch muscle fibers?

    A. mitochondria
    B. smooth endoplasmic reticulum
    C. free floating ribosomes
    D. lysosomes

181. In which of the following sports do athletes most likely rely primarily upon the phosphagen system for muscle contraction?

    A. tennis
    B. boxing
    C. diving
    D. cross-country skiing

182. Which of the following is depleted first in the body of an athlete performing maximal exercise?

    A. phosphocreatine
    B. ATP
    C. glycogen
    D. glucose

GO ON TO THE NEXT PAGE.

183. All of the following are true concerning the musculoskeletal system EXCEPT:

    A. Skeletal muscles function by pulling one bone toward another.
    B. Tendons connect muscle to muscle.
    C. Ligaments connect bone to bone.
    D. The biceps works antagonistically to the triceps.

184. Which of the following statements concerning bone is false?

    A. Bone functions as a mineral reservoir for calcium and phosphorous.
    B. Bone contains cells which differentiate into red and white blood cells.
    C. Bone acts as a thermostat for temperature control of the body.
    D. Bone provides a framework by which muscles can move the body.

STOP. IF YOU FINISH BEFORE TIME IS CALLED, CHECK YOUR WORK. YOU MAY GO BACK TO ANY QUESTION IN THIS TEST BOOKLET.

30-MINUTE IN-CLASS EXAM FOR LECTURE 9

Two schedules summarize the most important demographic information of a closed population: the *survivorship schedule* and the *fertility schedule*.

The survivorship schedule records the probability that an individual survive to a particular age, and is shown in Figure 1. There are three basic forms of survivorship curves in nature: type I; type II; and type III. Most species in nature exhibit a type III survivorship curve.

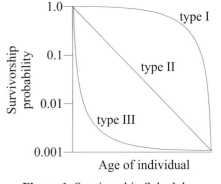

Figure 1 Survivorship Schedule

The *fertility schedule* records the average number of daughters that will be produced by all the females in the population at each particular age. The *net reproductive rate* is the average number of females produced in the lifetime of a single female. The net reproductive rate can be obtained from the survivorship and fertility schedules as the sum of the yearly products of the probability of survivorship and the average number of females born to a single female during life expectancy.

Any population reproducing in a constant environment (other than species breeding synchronously at a single age) will attain a stable age distribution. In such a population, the proportion of individuals at a given age will remain constant.

185. The annual adult mortality of white storks is approximately 21 percent regardless of age. What form of survivorship curve is exhibited by white storks?

A. type I
B. type II
C. type III
D. In order to predict the survivorship curve, the fertility schedule must be known.

186. Which of the following could be true concerning a population that has lived in a stable environment for hundreds of years and whose individuals have an average life expectancy of 20 years?

A. The number of individuals that are 13 years old doubles every 100 years.
B. If most of the individuals in the population are over 18 years old, then the population is aging.
C. Most individuals in the population live at least forty years.
D. If 60% of the population is 10 years old, then, in two years, 60% of the population will be 12 years old.

187. The net *reproductive rate* is most likely based only on females and not on males because:

A. males reproduce faster than females.
B. females live longer than males.
C. the maximum rate at which a female gives birth is not changed by the number of males in a population.
D. early population scientists designated this as the conventional standard.

188. Which form of survivorship curve would most likely be exhibited by a pure *r* strategist?

A. type I
B. type II
C. type III
D. *r* strategists do not exhibit typical survivorship curves.

189. Assuming a zero growth rate, the survivorship curve that most closely represents the population of humans in the U.S. during the late 20th century is:

A. type I
B. type II
C. type III
D. The survivorship curve for a zero growth rate population would be flat.

190. Which of the following is the LEAST important characteristic for the survival of a species living in a short-lived, unpredictable habitat?

A. large brood size
B. rapid development
C. early reproduction
D. efficient utilization of available resources

Passage II (Questions 191-196)

Isogenic mice are mice that have been inbred until they have nearly identical genotypes. In 1955, Eichwald and Silmser transplanted skin between isogenic mice. *Histocompatibility* is the acceptance of tissue grafts. A Y-linked gene located at the H-Y locus produces the H-Y antigen. Tissue grafts in isogenic mice may be rejected due to the H-Y antigen. In Eichwald and Silmser's experiment, only female mice rejected the skin grafts, and then only from male mice.

In 1984, McLaren bred mice in which the H-Y gene segregated separately from maleness. However, males which lacked the H-Y antigen had a defect in spermatogenesis. Then in 1987, Page mapped a Y-linked gene that coded for a factor important in the development of male genitalia, called the *testis-determining factor (Tdf)*, to a different region on the Y chromosome than that of the H-Y gene.

Not all animals have an X-Y chromosome makeup like that found in mammals. In birds, butterflies, moths, and some fish, the homozygous partner is male, while the heterozygous partner is female.

191. In chickens, nonbarred plumage is a sex linked recessive trait. What is the probability that a barred female and a nonbarred male produce a barred chic as their first female offspring?

- **A.** 0
- **B.** 25%
- **C.** 50%
- **D.** 100%

192. Which of the following mice do not produce H-Y antibodies?

- **A.** isogenic males of the Eichwald and Silmser experiment
- **B.** isogenic females of the Eichwald and Silmser experiment
- **C.** males of the McLaren experiment that lack the H-Y gene.
- **D.** wild female mice.

193. Which of the following is NOT true concerning the H-Y gene and the gene for *Tdf*?

- **A.** They obey the Mendelian Law of Segregation.
- **B.** They obey the Mendelian Law of Independent Assortment.
- **C.** They both code for proteins.
- **D.** They are normally found in males but not in females.

194. A human male may inherit all of the following EXCEPT:

- **A.** his mother's mother's X chromosome.
- **B.** his mother's father's X chromosome.
- **C.** his father's father's Y chromosome.
- **D.** his father's mother's X chromosome.

195. Colorblindness is a sex linked recessive trait. What are the possible genotypes of the mother of a colorblind female, if the mother's father was not colorblind?

- **A.** homozygous recessive only
- **B.** homozygous dominant only
- **C.** heterozygous only
- **D.** homozygous recessive or heterozygous

196. *Isogenic* mice are produced by breeding only siblings for many generations. Which of the following is true concerning the production of a population of isogenic mice?

- **A.** The Hardy-Weinberg law is not violated.
- **B.** It is an example of speciation.
- **C.** Isogenic mice are more likely to have autosomal recessive diseases than are normal mice.
- **D.** Isogenic mice have fewer chromosomes than normal mice.

GO ON TO THE NEXT PAGE.

Passage III (Questions 197-202)

Gregor Mendel, an Austrian monk, performed the first quantitative studies of inheritance. Using the garden pea, Mendel performed a series of hybridization experiments with lines differing in seven traits. His experiments followed these three basic steps:

1. Pea plants were allowed to self-fertilize for several generations and only plants which faithfully reproduced their original traits in each generation were used.

2. Mendel then cross-fertilized plants having different traits.

3. He allowed the hybrids to self-fertilize for several generations.

These steps had been done before by other scientists with the exception that, this time, Mendel counted and recorded the results.

By chance, each of the traits that Mendel studied followed the Law of Independent Assortment. Each trait also exhibited complete dominance.

197. If the traits that Mendel had studied had not followed the Law of Independent Assortment, how might Mendel's findings have changed?

   A. Alleles for the same trait would not have separated independently of each other.
   B. Alleles for different traits would not have separated independently of each other.
   C. Homologous chromosomes would exhibit alleles coding for different traits at the same loci.
   D. Phenotypes of a single trait would resemble both parental phenotypes on the same individual.

198. What was accomplished in step 1 of Mendel's experiment?

   A. Pure homozygous populations were created.
   B. Pure heterozygous populations were created.
   C. Any sick plants were naturally selected against.
   D. Recessive alleles were removed from the population.

199. If the red flower trait is dominant and the white flower trait is recessive in garden peas, what is the expected ratio of red flowered plants, pink flowered plants, and white flowered plants when step 3 is performed?

   A. 9, 3, 1
   B. 3, 0, 1
   C. 1, 1, 1
   D. 1, 2, 1

200. If the wrinkled pea trait is dominant to the smooth pea trait, which of the following is not possible?

   A. The offspring of two pea plants with wrinkled peas has smooth peas.
   B. The offspring of two pea plants with smooth peas has wrinkled peas.
   C. The offspring of a pea plant with smooth peas and a pea plant with wrinkled peas has smooth peas.
   D. The offspring of a pea plant with smooth peas and a pea plant with wrinkled peas has wrinkled peas.

201. Which of the following explains why a cross between a plant with green pea pods and a plant with yellow pea pods results in offspring of plants with either yellow or green pea pods, and does not result in plants with yellowish green pea pods?

   A. the Law of Segregation
   B. the Law of Independent Assortment
   C. the Hardy-Weinberg Principle
   D. pleiotropic separation

202. A dihybrid cross is made between individuals with the genotype BbFf. What is the ratio of the following genotypes BBFF, BBff, bbFF, bbff in the progeny?

   A. 9, 3, 3, 1
   B. 4, 3, 2, 1
   C. 1, 3, 3, 1
   D. 1, 1, 1, 1

266

203. All of the following habitats would probably favor an r strategist over a *K* strategist EXCEPT:

    A. the weedy cover of new clearings in forests.
    B. the mud surfaces of new river bars.
    C. the bottoms of nutrient-rich rain pools.
    D. a cave wall.

204. Healthy organisms living in the wild can typically mate and reproduce fertile offspring if they are members of the same:

    A. species.
    B. genus.
    C. family.
    D. phylum.

205. Colorblindness is a sex-linked recessive trait. If a colorblind man and a woman that is a carrier for the trait have two girls and two boys, what is the probability that at least one of the girls will be colorblind?

    A. 0 %
    B. 50 %
    C. 75 %
    D. 100 %

206. Sickle cell anemia is an autosomal recessive trait. The ability of homozygous recessives to survive and reproduce is greatly reduced. Which of the following must be true in order for the sickle cell gene to remain in the gene pool indefinitely?

    A. Homozygous dominates must have increased fitness over heterozygotes.
    B. Heterozygotes and homozygous dominates must have equal fitness.
    C. Heterozygotes have increased fitness over homozygous dominates.
    D. Heterozygotes must also have decreased fitness.

207. All of the following are examples of density dependent factors affecting population growth EXCEPT:

    A. competition for resources.
    B. catastrophic weather.
    C. predation.
    D. disease.

STOP. IF YOU FINISH BEFORE TIME IS CALLED, CHECK YOUR WORK. YOU MAY GO BACK TO ANY QUESTION IN THIS TEST BOOKLET.

STOP.

ANSWERS & EXPLANATIONS

FOR

30-MINUTE IN-CLASS EXAMINATIONS

ANSWERS TO THE 30-MINUTE IN-CLASS EXAMS

Lecture 1	Lecture 2	Lecture 3	Lecture 4	Lecture 5	Lecture 6	Lecture 7	Lecture 8	Lecture 9
1. C	24. C	47. A	70. B	93. B	116. A	139. A	162. B	185. B
2. A	25. C	48. A	71. C	94. C	117. B	140. B	163. A	186. A
3. C	26. A	49. B	72. D	95. A	118. A	141. B	164. D	187. C
4. C	27. C	50. C	73. B	96. C	119. A	142. A	165. D	188. C
5. A	28. B	51. B	74. D	97. C	120. B	143. D	166. A	189. A
6. B	29. A	52. C	75. D	98. A	121. D	144. A	167. C	190. D
7. A	30. D	53. C	76. C	99. D	122. A	145. D	168. B	191. A
8. A	31. A	54. C	77. B	100. D	123. B	146. D	169. B	192. A
9. C	32. A	55. D	78. B	101. A	124. C	147. B	170. B	193. B
10. A	33. D	56. C	79. C	102. D	125. D	148. D	171. D	194. D
11. A	34. A	57. B	80. B	103. A	126. D	149. B	172. A	195. C
12. A	35. A	58. A	81. B	104. B	127. A	150. B	173. A	196. C
13. D	36. A	59. B	82. B	105. C	128. C	151. C	174. D	197. B
14. B	37. B	60. D	83. D	106. B	129. A	152. A	175. C	198. A
15. A	38. C	61. C	84. B	107. B	130. C	153. C	176. D	199. B
16. B	39. A	62. B	85. A	108. C	131. B	154. C	177. C	200. B
17. B	40. D	63. C	86. B	109. B	132. C	155. C	178. C	201. A
18. B	41. C	64. A	87. C	110. D	133. B	156. B	179. B	202. D
19. D	42. D	65. D	88. C	111. B	134. C	157. D	180. A	203. D
20. C	43. D	66. A	89. D	112. C	135. A	158. A	181. C	204. A
21. A	44. C	67. D	90. C	113. C	136. B	159. A	182. A	205. C
22. C	45. D	68. A	91. B	114. A	137. C	160. B	183. B	206. C
23. C	46. B	69. D	92. A	115. A	138. B	161. A	184. C	207. B

MCAT BIOLOGICAL SCIENCES

Raw Score	Estimated Scaled Score
22-23	15
20-21	14
19	13
18	12
17	11
15-16	10
14	9
12-13	8

MCAT BIOLOGICAL SCIENCES

Raw Score	Estimated Scaled Score
11	7
9-10	6
8	5
6-7	4
5	3
3-4	2
1-2	1

EXPLANATIONS TO IN-CLASS EXAM FOR LECTURE 1

Passage I

1. **C is correct.** Protein disulfide isomerase is an enzyme. The function of any enzyme is to lower the activation energy of both the forward and reverse reactions. Enzymes cannot alter the equilibrium constant of a reaction; they can only increase the rate at which a reaction proceeds towards equilibrium.

2. **A is correct.** The first line of the passage says that folding is dependent upon amino acid sequence, which is the primary structure. The folding pattern itself is the tertiary structure. This question may seem ambiguous, but, especially with the directions "according to the passage", the best answer is A.

3. **C is correct.** Process of elimination is the best technique for this question. To help clarify a question, it is sometimes helpful to restate the question as a statement with "because", and add the answer choices. For example, "Attempts at predicting protein configuration based upon amino acid sequence have been unsuccessful because." Even if A were true, it does not answer the question. B is false because enzymes are catalysts. They do not determine the product; they only increase the rate of the reaction. D is generally false except in rare cases of neutral mutation. C is a direct response to the question and happens to be true.

4. **C is correct.** The folding of a peptide chain is called the tertiary structure of a protein. The passage states that chaperones assist in the "folding process". This is the best answer.

5. **A is correct.** Choice A describes exactly the role of chaperones as explained in the passage. A cell with more chaperones can synthesize proteins more easily, giving it a selective advantage. B is incorrect. Chaperones don't slow the process of protein folding. C is wrong for a number of reasons. For one, the energy for protein synthesis comes from ATP, not heat. For D, chaperones don't affect the rate of polypeptide synthesis; they affect polypeptide folding.

6. **B is correct.** Why does a protein denature in the presence of heat? The hydrogen bonds and other non-covalent interactions in the secondary and tertiary structure are disrupted. Another way to answer this question is to notice that all the bonds are covalent except hydrogen bonds. Although hydrogen bonds are the strongest type of intermolecular bond, they are much weaker than covalent bonds, and will be disrupted before covalent bonds when heat is applied.

7. **A is correct.** This question requires that you know that disulfide bond formation occurs in the synthesis of cystine, not proline, and that you know that an enzyme lowers the activation energy of a reaction. This question is pushing the envelope in how much detail is required by MCAT. In other words, this may be too detailed for MCAT. However, notice that even in this question, the amino acids chosen were ones with which you are likely to be familiar.

Passage II

8. **A is correct.** This question calls for some speculation. We must assume that carbohydrates do not react with SDS in the same way that proteins do. The passage explains that SDS gives proteins a charge "in approximate proportion to their size." It then says that the rate of movement is related to the molecular weight. The carbohydrates of a glycoprotein increase the mass without reacting to the SDS. Thus, they disrupt the relationship between mass and movement. Process of elimination can help. Acidic proteins are polar, so C can't be correct unless B is also correct. Enzymes include acidic proteins and polar proteins, so D can't be correct unless B and C are correct.

9. **C is correct.** According to the passage, coomassie blue is a dye added to the proteins after they have separated. Thus, it does not affect the results of the experiment. C is the only choice that does not affect the results.

10. **A is correct.** The passage states that SDS separates proteins based upon size only. The native charge is small compared to the SDS. The force on a protein is proportional to the charge, which is proportional to the size. So why wouldn't all proteins move at the same rate? The answer is because large proteins have difficulty fitting through the pores of the gel and thus move more slowly.

11. **A is correct.** The passage explains that SDS gives proteins a charge "in approximate proportion to their size." Only A offers an explanation why this is unnecessary with nucleic acids. C and D are simply false anyway.

12. **A is correct.** Primary structure is the order of amino acids, which is determined by covalent peptide bonds. The question says that SDS cannot disrupt covalent bonds.

13. **D is correct.** Because SDS breaks noncovalent bonds, and mercaptoethanol breaks disulfide bonds, the quaternary structure of a protein is disrupted. Once disrupted, the subunits separate according to their size as explained in the passage.

14. **B is correct.** Since SDS PAGE identifies proteins based upon size, two proteins with similar molecular weight would not be easily distinguished using SDS PAGE. C is true but is not a limitation in identifying the different proteins in a mixture. D is the very reason that SDS is used; to make charge correspond to size.

Passage III

15. **A is correct.** Aerobic respiration is the oxidation of glucose: glucose + oxygen = carbon dioxide + water. It is also combustion.

16. **B is correct.** Substrate-level phosphorylation is when an energy-rich intermediate transfers its phosphate group to ADP, forming ATP, without requiring oxygen. Oxidative phosphorylation is the production of ATP via the electron transport chain and ATP synthase.

17. **B is correct.** C and D are wrong because glycolysis is independent of oxygen. A has only one pyruvate, no NADH, and isn't balanced.

18. **B is correct.** The passage says that arsenate acts as a substrate for Glyceraldehyde 3-phosphate dehydrogenase. Thus, arsenic would prevent the reaction of glyceraldehyde 3-phosphate to 1,3-bisphosphoglycerate leading to a build up of the former.

19. **D is correct.** D directly explains the word committed. A contradicts the passage. B contradicts the diagram. C is not true and does not answer the question; reactions in chemical pathways do not have to create progressively lower energy molecules.

20. **C is correct.** The passage states that PFK activity is stimulated when cellular energy is low, and inhibited when energy is high. Low cellular energy corresponds to high ADP concentration because ATP has been used. ATP is, itself, an allosteric inhibitor of PFK, and an indicator that cellular energy is high, so D is wrong. (By the way, the concentration of ATP is held relatively constant within a cell, and is usually much higher than the concentration of ADP.) Glucagon is a peptide hormone that doesn't enter the cell, and therefore could not allosterically inhibit PFK, an enzyme of glycolysis, which takes place in the cytosol. Thus, B is wrong. Citrate is a intermediate in the Krebs cycle and actually acts as an allosteric inhibitor of PFK. This makes sense, because if there is an abundance of citrate, then there must be ample energy being produced in the cell. Thus, A is wrong.

21. **A is correct.** Arsenate competes for the active site; this is the definition of competitive inhibition. Arsenate would have to be a product of the glycolytic pathway in order for its action to be considered negative feedback.

Stand Alones

22. **C is correct.** A is wrong because not all enzymes require cofactors. B is wrong because not all cofactors are organic. D is wrong because cofactors aren't catalysts.

23. **C is correct.** The graph of reaction rate versus substrate concentration for enzymes is shown in Figure 1-8; enzyme activity is saturable.

EXPLANATIONS TO IN-CLASS EXAM FOR LECTURE 2

Passage I

24. **C is correct.** According to the passage, the *Pelomyxa* is a eukaryote and does not undergo mitosis. A is not true and not implied in the passage. Fungi lack centrioles and undergo mitosis. B is false. Prokaryotes do not undergo mitosis. They undergo binary fission. D is not true, nor does the passage imply it.

25. **C is correct.** In prophase, the nuclear membrane begins to disintegrate and chromosomes condense. Replication occurs during the S phase so A is wrong. There is no crossing over in mitosis so B is wrong. D is incorrect as well.

26. **A is correct.** The claim that *Pelomyxa* is a primitive eukaryote is based upon its apparently primitive method of nuclear division, and the fact that it has not acquired certain organelles such as mitochondria. Choice A would indicate that *Pelomyxa* underwent mitosis and contained mitochondria at one point and lost the ability and organelles through evolution. This would indicate that *Pelomyxa* was not a primitive eukaryote. B is wrong because it only tells us that Pelomyxa is phylogenetically older than diatoms. C is wrong because it is irrelevant. D is wrong because two species in the same phylum don't necessarily have similar phylogenic age.

27. **C is correct.** Anaphase is the separation of chromosomes. The formation of two daughter nuclei marks telophase. These two phases most closely resemble the separation of chromosomes and formation of the nuclear membrane between chromosomes in dinoflagellates.

28. **B is correct.** The passage states that the symbionts function like mitochondria. Mitochondria provide energy in the form of ATP from the metabolism of nutrients.

Passage II

29. **A is correct.** Choice A is the only choice that indicates that the poly-A tail is not coded for by DNA but is synthesized separately. It offers a mechanism by which synthesis may take place other than transcription. B seems to indicate that the poly-A tail is transcribed from DNA at the end of the gene. Even if C and D made sense, they would indicate that the poly-A tail was transcribed from a gene.

30. **D is correct.** The question is really asking, "Where does post-transcriptional processing take place?" The obvious answer is in the nucleus. Some (very little) post-transcriptional processing of mRNA does take place within prokaryotes, but snRNPs are not found in prokaryotes. Post-transcriptional processing of rRNA and tRNA commonly takes place in prokaryotes.

31. **A is correct.** Figure 2 is an electron micrograph of R looping. The passage explains that R looping is a technique where DNA and mature RNA are hybridized. Mature RNA has no introns. Therefore, the loops are parts of the DNA that correspond to the removed sections of the RNA; the loops are DNA introns.

32. **A is correct.** The DNA strand that is displaced is the template for the DNA strand shown.

33. **D is correct.** The question says that introns change while the genes that encompass them remain intact. This suggests that the intron plays little or no role in the phenotype. The phenotype is affected by selective pressure. Thus, selective pressure has no apparent mechanism by which to affect introns.

34. **A is correct.** The passage states that the 5′ end is capped forming a 5′-5′ triphosphate linkage. You must understand that the 5′ refers to the carbon number on the ribose. You don't have to know how the carbons are numbered, just that the same carbon on each ribose must be attached to the triphosphate group. This leaves only A or C. C has thymine, so it cannot be RNA. Note: Sometimes, carbon 2 on the first two nucleotides after guanosine has an O-methyl group instead of a hydroxyl group.

Passage III

35. **A is correct.** The passage says that glycosylases recognize uracil in DNA as a product of deamination of cytosine. After recognizing uracil, the glycosylases change it to cytosine. If DNA naturally contained uracil, glycosylases would have no way to distinguish the good uracil from the uracil that is produced by the deamination of cytosine. B indicates an advantage of having uracil, not a disadvantage. Who knows if C is true or not? If it is, it is knowledge that is beyond that required by the MCAT, and, therefore, cannot be the correct answer. D is a false statement.

36. **A is correct.** Only answer choice A gives a logical reason. The passage says that much DNA damage is not permanent if repaired before replication. The time between S phases (replication) is shorter for rapidly reproducing cells. Cancer cells reproduce rapidly; thus, cancer cells cannot repair the damage as thoroughly as normal cells. B is not true, and we are given no reason to believe that it is. C is not true, and we are given no reason to believe that it is. D is very unlikely; if tumor cells outnumbered normal cells, the patient would have to be dead or near death. Anyway, there is no strong scientific reason that we would know if D is true or not, whereas A is supported by the passage and our knowledge of the cellular life cycle.

37. **B is correct.** The passage states that neoplastic diseases arise from mutations in somatic cells. Thus, they cannot be passed on to offspring. This is not the same thing as saying that susceptibility to neoplastic diseases cannot be inherited. For instance, breast cancer is a neoplastic disease. Susceptibility to breast cancer is inheritable; breast cancer is not.

38. **C is correct.** C is the only choice that answers the question and is true. A and B are false statements; prokaryotes must possess a ligase to connect DNA molecules because they replicate DNA in a very similar fashion to the way eukaryotes replicate; they make Okazaki fragments. Prokaryotic DNA is double stranded. D does not answer the question of why eukaryotes <u>can</u> repair the breaks.

39. **A is correct.** We are looking for two pyrimidines that are both found in DNA. Only A meets this criterion. B and C do not contain two pyrimidines, and D must be RNA.

40. **D is correct.** This is tricky. The sense strand might code for an antisense strand where bromouracil substitutes for T. The next replication, the new antisense strand with bromouracil would code for a guanine at that spot. Thus the new sense strand would be identical to the original except that guanine would replace adenine. (Sense and antisense are ambiguous terms, but their ambiguity is irrelevant to the answer to this question.)

$$\text{original sense strand} = 5'\text{-GGCGTACG-}3'$$
$$\text{new antisense strand in presence of Br} = 3'\text{-CCGCA}\underline{B}\text{GC-}5'$$
$$\text{new sense strand} = 5'\text{-GGCGTGCG-}3'$$

Since there is no answer choice where bromouracil might be present in the second replication, we don't worry about that possibility.

41. **C is correct.** The passage states that a tautomeric shift causes a base pair mismatch. Although this may lead to a deletion upon repair by glycosylases, in their absence, the resulting mutation is a substitution. Choice A requires a deletion or insertion. Choice D is the result of an improper number of chromosomes. There is no mechanism by which a tautomeric shift would result in more or fewer chromosomes.

42. **D is correct.** D is a summary of the last paragraph. Rapidly reproducing cells have less time to repair DNA between replications. A is wrong because there is the basal mutation rate which the rate of mutations in the absence of mutagens. B is wrong because the environmental mutagens simply add to the basal mutation rate, they don't change it. C is wrong because hypoxanthine results from a mutagen; it is not a mutagen itself.

Stand Alones

43. **D is correct.** DNA replication takes place during the S stage of interphase. You should memorize the stages of the cell life cycle.

44. **C is correct.** Both children are female because the genotype of Turner's syndrome, as per the question, must be XO. The O came from the parent within whom nondisjunction occurred; the X came from the other parent. The mom only has healthy Xs to pass on, the father has only one colorblind X to pass on. Thus the colorblind child got her X from Dad and her O from Mom, and the healthy child got her X from Mom and her O from Dad.

45. **D is correct.** The primary spermatocyte is the spermatogonium just after DNA replication. Humans never have more than 46 chromosomes. The primary spermatocyte has 92 chromatids but only 46 chromosomes.

46. **B is correct.** Memorize this for the MCAT.

EXPLANATIONS TO IN-CLASS EXAM FOR LECTURE 3

Passage I

47. **A is correct.** The passage says that the death rate is exponential. This means that it decreases by the same fraction at even time intervals. In the first three minutes the population dropped to 1/25 of its original size. In the second three minutes it should do this again. $2.5 \times 10^{11}/25 = 1.0 \times 10^{10}$.

48. **A is correct.** The *z* value for *Salmonella* is 5°C. This means that for each increase of 5°C, the *D* value is reduced by a factor of 10. Table 1 reports that the *D* value for *Salmonella* at 60°C is 0.4 minutes or 24 seconds. 70° is 10 degrees more than 60°. 10 degrees is two *z* values for *Salmonella*. Thus, we divide 24 seconds by 10 twice. We get 0.24 seconds.

49. **B is correct.** In order to go from 10^{12} to 1 ($1 = 10^0$), we divide by 10 twelve times. This is 12 *D* values (a 90% reduction is the same as dividing by 10). One *D* value is 0.2 minutes, so 12 *D* values is 2.4 minutes. While answer D may be appealing mathematically, it is logically ridiculous. It is equivalent to 380,500 years. These regulations would be a little stiff.

50. **C is correct.** Exponential growth on a logarithmic scale is a straight line. Get used to reading log scales. They may be on the MCAT.

51. **B is correct.** The passage states that disinfection is the killing or inhibition of pathogens. Pathogens are disease causing microbes (stated in the passage). Disinfection methods (including the controlling methods discussed in the passage) are likely to destroy some nonpathogens as well as pathogens.

52. **C is correct.** The three *D* values for *S. aureus* vary according to substrate. This supports statement C. A is not true, nor is there sufficient evidence in the table to support it as strongly as C. D is false, and contradicts Table 1. There doesn't appear to be any evidence for B.

53. **C is correct.** *C. botulinum* can be killed by boiling water. However, it takes a long time. Table 1 predicts a *D* value of 20 minutes for *C. botulinum* at 101°C. A is not true because *C. botulinum* is an obligate anaerobe. B is not true because *C. botulinum* is a gram positive bacterium. D is false because, according to the passage, horses make antibodies to *C. botulinum* toxin.

Passage II

54. **C is correct.** The passage says that the ends of the strands are reverse compliments of each other. The complement sequence is 5′-TGACAATTC-3′. The reverse of this is 5′-CTTAACAGT-3′.

55. **D is correct.** D is the only choice that is consistent with both viruses and transposons. A, B, and C are all unique to viruses. It should be clear from the passage that transposons are not viruses, they are a form of TGE, which is a vehicle for genetic shuffling.

56. **C is correct.** Binary fission typically produces identical daughter cells. It is not a method of genetic recombination in prokaryotes.

57. **B is correct.** Conjugation is genetic recombination between prokaryotic individuals, which typically involves the F plasmid or some other plasmid. D is not genetic recombination. C does not take place in prokaryotes. A does not typically involve plasmids.

58. **A is correct.** Conjugation is a one way transfer of genetic information from the initiator or F+ to the receiver or F-. The passage says that replicative transposition is when a copy of the gene is transferred. In this case, bacterium A lost the ability to live on histidine lacking medium. Thus the gene must have been completely removed and transferred; conservative transposition.

Passage III

59. **B is correct.** A virion is the inert form of a virus that exists outside the host cell. A prophage is the name used to describe the virus while it is incorporated into the host cell DNA.

60. **D is correct.** The passage states that λ phage is a bacteriophage. Bacteria don't have nuclei, mitochondria, or endoplasmic reticulum. Their DNA is in the cytoplasm.

61. **C is correct.** You should know that UV light causes DNA damage. The passage states that DNA damage leads to the lytic cycle. A is wrong because the genes of a second λ phage would be repressed by the λ repressor of the first λ phage. B is wrong because the lysogenic stage can last hundreds of thousands of generations. D is wrong because bacteria don't undergo mitosis.

62. **B is correct.** All viruses need a specific protein to bind in order to attach to and infect a host cell. The host cell does not have to be weakened. Mutagens are an entirely separate topic from viral infections. Bacteria don't have nuclei.

63. **C is correct.** A is wrong because DNA is not translated. B is wrong because only some RNA viruses contain reverse transcriptase and lambda phage isn't one of them. D is wrong because bacteria have no nuclei and viruses are never injected into a nucleus.

64. **A is correct.** The passage states that lambda phage is a bacteriophage. This means it attacks bacteria.

65. **D is correct.** No virus has both DNA and RNA.

Stand Alones

66. **A is correct.** This question should be easy. Bacteria have no complex membrane bound organelles. The MCAT is likely to leave out the 'complex' part.

67. **D is correct.** Yeast is probably the only member of the Fungi kingdom about which you must know anything specific.

68. **A is correct.** All viruses require the host cell in order to replicate.

69. **D is correct.** Only eukaryotes have centrioles. Prokaryotes have peptidoglycan cell walls, and some eukaryotes have cellulose or chitin walls. Of course, both have RNA and ribosomes, so that they can synthesize protein. All living organisms have both DNA and RNA. (Viruses are not living.) By the way, since ribosomes are made from RNA, if you chose B, then C would also have to be true.

EXPLANATIONS TO IN-CLASS EXAM FOR LECTURE 4

Passage I

70. **B is correct.** Chromaffin cells are modified postganglionic sympathetic neurons, so they most likely secrete norepinephrine and epinephrine.

71. **C is correct.** The diaphragm is skeletal muscle and is not innervated by the autonomic nervous system. You may be able to eliminate A, B, and D from memory. By the way, sympathetic activity does not innervate skeletal muscle, but its effects increase glycogenolysis in skeletal muscle.

72. **D is correct.** Cardiac cells are innervated by both autonomic nervous systems. Sympathetic neurons release epinephrine onto cardiac muscle cells. A and B would be expected to have similar receptors because chromaffin cells are modified postganglionic sympathetic neurons. All autonomic preganglionic neurons release acetylcholine onto postganglionic neurons.

73. **B is correct.** You should know that A, C, and D are autonomic responses. A and C are autonomic because they have to do with smooth muscle. Sweating is controlled by both the sympathetic and the parasympathetic as well. Shivering is skeletal muscle contraction, which you should know is not controlled by the autonomic nervous system.

74. **D is correct.** The somatic nervous system governs skeletal muscles, which are involved in the simple reflex arc.

75. **D is correct.** The nervous system acts directly and immediately (in seconds or less), while hormones are indirect and require more time.

76. **C is correct.** Epinephrine is a sympathetic neurotransmitter. The sympathetic nervous system dilates the pupils (so you can hunt in the dark [memory aid]).

Passage II

77. **B is correct.** From the passage, we know that one function of the smooth ER is hormonal synthesis. Thus, B is the best answer.

78. **B is correct.** The passage states that all phospholipid synthesis takes place on the cytosol side of the smooth ER.

79. **C is correct.** The passage states that ingestion of phenobarbital leads to greater production of smooth ER and mixed-function oxidases but not other enzymes. More mixed-function oxidases means more efficiency in degrading phenobarbital. This is why people that take sleeping pills must take increasingly greater doses and are less responsive to antibiotics.

80. **B is correct.** The passage states that phospholipid translocators flip phospholipids from one side of the membrane to the other.

81. **B is correct.** Proteins are synthesized on the rough ER or on ribosomes in the cytosol.

82. **B is correct.** Since oxygen is <u>reduced</u> to water, the iron must be <u>oxidized</u>.

83. **D is correct.** Fat is stored energy.

Passage III

84. **B is correct.** Although the calcium channels begin opening immediately, they are slow to open and slow to close. The result is the extended plateau of section 2. The sodium channels close very quickly and are the major contributors to depolarization or section 1.

85. **A is correct.** Depolarization is the initial influx of sodium ions.

86. **B is correct.** Acetylcholine is the neurotransmitter used by the parasympathetic nervous system, particularly the vagus nerve innervating the heart. The acetylcholine binds to muscarinic receptors, which stimulate the opening of K^+ channels and thus inhibit depolarization. You should know that the time between heartbeats is increased by acetylcholine. Thus, section 4 lengthens.

87. **C is correct.** Near the end of the action potential potassium channels are slow to close while potassium ions exit the cell. Calcium ions are being pumped back into the sarcoplasmic reticulum. Sodium channels are closed.

88. **C is correct.** Na^+ voltage gated channels contribute to the action potential but not to the resting potential. They are only open during an action potential. If you removed them, the resting potential would not be affected.

89. **D is correct.** Only cells that experience action potentials contain Na^+ voltage gated channels. Muscle cells and neurons are two examples. Since the passage is about action potentials in muscle, this question should be deducible even without this information.

90. **C is correct.** You should know that Na^+ channels are shut at the end of section 1 and the beginning of section 2 of the graph in Figure 1. From the passage, you should deduce that they are inactivated not closed.

Stand Alones

91. **B is correct.** You should know that Na^+ channels are more sensitive than K^+ channels. That's why they open first.

92. **A is correct.** Choice A describes saltatory conduction.

EXPLANATIONS TO IN-CLASS EXAM FOR LECTURE 5

Passage I

93. **B is correct.** Cortisol is a steroid. The passage states, and you should know, that steroids diffuse through the membrane and bind to a receptor in the cytosol, where they are carried to the nucleus.

94. **C is correct.** Caffeine inhibits phosphodiesterase, which leads to an increase in cyclic AMP. This much is derived from the question. Cyclic AMP activates phosphorylase, resulting in breakdown of glycogen to glucose. This is from the last sentence in the passage.

95. **A is correct.** From Figure 1, the only missing link in the chain reaction from hormone to cyclic AMP as explained in the passage is the G_s-protein. G_s-protein does not activate adenylyl cyclase; it inhibits, so B must be incorrect. C is incorrect as epinephrine binds to plasma membrane receptors. Nowhere in the passage is D discussed; protein kinase A is downstream of the receptor, G_s-protein, and adenylyl cyclase.

96. **C is correct.** C describes a mechanism by which the same hormone can stimulate different processes in different cells through increasing cyclic AMP levels. A is wrong because liver cells must contain a G_s-protein, since addition of cyclic AMP activates phosphorylase. B is wrong because from the passage we know that liver cells are capable of production of cyclic AMP. D is wrong because the passage says that glucagon works through cyclic AMP levels.

97. **C is correct.** Cyclic AMP is in the cytosol. Look at Figure 1. Cyclic AMP could not be a part of the membrane presence because it is not part of the membrane. It must be the right answer.

98. **A is correct.** Epinephrine is a sympathetic hormone and stimulates the heart while inhibiting smooth muscle of the gut. The answer choices only give the option of a mechanism differing in G_s-proteins and G_i-proteins. The answer choice must coincide with epinephrine stimulating heart muscle and inhibiting smooth muscle.

99. **D is correct.** Since the Gs-protein can't hydrolyze GTP, it can't turn off. This leads to increased cAMP. A is a bad answer because the passage states that the G-protein must hydrolyze GTP in order to be inactivated. You have no way of knowing about electrolyte concentration so B must be wrong. C is wrong because nothing in the passage mentions any connection between G-proteins and hormone binding to receptors on the outside of the cell.

100. **D is correct.** Aldosterone is a steroid. Its effect is at the level of transcription. It increases protein production.

Passage II

101. **A is correct.** These two hormones even use the same receptor protein as stated in the passage.

102. **D is correct.** Progesterone prepares the uterus for pregnancy.

103. **A is correct.** The passage states that HCG uses a membrane bound receptor; therefore, HCG is a peptide hormone. It is really a glycoprotein like LH and FSH.

104. **B is correct.** The remaining part of the follicle after it bursts to release the egg is the corpus luteum. The corpus luteum secretes estradiol and progesterone until it degrades into the corpus albicans. You should know this for the MCAT.

105. **C is correct.** Since HCG is produced by the placenta, it is only found when there is a pregnancy. All other hormones normally occur during the menstrual cycle.

106. **B is correct.** Only estrogen is a steroid like all the cortical hormones, so only estrogen could act as a substrate.

Passage III

107. **B is correct.** Know your mitosis. You should also know that at birth the oocytes are arrested in prophase of meiosis I until puberty.

108. **C is correct.** These two cells arise from the interstitial cells and secret steroids. They are phylogenetically related.

109. **B is correct.** Androgens are male hormones like testosterone. From the passage, testosterone inhibits GnRH. GnRH stimulates FSH and LH which are required for gamete production. Choice A is wrong because, although there would be a decrease in testosterone production, the exogenous production more than makes up for this decrease. Otherwise, there would be no point in taking androgens. C and D are wrong for the reasons that B is correct.

110. **D is correct.** The only steroid choice. Steroids act in the nucleus. Peptides don't enter the cell. Woman produce some testosterone from the adrenal cortex.

111. **B is correct.** Osteoclasts breakdown bone, osteoblasts build bone. This will be covered in Lecture 8.

112. **C is correct.** The follicle bursts on ovulation and remains behind as the corpus luteum. The secondary oocyte and corona radiata are released into the body cavity and swept into the fallopian tubes by fimbriae. The corona radiata consists of the zona pellucida and some granulosa cells.

113. **C is correct.** The LH surge causes ovulation. Yes, this question seems to have more than one answer. Couldn't estrogen also be true? However, read the question closely. It says MOST. LH alone will cause ovulation; estrogen alone will not.

114. **A is correct.** FSH blockage would prevent spermatogenesis by interfering with Sertoli cells and would not interfere with Leydig cells, which produce androgens. The others would affect testosterone production.

Stand Alones

115. **A is correct.** All hormones work through negative feedback. The negative feedback begins when the effector is overproducing. The effector of TSH is the thyroid. In this case, the effector would be under producing. TSH production would increase to try to correct this.

EXPLANATIONS TO IN-CLASS EXAM FOR LECTURE 6

Passage I

116. **A is correct.** The volume of a sphere is $4/3\ \pi r^3$, and the surface area of a sphere is $4\pi r^2$. These equations tell us that while the surface area of a sphere is proportional to the square of the radius, the volume is proportional to the cube of the radius. Since $4/3\ \pi$ and $4\ \pi$ in these equations are just constants, and we do not want to know the exact values of surface area and volume, we can ignore them. For the original fat globule, we will arbitrarily assume its radius = 2; using this radius the volume is 8 while the surface area is 4 (ignoring constants). The problem states that the bile is reducing the diameter (and thus the radius) by a factor of 2. Decreasing the radius by a factor of 2 (new radius = 1) makes the volume = 1, and surface area = 1. Notice that the volume decreased from 8 to 1, while the surface area decreased from 4 to 1. Now you must realize that no fat was lost in this process, so a decrease in volume by a factor of 8 while maintaining the same amount of fat means that the single fat droplet must be divided into 8 smaller fat droplets (each with a volume = 1 and surface area = 1). 8 new fat globules, each with a surface area 4 times as small as the original add up to a total cumulative surface area of 8 (versus the original surface area of 4). The surface area increased by a factor of 2.

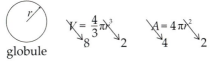

117. **B is correct.** You should narrow this down to A or B from the passage. The thoracic duct delivers lymph to the venous circulation from the lower part of the body and the left arm. This question requires that you know either that the thoracic duct delivers the fat to the blood, or that the lymphatic ducts empty their contents into the veins, not the arteries. This question is on the trivial side for an MCAT question, but it is not impossible that they would ask it.

118. **A is correct.** Only A shows that serum cholesterol level alone might not indicate a health risk to the patient. The HDLs might cause a high serum cholesterol but indicate a healthy patient.

119. **A is correct.** From the passage, lingual lipase works in the stomach.

120. **B is correct.** You should know that trypsin is activated by enterokinase and then activates the other pancreatic enzymes. C and D must be wrong because the duodenum is at a pH of 6.

121. **D is correct.** You should know the word emulsification.

122. **A is correct.** You should know the major pancreatic enzymes, trypsin (works on proteins), chymotrypsin (works on proteins), lipase (works on lipids), and amylase (breaks starch into disaccharides). Disaccharides are broken down by intestinal enzymes. Lactase breaks down lactose, a disaccharide in milk.

Passage II

This passage offers an opportunity to see some topics that are not required by the MCAT but are likely to be used in a passage. Glomerular filtration rate (GFR) is the rate at which filtrate enters Bowman's capsule. The rate of filtration is affected by: 1) the oncotic pressure difference between the blood and Bowman's capsule (oncotic pressure is osmotic pressure that tends to move fluid back to the blood); 2) the hydrostatic pressure difference between the blood and Bowman's capsule (overpowers the oncotic pressure and moves fluid into Bowman's capsule); and 3) the rate of blood flow (the faster, the greater the GFR). Conventional thinking (MCAT) says that the change in hydrostatic pressure is the regulator of GFR. If we multiply the plasma concentration of solute z (P_z) times the GFR, this should tell us the rate at which the solute is excreted. If there is no resorption or secretion, this number should equal the solute concentration in the urine (U_z) times the urine volume (V). GFR $\times$ P$_z$ = U$_z$ $\times$

V. If we take into account, secretion, resorption, and change in volume of total filtrate, we can change GFR in the equation to C_z. $C_z \times P_z = U_z \times V$. This is the first equation in the passage. Another way to look at clearance is as a type of comparison between the concentration in the urine to the concentration in the plasma. $C_z = (U_z/P_z) \times V$. RBF is the rate at which blood flows through the glomerulae of the kidney. RPF is the rate at which plasma flows through the glomerulae of the kidney. Still another way to look at clearance is as the volume of plasma that must be filtered in order to produce the solute by filtrate alone (i.e. without secretion or resorption).

123. **B is correct.** The GFR is the volume of plasma filtered each minute. The renal clearance of a substance is the minimum volume of plasma needed to be filtered in order to produce that much substance by filtration alone. If clearance is greater than filtration, then the urine must be receiving an additional supply of the substance from some other source. Secretion is the only answer.

124. **C is correct.** To accurately measure GFR you need a substance that is cleared from the plasma solely by glomerular filtration in the absence of any complicating factors that might cause a difference in the concentration of the substance in the filtrate vs. its concentration in the plasma. These substances should not be reabsorbed or secreted by the tubules. It should also not be destroyed, synthesized or stored by the kidneys and pass through the glomerular filtration membrane unhindered. In summary, the substance must not be affected by the nephron (except to be filtered) filtrate should be equal to its concentration in the plasma barring water resorption by the tubules. Choices A and B describe a situation where inulin is affecting the physiology of the kidney, so they must be incorrect. D is incorrect because the passage states that almost no solute is completely filtered in one pass through the renal corpuscles, yet it is possible to measure GFR.

125. **D is correct.** The passage states that inulin is a polysaccharide, so it must be larger than glucose. You should know that a red blood cell is too large to be filtered, so inulin must be smaller. Albumin is just small enough to be filtered but its negative charge prevents filtration. (This knowledge concerning albumin is not required for the MCAT or this question since you know that D must be true of any polysaccharide.)

126. **D is correct.** The renal clearance must be affected by the ability of a substance to filter into Bowman's capsule. This is dependent upon size and charge. The renal clearance is also affected by the filtration rate, which is dependent upon the hydrostatic pressure difference between the glomerulus and Bowman's capsule, and the oncotic pressure difference. PAH is a solute used to measure the plasma flow, and should not significantly affect the clearance of another solute or else its accuracy would be diminished.

127. **A is correct.** You should know that glucose is completely resorbed in the proximal tubule of a healthy adult.

128. **C is correct.** A close reading of the question reveals that it is asking for the filtered fraction. This is equal to GFR/RPF. The clearance of inulin is the GFR, and the clearance of PAH is the RPF. This equals $125/625 = 1/5 = 0.2 = 20\%$

129. **A is correct.** You must see from the passage that a normal GFR is 125 ml/min. Then you must recognize that creatinine is just like inulin, since neither is absorbed nor secreted. Thus, the clearance of creatinine is equal to the GFR. Now use the equation:

$$GFR = C = \frac{U \times V}{P}$$

$$\therefore \ P = \frac{U \times V}{GFR} = \frac{2.5 \ mg/ml \times \ 1 \ ml/min}{125 \ ml/min}$$

130. **C is correct.** A hematocrit level of 50% means that 50% of the blood by volume is red blood cells and the other 50% is plasma. The renal blood flow, then, is twice the renal plasma flow. The renal plasma flow is equal to the PAH clearance.

Passage III

131. **B is correct.** Because amino acids and sodium are reabsorbed together in the proximal tubule, less sodium reaches the macula densa cells in the distal tubule. Renin is released, leading to decreased resistance in the afferent arterioles, increased resistance in the efferent arterials, and increased renal blood flow. The increased renal blood flow increases the GFR.

132. **C is correct.** The renin-angiotensin system causes increased systemic blood pressure as per the passage. The renal blood pressure does not increase because blood flow to the kidneys is impeded by the stenosis.

133. **B is correct.** Aldosterone comes from the adrenal cortex; ADH is from the posterior pituitary; and the sympathetic nervous system is part of the autonomic nervous system. The thyroid is not mentioned.

134. **C is correct.** The passage states that renin is an enzyme. An enzyme is a catalyst. A catalyst lowers the energy of activation of a reaction without being permanently altered.

135. **A is correct.** Urine volume is reduced by ADH secretion among other things. Aldosterone causes B, C, and D.

136. **B is correct.** Angiotensin II is a peptide (from the passage). Peptide hormones act via second messenger.

Stand Alones

137. **C is correct.** Urine is concentrated in the collecting ducts, but the loop of Henle plays the major role in allowing that to happen by establishing a concentration gradient between the collecting duct and the medulla.

138. **B is correct.** The small intestine is the major site for absorption of nutrients.

EXPLANATIONS TO IN-CLASS EXAM FOR LECTURE 7

Passage I

139. **A is correct.** Since antibodies are produced, this is an example of humoral immunity. Humoral immunity is directed against an *exogenous antigen* (one found outside the cell) such as fungi, bacteria, viruses, protozoans, and toxins. Cell mediated immunity (T-cells) works against infected cells, cancerous cells, skin grafts, and tissue transplants.

140. **B is correct.** A nude mouse lacks a thymus. T-cells require a thymus for maturation. An antibody is not a cell, so A is wrong. The MCAT may be misleading in this fashion, so read the question closely.

141. **B is correct.** The experiment begins with the premise that nude mice do not produce antibodies or T cells. They do produce B-cells. Since antibodies are produced only after exposure to the donor T cells, either the T-cells or the B-cells are producing antibodies. Since the B-cells continue to produce antibodies after the T cells are removed, B-cells produce antibodies only after exposure to helper T-cells.

142. **A is correct.** Interleukins are protein hormones, so they act through a second messenger system at the membrane surface.

143. **D is correct.** You should know that all blood cells arise from the same stem cells in the bone marrow.

144. **A is correct.** A foreign particle capable of provoking an immune response is the definition of an antigen. Choices B, C, and D are all molecules being produced by the body (recognized as self) and should not illicit an immune response.

145. **D is correct.** Each B-cell is capable of reproducing only one type of antibody specific to one antigen. When many memory cells are produced, the body will start with many more B-cells that make antibodies against the same antigen, allowing for a faster response in case of a second infection.

Passage II

146. **D is correct.** Quaternary structure consists of the joining of separate polypeptide chains. From the Figure 1, you should see that the quaternary structure is disrupted when disulfide bonds are broken. You should know that disulfide bonds are also involved in tertiary structure.

147. **B is correct.** If antigens are released within a healthy individual, that individual will make memory B cells for a secondary immune response, often times making the individual immune to infection.

148. **D is correct.** The light chain is the smaller polypeptide. If we break apart the antibody shown in Figure 1 at its disulfide bonds, the passage says that we are left with two heavy and two light chains. Since the light chain contains no part of the F_c region, E and F must mark the light chain. The passage says that the F_c region is constant from antibody to antibody; therefore, it must contain at least, A and B. Since the light chain contains no part of the F_c region, E cannot be in the F_c region.

149. **B is correct.** The variable and hypervariable regions are responsible for antigen binding. Different antibodies have different variable and hypervariable regions that make each antibody specific for particular antigens.

150. **B is correct.** Proteins for secretion are produced at the rough endoplasmic reticulum.

151. **C is correct.** Plasma cells arise from B-cells, not T-cells. A is a true statement. Remember how the vaccine for chickenpox was found through cowpox. This does not contradict the passage. The pathogens or disease carriers may be different but carry similar antigens.

152. **A is correct.** From the passage, the F_c region is constant and not part of the light chain. (See the answer to question 148.)

Passage III

153. **C is correct.** We want to know why hemoglobin evolved to have less affinity for CO than the heme group has for CO (as explained in the passage). This would only happen if there was some advantage to not binding with CO, and there would only be an advantage if CO were present. So what explains the presence of CO in the cell? Since the passage tells us that the break down of heme produces CO, this is a logical source. Thus **the production of CO by the breakdown of heme, favored the selection of a form of oxygen carrier that had less affinity for CO.** For choice D, there is no simple mechanism that an MCAT test-taker should know that would result in CO from carbonic acid in the blood. Choice A is a far too recent event to account for such broad evolutionary change, and, of course, hemoglobin existed before the Industrial Revolution. For B there is no CO involved.

154. **C is correct.** They differ in their amino acid sequence (primary structure) but have a similar three-dimensional shape (tertiary). Hemoglobin is made from four polypeptides while myoglobin is made from one (quaternary structure).

155. **C is correct.** The passage states that M has a greater affinity for O_2. Thus, at a given pressure the M saturation will always be higher. It also states that BPG (in red blood cells) affects the characteristic sigmoidal oxygen dissociation curve for hemoglobin. The M curve should not be sigmoidal because it is not exposed to BPG.

156. **B is correct.** Lactic acid lowers pH. The passage says that only hemoglobin responds to low pH.

157. **D is correct.** The passage states that BPG is found in red blood cells. A and C are not red blood cells, so they must be wrong. As per the passage, BPG affects the characteristic sigmoidal oxygen dissociation curve for hemoglobin. Thus at low pressures of O_2 the BPG has the greatest effect. Low levels of O_2 would result in the capillaries of skeletal muscle not the lungs.

158. **A is correct.** Since the passage states that the tertiary structure (or three dimensional shape of the polypeptide) does not vary significantly between species it is likely that this shape is fundamental to oxygen transport.

159. **A is correct.** High pH means low H^+ concentration and slower breathing so that one doesn't lose too much CO_2 (an acid).

Stand Alones

160. **B is correct.** The lymphatic system is an open system, meaning something goes in one end and out the other. A closed system is like the blood, where fluid doesn't exit or enter the system.

161. **A is correct.** Platelets do not contain a nucleus and they are not cells. Human erythrocytes contain no organelles.

EXPLANATIONS TO IN-CLASS EXAM FOR LECTURE 8

Passage I

162. **B is correct.** Myosin hydrolyzes ATP to ADP.

163. **A is correct.** From the passage you know that Ca^{2+} is bound to calsequestrin in the sarcoplasmic reticular lumen. This lowers the concentration inside the lumen, which weakens the gradient that the calcium pumps must work against.

164. **D is correct.** From the passage, we know that the red color comes from myoglobin and cytochromes within the mitochondria. These supply ATP to muscle. Muscle with only small amounts of these, such as white muscle, is capable of only short periods of contraction.

165. **D is correct.** The passage states that ATP levels remain constant. Ca^{2+} concentrations change to create muscle contraction.

166. **A is correct.** From the passage, the function of creatine kinase is to allow phosphocreatine to give or receive a phosphate group to or from ATP. This is what maintains ATP levels during muscle contraction. The passage states that cellular respiration is not fast enough to maintain ATP levels.

167. **C is correct.** T-tubules are invaginations of the sarcolemma which deliver the action potential directly to the sarcoplasmic reticulum along the center of each sarcomere.

168. **B is correct.** This question is a little bit trivial for the MCAT, but is still within the realm of possibility. You should be aware that some cell types in the human body, such as muscle cells and neurons, are so specialized that they have lost the ability to undergo mitosis.

169. **B is correct.** The mechanism involves an integral protein of the sarcoplasmic reticulum not the sarcolemma. The passage says that uptake is active, so it requires ATP and can occur against the concentration gradient of Ca^{2+}.

Passage II

170. **B is correct.** This knowledge may be too trivial to be required on the MCAT. Only one or two questions that require this much detail will be on any given MCAT.

171. **D is correct.** As osteoblasts release matrix materials around themselves, they become enveloped by the matrix and differentiate into osteocytes. This knowledge is required by the MCAT.

172. **A is correct.** The passage mentions strength and flexibility when talking about structural proteins and collagen. This is a hint. The passage also says that collagen is a protein; another hint. You should have some idea that collagen is a structural protein.

173. **A is correct.** Parathyroid increases blood calcium by breaking down bone via stimulation of osteoclasts.

174. **D is correct.** Bone is living tissue containing vascular connective tissue (blood), and nerves.

175. **C is correct.** This question is reading comprehension. The passage says that osteoclasts are differentiated from phagocytotic blood cells. All blood cells differentiate from the same precursor.

Passage III

176. **D is correct.** These are in the same ratio as the table. Since ATP concentration remains relatively constant within the cell, it makes sense that ATP production rate would mirror muscle power for these systems.

177. **C is correct.** The body must reconstitute the phosphagen system, which requires ATP to make phosphocreatine. Oxygen is required to make ATP. This is the very heavy breathing following exercise. The heavier than normal breathing that follows for approximately 1 hour takes in oxygen to convert most of the lactic acid produced by the glycogen-lactic acid system back into glucose. (This is done principally in the liver.) Choice A is ridiculous because the time frame is to short to make new hemoglobin, and the increase from 2.5 to 11.5 is way too high for the change to be due to hemoglobin. B is not an explanation of why. D is wrong because the oxygen used during aerobic respiration is breathed not stored; the passage states that the aerobic contractions can continue indefinitely or as long as nutrients last. Nutrients are not replenished with oxygen, but with food.

178. **C is correct.** The passage states that glycogen is replenished in 2 days for an individual on a high carbohydrate diet.

179. **B is correct.** The liver makes glucose from lactic acid.

180. **A is correct.** Slow twitch must rely upon aerobic respiration.

181. **C is correct.** Only C is an event that lasts for a period of 10 seconds or less.

182. **A is correct.** ATP levels remain nearly constant. Phosphocreatine is the first thing used to maintain those levels.

Stand Alones

183. **B is correct.** Tendons connect muscle to bone. Muscle pulls on bone not other muscles.

184. **C is correct.** Bone does not act as a thermostat for the body.

EXPLANATIONS TO IN-CLASS EXAM FOR LECTURE 9

Passage I

185. **B is correct.** The best way to answer this is to imagine a simpler example. Imagine 50% died each year. (This is simply the half-life curve. You may already know that a half-life curve is a straight line on a semilog plot.) Now start with a population of 100, record the results for 5 years, and plot them. The amount of adults left alive each year would be: 50, 25, 12.5, 6.25, 3.125. The probability of an individual living to a certain age is simply the number of individuals living to that age divided by total born. This leaves us with 0.5, 0.25, 0.125, 0.0624, and 0.03125. Plotting this on the graph in Figure 1 gives:

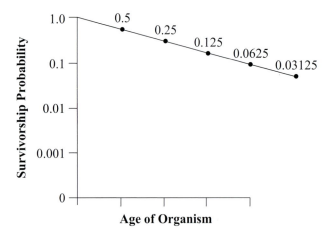

186. **A is correct.** As the answer explanation explains, the last paragraph is key; according to the last paragraph, the population in a constant environment attains a stable age distribution (a bell curve) where the proportion of individuals at a particular age remains constant (achieves a steady state). This does not mean that the population is not growing or shrinking; only that the relative number of each age group remains constant. The only way for a particular age class to maintain itself with respect to other age groups is for the birth and death rate to be equal. Choice B does not describe this; instead it describes a situation where the birth rate has plummeted and the population ages due to no great influx of newborns. Remember that the last paragraph said that the proportion of individuals at a particular age remains constant; that means that if 60% of the population is 10 years old now, then 60% of the population should be 10 years old two years from now. Choice D violates this because choice D assigns the same proportion of individuals changing in a two year span of time. Choice C is wrong because if more than half the individuals live to age 40, the average life expectancy must be over 20. A is true only if individuals of all different ages increases proportionally over the same time, and thus could be true, as the question asks.

187. **C is correct.** Imagine one female human. In 90 months she could not have more than 10 babies regardless of how many males were in the population. Now imagine one male human. In 90 months, the number of babies he could produce depends only upon the number of females.

188. **C is correct.** Remember that an *r* strategist has many babies to ensure the survival of only a few. Thus, many die early on, and type III is the most likely curve.

189. **A is correct.** The population growth rate doesn't change the survivorship curve. Humans in the U.S. are expected to live long lives with an increasing expectancy of death later in life.

190. **D is correct.** A short-lived, unpredictable habitat is best suited for an *r* strategist. D describes a *K* strategist.

191. **A is correct.** The passage states that male birds are the homozygous partner in the sex chromosomes meaning that they are XX. Thus:

$$X^bX^b \qquad X^BY$$
male $\qquad$ female

Parents

	X^B	Y
X^b	X^BX^b barred male	X^bY unbarred female
X^b	X^BX^b barred male	X^bY unbarred female

192. **A is correct.** Those that produce the H-Y antigen will not make the H-Y antibodies. These are the same individuals that have the H-Y gene.

193. **B is correct.** The law of independent assortment does not work for genes on the same chromosome. It says that alleles will separate independently (in a random fashion) during meiosis. The law of segregation says that homologous chromosomes will segregate during meiosis.

194. **D is correct.** D cannot happen because the father cannot pass an X to his son. The Y must come from the father.

195. **C is correct.** If the mother's father was not colorblind, then he gave her one dominant X. If the girl is colorblind, then she inherited a recessive X from her mom. Thus the mother is heterozygous.

196. **C is correct.** Since siblings are more likely to have recessive alleles at the same locus, recessive diseases are more likely.

Passage III

197. **B is correct.** The Law of Independent Assortment states that alleles for different traits will sort independently of each other. This is only true if the traits exist on separate chromosomes.

198. **A is correct.** Any heterozygotes would have produced both phenotypes when self fertilized. Bb x Bb. When they produced the phenotype opposite to their own, they were removed from the population. Eventually, only homozygotes remained.

199. **B is correct.** This is the Mendelian ratio. BB x bb

200. **B is correct.** bb x bb cannot produce anything but bb. A wrinkle phenotype might be Bb or BB.

201. **A is correct.** The law of segregation states that homozygous alleles segregate independently and do not blend, but show complete dominance.

202. **D is correct.** You can do a Punnett square to figure this out, but why would one genotype be more likely than another? By the way, the phenotypic ratio is 9, 3, 3, 1.

Stand Alones

203. **D is correct.** We are looking for a habitat that does not change very often. One in which a species would have to make efficient use of the resources in order to survive. *r* strategists reproduce quickly to take advantage of rapidly changing, short-lived habitats. In longer lived habitats, *K* strategists gain the advantage. The first three answer choices are short-lived habitats.

204. **A is correct.** Members of the same wild species can <u>typically</u> reproduce fertile offspring. There are many exceptions to this rule, but the MCAT tests for the rules, not the exceptions. A wolf and a dog are different species. They can breed and produce fertile offspring. There are cabbage, fish, tobacco plants, and millions of other examples. The rule is "separate species don't <u>normally</u> reproduce fertile offspring in the wild due to a variety of factors."

205. **C is correct.** There is a 50% probability that each girl will be colorblind. This means that there is a 25% probability that both will be colorblind and 25% that neither will be colorblind, leaving 50% that either one or the other might. The boys are irrelevant. We want all situations where at least one is colorblind. 50 + 25 = 75.

206. **C is correct.** If the gene is actively selected against in one form, it must be actively selected for in another form or it will be eliminated from the population.

207. **B is correct.** Density independent factors alter birth, death, or migration rates without having the effects influenced by population density. A storm destroying one half of an island affects all species in the same proportions regardless of density. This is density independent.

ANSWERS & EXPLANATIONS

FOR

QUESTIONS IN THE LECTURES

ANSWERS TO LECTURE QUESTIONS

Lecture 1	Lecture 2	Lecture 3	Lecture 4	Lecture 5	Lecture 6	Lecture 7	Lecture 8	Lecture 9
1. B	25. C	49. A	73. D	97. D	121. C	145. D	169. D	193. B
2. B	26. C	50. D	74. B	98. C	122. D	146. C	170. C	194. A
3. A	27. D	51. D	75. A	99. B	123. D	147. A	171. C	195. C
4. B	28. C	52. A	76. A	100. C	124. A	148. A	172. D	196. B
5. C	29. A	53. D	77. B	101. D	125. A	149. D	173. B	197. A
6. D	30. D	54. A	78. D	102. B	126. D	150. A	174. C	198. B
7. C	31. B	55. D	79. A	103. B	127. A	151. A	175. A	199. D
8. A	32. D	56. C	80. A	104. B	128. C	152. D	176. D	200. C
9. B	33. B	57. D	81. B	105. C	129. B	153. B	177. B	201. B
10. C	34. C	58. C	82. D	106. C	130. D	154. C	178. C	202. C
11. C	35. D	59. B	83. A	107. D	131. D	155. A	179. A	203. D
12. C	36. C	60. B	84. D	108. A	132. A	156. B	180. C	204. C
13. B	37. D	61. D	85. A	109. D	133. B	157. C	181. A	205. C
14. A	38. B	62. B	86. B	110. A	134. B	158. B	182. D	206. A
15. D	39. B	63. C	87. C	111. C	135. B	159. A	183. C	207. D
16. C	40. D	64. B	88. B	112. D	136. A	160. D	184. A	208. D
17. D	41. B	65. D	89. B	113. C	137. C	161. C	185. C	209. B
18. B	42. B	66. D	90. A	114. A	138. D	162. B	186. A	210. C
19. B	43. A	67. D	91. A	115. A	139. B	163. D	187. C	211. D
20. C	44. A	68. B	92. C	116. C	140. D	164. D	188. A	212. C
21. C	45. D	69. A	93. D	117. D	141. C	165. D	189. C	213. B
22. D	46. C	70. C	94. A	118. C	142. B	166. A	190. A	214. C
23. B	47. C	71. B	95. C	119. C	143. B	167. D	191. B	215. C
24. C	48. B	72. B	96. A	120. D	144. A	168. C	192. B	216. B

EXPLANATIONS TO QUESTIONS IN LECTURE 1

1. **B is correct.** Answers A and C are anabolic reactions. Hydrolysis is used to break down triglycerides, proteins, carbohydrates, and is involved in nucleotide catabolism.

2. **B is correct.** DNA is a nucleotide polymer. A nucleotide is a ribose sugar, a phosphate group, and a nitrogenous base. The nucleotides in DNA are held together by phosphodiester bonds.

3. **A is correct.** Plants store carbohydrates as starch. Animals store carbohydrates as glycogen. Glucose is not a polymer. Cellulose is found in plant cell walls and is not digestible by humans.

4. **B is correct.** This question simply requires that you recognize that protein is the only major nutrient containing nitrogen.

5. **C is correct.** The bending of the polypeptide chain is the tertiary structure of a protein.

6. **D is correct.** Fats are a more efficient form of energy storage than carbohydrates and proteins. The phospholipid bilayer membrane is a fatty component of cell structure. Fats such as prostaglandins behave as hormones.

7. **C is correct.** DNA is double stranded with A, C, G, and T, while RNA is single stranded with uracil (U) replacing T. The 'D' in DNA stands for deoxy-, meaning that DNA lacks a hydroxyl group possessed by RNA at its second pentose carbon atom.

8. **A is correct.** Both alpha and beta linkages in polysaccharides are hydrolyzed by adding water. The question asks for a reactant; choices C and D are enzyme catalysts which are never reactants.

---

9. **B is correct.** Enzymes function by binding the substrates on their surfaces in the correct orientation to lower the energy of activation. The change in free energy for the reaction, ΔG, is the difference in energy between reactant and products and is not changed by enzymes.

10. **C is correct.** Decreasing the temperature always decreases the rate of any reaction. Lowering the concentration of a substrate will only lower the rate of an enzymatic reaction if the enzyme is not saturated. Adding a noncompetitive inhibitor will definitely lower the rate of a reaction because it lowers V_{max}. Changing the pH will increase or decrease the rate of an enzymatic reaction depending upon the optimal pH.

11. **C is correct.** A high temperature would denature the enzyme.

12. **C is correct.** Feedback inhibition works by inhibiting enzyme activity and preventing the build up and waste of excess nutrients. Nonenzymatic feedback mechanisms also exist, like the action potential in the neuron.

13. **B is correct.** Noncompetitive inhibition changes the configuration of the enzyme.

14. **A is correct.** The implanted, fertilized embryo produces human chorionic gonadotropin, which stimulates the corpus luteum to continue producing progesterone in a positive feedback mechanism; feedback enhancement is made up.

15. **D is correct.** Peptidases that function in the stomach work at a low pH. Once chyme enters the small intestine, it encounters an alkaline environment, meaning high pH or low hydrogen ion concentration.

16. **C is correct.** A competitive inhibitor may be overcome by increasing the concentration of substrate.

---

17. **D is correct.** Oxygen accepts the electrons (along with protons) to form water.

18. **B is correct.** The Krebs cycle occurs within the mitochondrial matrix in all eukaryotic cells.

19. **B is correct.** As electrons flow, the carriers pass along one or two electrons, and are reduced (gain electrons) then oxidized (lose electrons) until the last carrier donates electrons to oxygen.

20. **C is correct.** The electrons from NADH drive protons outward across the inner mitochondrial membrane.

21. **C is correct.** Glycolysis occurs in aerobic and anaerobic respiration.

22. **D is correct.** ATP is a product. Two ATPs enter the reaction to "prime the pump", and four ATPs are produced. Glucose is a reactant and pyruvate is a product. Oxygen plays no role in glycolysis.

23. **B is correct.** The process of fermentation includes glycolysis, which produces two ATPs.

24. **C is correct.** This answer can most easily be found by process of elimination. Choice A is incorrect because ATP synthase is on the *inner* mitochondrial membrane. Choice B and D are poor answers because they mention the specific processes, Glycolysis and Krebs cycle, with which you should be familiar. A change in these processes would indicate a completely different process. Choice C, on the other hand, refers to membrane transport in a more general way allowing for the possibility that a specific mechanism of transport may differ in heart and liver cells.

EXPLANATIONS TO QUESTIONS IN LECTURE 2

25. **C is correct.** Since A always binds with T and G always binds with C, both the ratio of A/T and the ratio of G/C equal one.

26. **C is correct.** The introns (intervening sequences) are removed during posttranscriptional modification.

27. **D is correct.** This question requires no knowledge of PCR. It requires only that you know that a DNA polymerase replicates from 5′ to 3′, and that you know the complementary bases. (Complementary bases will be covered in this the next section of this lecture. Since DNA is replicated from 5′ to 3′, the primer must be the complement of the 3′ end of the DNA fragment. In other words, the DNA polymerase can only read from 3′ to 5′, so it must start at the 3′ end of the DNA fragment. The complement of the 3′ end of the DNA fragment is answer choice D. (By the way, the primer does not have to start exactly at the end of a DNA fragment in PCR, and a primer is longer than 3 nucleotides.)

28. **C is correct.** DNA replication is semiconservative, which means that both strands are replicated, and each old strand is combined with a new strand.

29. **A is correct.** Introns are removed from the primary transcript during posttranscriptional processing. The number of nucleotides in the mature mRNA would have to be less than the number of base pairs of the gene.

30. **D is correct.** Note that the question asks about complementary strands, which are the two strands in a double strand of DNA.

31. **B is correct.** You should know that mRNA leaves the nucleus in its finished form and that the process of RNA production is called transcription.

32. **D is correct.** Electrophoresis uses an electrolytic cell with a positively charged anode and negatively charged cathode. The phosphate group of the DNA fragment gives it a negative charge that is attracted to the positively charged anode.

33. **B is correct.** The start codon is AUG. mRNA is translated 5′→3′. (Note: You did not need to know that to answer this question correctly.) We are looking for AUG 5′→3′. Only A, B and C have an AUG sequence. However, if there are three codons in C, they must be: AAU, GCG, and GAC. The three in A must be: GAU, GCC, and GGA.

34. **C is correct.** Translation does not take place within the nucleus.

35. **D is correct.** There are 4^3 possible different codons. There are more codons than amino acids (used in proteins). This means that any amino acid could have several codons. The genetic code is evolutionarily very old, and almost universal. Only a few species use a slightly different genetic code.

36. **C is correct.** The ribosome is made in the nucleolus from rRNA and protein. It does not have a membrane.

37. **D is correct.** The complementary sequence to 5'-AUG-3' is 5'-CAU-3'. Only D contains this sequence in any order. Remember, thymine is only found in DNA, not RNA, so B and C must be wrong.

38. **B is correct.** Only choice B is both true and concerns translation. Ribosomes contain two subunits for both eukaryotes and prokaryotes, so choice A is incorrect. Prokaryotes do contain ribosomes, so choice C is wrong. Translation does not concern DNA, so choice D is incorrect.

39. **B is correct.** The P stands for peptidyl site, where the growing peptide chain attaches to the tRNA.

40. **D is correct.** Signal peptides attach to SRPs to direct the ribosome to attach to a membrane such as the endoplasmic reticulum. The signal peptide is usually removed during translation.

---

41. **B is correct.** A primary spermatocyte has finished the S stage of interphase but not the first meiotic division. Thus, it has 46 chromosomes.

42. **B is correct.** Replication takes place only during the synthesis phase.

43. **A is correct.** In normal meiosis, the only change in the nucleotide sequence of the third chromosome will occur during crossing over. Crossing over occurs in prophase I.

44. **A is correct.** Only germ cells undergo meiosis.

45. **D is correct.** In metaphase I we see tetrads.

46. **C is correct.** Centrioles migrate in prophase of mitosis. Chromosomes align in metaphase; centromeres split in anaphase; cytokinesis usually occurs during telophase.

47. **C is correct.** In prophase I, a tetrad will form and genetic recombination will occur; a spindle apparatus will always form; BUT chromosomal migration describes anaphase.

48. **B is correct.** The life cycle of all oocytes is arrested at the primary oocyte stage until puberty.

EXPLANATIONS TO QUESTIONS IN LECTURE 3

49. **A is correct.** A retrovirus contains RNA which is reverse transcribed to DNA and then incorporated into the host cell genome.

50. **D is correct.** A virus cannot contain both DNA and RNA. Many viruses contain proteins.

51. **D is correct.** The first step in the infection of a host is attachment of the phage tail to a specific receptor on the host cell membrane. The capsid on the bacteriophage does not enter the host cell.

52. **A is correct.** Animal viruses attach by recognizing a receptor protein and entering through endocytosis.

53. **D is correct.** Viruses are not living and do not carry out any type of respiration. They require no nutrients, using energy from their host cell. They cannot reproduce inside nonliving organic matter. Viruses reproduce at the expense of a host. Thus, they most closely resemble parasites.

54. **A is correct.** In the lysogenic cycle of viral infection, a cell harbors inactive viral DNA in its genome.

55. **D is correct.** A bacteriophage has a tail and fibers.

56. **C is correct.** Viruses do not carry ribosomes.

---

57. **D is correct.** Prokaryotes have a cell wall that contains peptidoglycan, ribosomes, and a plasma membrane without cholesterol.

58. **C is correct.** Because DNA is acquired directly from the medium, this is transformation. Transduction is the transfer of DNA via a virus. There is no sexual reproduction in bacteria. Conjugation occurs between two bacteria.

59. **B is correct.** Transduction is the transfer of DNA via a virus.

60. **B is correct.** You should arrive at this answer by process of elimination. You should know that gram negative bacteria do not retain gram stain, so 'A' is wrong. You should know that the membrane is made from phospholipids, so C is wrong. (Remember, archaebacteria are not on the MCAT.) D is wrong because fimbriae allow a bacterium to hold to solid objects. Finally, you should know that a bacterium with an outer membrane is gram negative and protected against certain antibiotics such as penicillin.

61. **D is correct.** The exponential growth in bacteria is due to binary fission—asexual reproduction.

62. **B is correct.** Bacilli are rod-shaped; spirilli are rigid helixes; spirochetes are not rigid; AND cocci are round.

63. **C is correct.** Although this is a simple question, it is a reminder that transduction, transformation, and conjugation are not methods of reproduction in bacteria. They are methods of genetic recombination, which is associated with sexual reproduction in eukaryotics, but is not necessarily associated with reproduction in prokaryotes.

64. **B is correct.** Bacterial plasma membranes are a phospholipid bilayer. Bacteria do not have a nucleus. Ribosomes do not contain peptidoglycans. Bacterial cell walls are made from peptidoglycan.

---

65. **D is correct.** Fungi are unique because they are immotile and have a cell wall (like most plants), but are heterotrophic and not photosynthetic (like most animals).

66. **D is correct.** Hyphae are haploid and lengthen through mitosis.

67. **D is correct.** Fungi are saprophytic. Saprophytes are organisms that break down the dead remains of living organisms.

68. **B is correct.** Fungi are heterotrophs, not autotrophs.

69. **A is correct.** Haploid spores can form and spread faster and more efficiently than diploid zygotes because they don't undergo meiosis.

70. **C is correct.** Because fungus is more like human cells, drugs that attack fungi are more likely to affect human cells.

71. **B is correct.** Similar to the plant kingdom the fungi kingdom is divided into divisions.

72. **B is correct.** Fungi are exodigesters. They put enzymes into their food while it is outside their bodies and then absorb the nutrients. Although dead matter is more susceptible to fungal attack, fungi may attack living or dead matter. Meiosis is associated with sexual reproduction. Yeast is an example of a facultative anaerobe.

EXPLANATIONS TO QUESTIONS IN LECTURE 4

73. **D is correct.** The flagella of bacteria are made from the protein flagellin.

74. **B is correct.** The nucleolus is the site of rRNA transcription not translation. It is not membrane bound and should not be confused with the nucleoid of prokaryotes.

75. **A is correct.** Anytime a compound moves against its electrochemical gradient across a membrane, it is active transport. The sodium electrochemical gradient was established by the expenditure of ATP, making this secondary active transport.

76. **A is correct.** We are looking for the cell that is most active in detoxification, one of the jobs of smooth ER. That would be the liver.

77. **B is correct.** I and II are true, but desmosomes are anchored to the cytoskeleton and are stronger than tight junctions.

78. **D is correct.** Ribosomes are made of RNA and protein. They do not have a phospholipid bilayer.

79. **A is correct.** The nucleus runs the cell and makes nucleic acids; the Golgi body packages materials for transport. The rough endoplasmic reticulum makes proteins for use outside the cell. The smooth endoplasmic reticulum helps to detoxify alcohol in the liver.

80. **A is correct.** The hydrolytic enzymes of lysosomes are activated by a low pH achieved by pumping protons into the interior.

---

81. **B is correct.** A signal is typically transmitted to the dendrites to the cell body and then down the axon; however, synapses are found all along the neuron and a signal may begin anywhere on the neuron. Although an action potential moves in all directions along an axon, the cell body and dendrites do not normally contain enough sodium channels to conduct the action potential for any length.

82. **D is correct.** This question is testing your knowledge of an action potential. The major ions involved in the action potential are sodium and potassium. Blocking sodium channels is the only way given that would block an action potential.

83. **A is correct.** The sodium/potassium pump moves potassium inside the membrane. Potassium is positively charged making the inside of the membrane more positive. The resting potential is measured with respect to the inside.

84. **D is correct.** Acetylcholinesterase is an enzyme that degrades acetylcholine. You should gather this from the name. If this enzyme is inhibited, then acetylcholine will not be catabolized as quickly, and it will bind and release repeatedly with postsynaptic receptors.

85. **A is correct.** White matter is composed primarily of myelinated axons.

86. **B is correct.** This is a knowledge based question. You should know this term.

87. **C is correct.** A negative potential is created inside the cell so excess positive charge is pumped out of the cell.

88. **B is correct.** Invertebrates do not have myelinated axons to accelerate nervous impulse transmission. Instead, they rely upon increased size. Vertebrata is a subphylum of Chordata which is characterized by a dorsal nerve chord at some point in their development.

---

89. **B is correct.** The cerebellum controls finely coordinated muscular movements, such as those that occur during a dance routine. Involuntary breathing movements are controlled by the medulla oblongata. The knee-jerk reflex is governed by the spinal cord.

90. **A is correct.** Every type of synapse in the peripheral nervous system uses acetylcholine as its neurotransmitter except the second (the neuroeffector) synapse in the sympathetic nervous system. You may not have known what a neuroeffector synapse was, but you should have been able to reason that it is an end-organ synapse. An effector is an organ or a muscle, something that responds to neural innervation by making something happen in the body.

91. **A is correct.** Parasympathetic stimulation results in "rest and digest" responses, or responses that are not involved in immediate survival or stress. B is a sympathetic response, as is C. D is mediated by skeletal muscles, which do not receive autonomic innervation.

92. **C is the right answer.** Pressure waves, or sound, are converted to neural signals by hair cells in the organ of Corti in the cochlea.

93. **D is correct.** In order to prevent conflicting contractions by antagonistic muscle groups, reflexes will often cause one muscle group to contract while it sends an inhibitory signal to its antagonistic muscle group. Motor neurons exit ventrally from the spinal cord, not dorsally, so A is out. Reflex arcs (at least somatic ones) are usually confined to the spinal cord; they do not require fine control by the cerebral cortex. This eliminates B. Reflex arcs may be integrated by an interneuron in the spinal cord. C is out as well.

94. **A is correct.** The central nervous system is comprised of the brain and spinal cord. An effector is organ or tissue affected by a nervous impulse.

95. **C is correct.** This is a knowledge based question. The cerebrum is also called the cerebral cortex.

96. **A is correct.** The question describes a simple reflex arc which does not involve neurons in the brain.

EXPLANATIONS TO QUESTIONS IN LECTURE 5

97. **D is correct.** Aldosterone, as we can tell by its name, is a steroid (any hormone whose name ends in "sterone" or something similar is a steroid). This allows us to eliminate choice A, because steroid hormones do not need cell membrane receptors or second-messenger systems. They simply diffuse across the cell membrane. We can eliminate B because the adrenal cortex is aldosterone's source, not its target tissue. C describes events at a synapse. Aldosterone actually exerts its effect by doing what D says, increasing the production of sodium-potassium pump proteins.

98. **C is correct.** This is a negative feedback question. If another source of aldosterone exists in the body besides the adrenal cortex, negative feedback (through the renin-angiotensin system and increased blood pressure) would suppress the level of aldosterone secreted by the adrenal cortex. A is out because the levels of renin in the blood would decrease, not increase; aldosterone release would increase blood pressure, and renin is released in response to low blood pressure. Oxytocin plays no role in blood pressure (vasopressin does) and would not be affected by this tumor. Now, to choose between C and D: we know that aldosterone from the adrenal cortex would respond to negative feedback, but we aren't sure whether the tumor would. At this point, we know enough to go with C. If we're sure C is a correct response, that must mean D is an incorrect response, and we can eliminate it. In fact, this is a good choice because normally, hormone-secreting tumors will not respond to negative feedback.

99. **B is correct.** All hormones bind to a protein receptor, whether at the cell membrane, in the cytoplasm, or in the nucleus of the cell. Steroids and thyroxine require a transport protein to dissolve in the aqueous solution of the blood. Steroids are derived from cholesterol, not protein precursors.

100. **C is correct.** Acetylcholine acts through a second messenger system, and is not a second messenger itself.

101. **D is correct.** Steroids are lipid soluble. Different steroids may have different target cells. For instance, estrogens are very selective while testosterone affects every, or nearly every, cell in the body. Steroids act at the transcription level in the nucleus, and are synthesized by the smooth endoplasmic reticulum.

102. **B is correct.** Exocrine function refers to enzyme delivery through a duct.

103. **B is correct.** Steroids act at the level of transcription by regulating the amount of mRNA transcribed.

104. **B is correct.** You should know that T_3 and T_4 (thyroxine) production are controlled by a negative feedback mechanism involving TSH (thyroid stimulating hormone) from the anterior pituitary; parathyroid hormone production is not be affected by thyroxine levels.

---

105. **C is correct.** Epinephrine release leads to "fight or flight" responses, as does sympathetic stimulation. A is out because insulin causes cells to take up glucose. It is not involved in "fight or flight" responses. B is out because acetylcholine is a neurotransmitter; it has few, if any, known hormonal actions. D is out because aldosterone is involved in sodium reabsorption by the kidney; it has no role in "fight or flight" responses.

106. **C is correct.** The nervous and endocrine systems are, in general, the two systems that respond to changes in the environment. In general, the endocrine system's responses are slower to occur but last longer.

107. **D is correct.** The only important thing to recognize from the question is that high insulin levels exist. Then go to the basics; insulin decreases blood glucose.

108. **A is correct.** This is an important distinction to be made. The hormones of the posterior pituitary are synthesized in the bodies of neurons in the hypothalamus, and transported down the axons of these nerves to the posterior pituitary.

109. **D is correct.** Calcitonin builds bone mass. Menopause contributes to osteoporosis by reducing estrogen levels leading to diminished osteoblastic activity. You are not required to know this, and may have had difficulty in eliminating this answer. Instead, you should answer this question by realizing that D was the exception.

110. **A is correct.** Thyroxine (T_4) is produced by the thyroid gland.

111. **C is correct.** Glucagon increases blood sugar, a good thing if you are running a marathon. An increased heart rate and sympathetic blood shunting might similarly be expected in someone who had just run 25 miles.

112. **D is correct.** Parathyroid hormone stimulates osteoclast (bone resorbtion) activity. It also works in the kidney to slow calcium lost in urine. It controls blood calcium levels via these two mechanisms.

---

113. **C is correct.** A looks good (testosterone does stimulate the testes to descend) until you notice that we're dealing with a physically mature male. The testes normally descend during late fetal development. B is out because increased testosterone would cause puberty to occur early, and would not change the timing of puberty if it's already happened (we are, after all, dealing with a physically mature male). D is out because we don't know of any direct mechanism by which testosterone increases body temperature.

114. **A is correct.** Increased secretion of estrogen sets off the luteal surge, which involves increased secretion of LH and leads to ovulation.

115. **A is correct.** The epididymis is where the sperm goes to mature and be stored until ejaculation. Testosterone is secreted by the seminiferous tubules.

116. **C is correct.** Decreased progesterone secretion results from the degeneration of the corpus luteum, which occurs because fertilization of the egg and implantation didn't happen. A is out because thickening of the endometrial lining occurs while estrogen and progesterone levels are high, not while progesterone secretion is decreasing. B is out because increased estrogen secretion causes the luteal surge, and because the luteal surge occurs earlier in the cycle. D is out because, while the flow phase does follow decreased progesterone secretion, it does not occur as a result of increased estrogen secretion.

117. **D is correct.** The layer of cilia along the inner lining of the Fallopian tubes serves to help the egg cell move towards the uterus, where it will implant if it has been fertilized. (Fertilization usually happens in the Fallopian tubes.) A describes what the ciliary lining in the respiratory tract does. B may sound good, but the Fallopian tubes are far enough away from the external environment that protection from its temperature fluctuations is not an issue. C would seem to gum up the whole "continuation of the species" plan. It's not a good answer.

118. **C is correct.** The adrenal cortex makes many other steroid based hormones, as well as testosterone.

119. **C is correct.** Mammalian eggs undergo holoblastic cleavage where division occurs throughout the whole egg. At first glance, this question appears to ask for somewhat obscure knowledge about meroblastic cleavage. However, you should be able to eliminate A, B, and D quite easily as being part of human embryonic cleavage, so it is unnecessary to know meroblastic or holoblastic cleavage.

120. **D is correct.** Generally, the inner lining of the respiratory and digestive tracts, and associated organs, come from the endoderm. The skin, hair, nails, eyes and central nervous system come from ectoderm. Everything else comes from the mesoderm. The gastrula is not a germ layer.

Lecture Question Expls.

EXPLANATIONS TO QUESTIONS IN LECTURE 6

121. **C is correct.** Pepsin, whose optimum pH is around 2.0, denatures in the environment of the small intestine, whose pH is between 6 and 7. A is wrong because pepsin isn't working at all in the small intestine; it won't be working synergistically with trypsin. B is wrong because pepsinogen is activated in the stomach by low pH. D is wrong because pepsin is a catalyst, which makes it a protein; amylase digests starch.

122. **D is correct.** The best answer is D because only D is both true and reveals a benefit for enzymes to be inactive while in the pancreas. A may seem logical but would only apply to lipase. B is false. C is true but is not an adequate explanation.

123. **D is correct.** The stomach doesn't digest carbohydrates. If stomach acid secretion is obstructed then: A) fewer bacteria will be killed in the stomach; B) less pepsinogen will be activated; C) the pH will rise.

124. **A is correct.** All macronutrients are digested through hydrolysis, or the breaking of bonds by adding water.

125. **A is correct.** The large intestine absorbs water. Fat is digested in the small intestine. Urea is secreted by the kidney.

126. **D is correct.** Amylases digest sugars; lipases digest fats; and proteases digest proteins.

127. **A is correct.** Pancreatic exocrine function includes enzymes made in the pancreas and secreted through a duct. Bile is not an enzyme. It is made in the liver and stored in the gallbladder.

128. **C is correct.** Most chemical digestion occurs in the first part of the small intestine, the duodenum.

---

129. **B is correct.** Parietal cells secrete HCl. Goblet cells secrete mucus. Chief cells secrete pepsin. G cells secrete gastrin into the blood.

130. **D is correct.** Gluconeogenesis is the production of glycogen from noncarbohydrate precursors. This function is performed mainly in the liver. Glycolysis can be performed by any cell. Fat storage takes place in adipocytes. Protein degradation occurs in all cells.

131. **D is correct.** Most fat digestates enter the lymph as chylomicrons via lacteals. Smooth endoplasmic reticulum synthesizes triglycerides.

132. **A is correct.** 'Essential' means that the body cannot synthesize them. Nonessential amino acids are synthesized by the liver. Amino acids are absorbed by facilitated and active transport. Urea is the end product of amino acid deamination in the liver.

133. **B is correct.** Glucose is absorbed in a symport mechanism with sodium. Sodium is an electrolyte. The absorption of glucose increases the absorption of sodium. Glucose is not an electrolyte. Glucose does not stimulate the secretion of amylase.

134. **B is correct.** Insulin decreases blood sugar levels in several ways. One of the ways is by inhibiting glycogenolysis.

135. **B is correct.** This is a knowledge based question. You should know that blood is an aqueous solution.

136 **A is correct.** Macromolecules are broken down into their basic nutrients.

---

137. **C is correct.** The only process available for the removal of wastes by the Bowman's capsule is diffusion, aided by the hydrostatic pressure of the blood.

138. **D is correct.** Glucose is normally completely reabsorbed from the filtrate and thus does not appear in the urine. When glucose does appear in the urine, the glucose transporters in the PCT (not in the loop of Henle, as stated in answer choice A) are unable to reabsorb all of the glucose from the filtrate. C is wrong because the proximal tubule does not secrete glucose.

139. **B is correct.** The purpose of the brush border is to increase the surface area available to reabsorb solutes from the filtrate. The brush border is made from villi, not cilia, and so has little or no bearing on the direction or rate of fluid movement.

140. **D is correct.** Renin secretion catalyzes the conversion of angiotensin I to angiotensin II, which increases the secretion of aldosterone. If renin is blocked, then aldosterone cannot cause increased synthesis of sodium absorbing proteins, and sodium absorption decreases. Without renin secretion, production of angiotensin II would decrease. Blood pressure would decrease, not increase; A is wrong. Platelets are irrelevant; B is wrong.

141. **C is correct.** This is the correct order of structures.

142. **B is correct.** Vasopressin is antidiuretic hormone increasing water retention. ADH levels will be rise in response to dehydration. Aldosterone is a mineral corticoid released by the adrenal cortex in response to low blood pressure. In a severely dehydrated person, blood volume would be low, likely resulting in diminished blood pressure. Aldosterone levels will rise in response to the low blood pressure.

143. **B is correct.** High blood pressure would result in more fluid being forced into Bowman's capsule.

144. **A is correct.** The loop of Henle concentrates the medulla via a net loss of solute to the medulla. This process is critical to the function of other parts of the nephron; a medulla with a high concentration of solute allows for the passive absorption of water from the filtrate in other areas of the nephron.

EXPLANATIONS TO QUESTIONS IN LECTURE 7

145. **D is correct.** The atrioventricular node, which sits at the junction between the atria and the ventricles, pauses for a fraction of a second before passing an impulse to the ventricles.

146. **C is correct.** Stroke volume must be the same for both ventricles. If it weren't, we'd have a never-ending backlog of blood in one or the other circulations, ending with the faster circulation running dry. To keep the whole system running smoothly, both halves of the circulation must pump the same quantity of blood with each stroke.

147. **A is correct.** The cardiac action potential is spread from one cardiac muscle cell to the next via ion movement through gap junctions.

148. **A is correct.** Less oxygenated blood will reach the systemic system because some oxygenated blood will be shunted from the aorta to the lower pressure pulmonary arteries. The pulmonary circulation will carry blood that is more oxygenated than normal, since highly oxygenated blood from the aorta is mixing with deoxygenated blood on its way to the lungs. The entire heart will pump harder in order to compensate by with more blood to the tissues.

149. **D is correct.** Oxygenated blood returning from the lungs feeds into the heart at the left atrium. From there, it flows into the left ventricle, which pumps it to the systemic circulation. The right atria and the right ventricle pump deoxygenated blood to the lungs.

150. **A is correct.** Blood loss is likely to be more rapid during arterial bleeding due the greater blood pressure in the arteries.

151. **A is correct.** Don't let the not-so-subtle physics reference fool you. Bernoulli's equation, which would indicate a greater pressure at the greater cross-sectional area, doesn't work here. You should memorize that blood pressure in a human is greatest in the aorta and drops until the blood gets back to the heart.

152. **D is correct.** The capillaries are one cell thick and blood moves slowly to allow for efficient oxygen exchange.

---

153. **B is correct.** Hyperventilation results in loss of CO_2, leading to lower concentrations of carbonic acid in the blood, and an increase in pH. Hypoventilation would result in the reverse. Breathing into a bag would increase the CO_2 content of the air and lead to acidosis. <u>Excess</u> aldosterone may lead to metabolic alkalosis due to hydrogen ion exchange in the kidney.

154. **C is correct.** Carbonic anhydrase is a catalyst. Catalysts increase the rate of a reaction. If the catalyst is inhibited, the rate decreases. Since the reaction moves in one direction in the lungs, and the opposite direction in the tissues, answer choice A is ambiguous. Unless one believes that a carbonic anhydrase inhibitor will affect transcription or degradation of hemoglobin, choices B and D are equivalent. If one were true, the other should also be true.

155. **A is correct.** Cellular respiration produces carbon dioxide, which, in turn, lowers blood pH. During heavy exercise, capillaries dilate in order to deliver more oxygen to the active tissues. Nitrogen is irrelevant to respiration.

156. **B is correct.** The increased hemoglobin concentration in the blood after reinjection increases the blood's ability to deliver oxygen to the tissues, often a limiting factor in endurance competitions. A is wrong because, while this may happen, it is not the primary benefit of blood doping. B is a much better answer. C is wrong because, since we are only reinjecting blood cells, we are not increasing the body's hydration status. D is wrong because, while blood is less viscous with fewer red blood cells, we are removing whole blood, not just red blood cells, so we won't be changing the blood's viscosity.

157. **C is correct.** Carbon dioxide is produced in the tissues. It is transported by the blood to the lungs, where it is expelled by diffusing into the alveoli. Since the concentration gradient carries CO_2 into the capillaries from the tissues and from the blood into the alveoli, we can reason that there is a higher concentration of CO_2 in the tissues than in the alveoli. A is wrong because blood CO_2 concentration will not change in the veins; it has nowhere to go. B is wrong because CO_2 is expelled into the lungs at the pulmonary capillaries, so CO_2 that was present in the pulmonary arteries (before the capillaries) will largely be gone in the pulmonary veins (after the capillaries). D describes the opposite of the concentration gradient that actually exists for CO_2 between the systemic tissues and the systemic capillaries.

158. **B is correct.** The person would need increased vascularity to deliver more blood to the tissues because the blood would carry less oxygen.

159. **A is correct.** Constricted air passages is the clue. The bronchioles are surrounded by smooth muscle and small enough to constrict. Cartilage does not constrict, muscle does. The skeletal muscle in the thorax does not constrict the air passages. The alveoli are not part of the air passages.

160. **D is correct.** Heavy exercise manifests increased carbon dioxide production that leads to increased carboxyhemoglobin.

---

161. **C is correct.** Hemolytic anemia can result from abnormalities of the red bloods cells that make them fragile and more susceptible to rupture when they are processed by the spleen. You should know for the test that the spleen destroys old, worn out red blood cells.

162. **B is correct.** "Swollen glands" are actually swollen lymph nodes that are bulging with immune cells gearing up to fight the invasion. A is simply not what's going on here. C is out because an inflammatory response would draw fluid into the inflamed area, not drain it away. D is also just a goofy answer.

163. **D is correct.** Immunoglobulins, or antibodies, are involved in the humoral immune system, or the B cell system. A is out because cytotoxic T cells work in cell mediated immunity. B is out because stomach acid plays a role in the nonspecific innate immunity; humoral immunity is specific and acquired. C is irrelevant.

164. **D is correct.** Antibodies bind to antigens through interactions between the antibody's variable region and the antigen. Antibodies do not phagocytize anything, so A is out. Antibodies are produced by plasma cells, they don't normally bind to them; B is out. Plasma cells are derived from stem cells in the bone marrow. Antibodies will not usually prevent their production. C is out.

165. **D is correct.** Fluid that is picked up by the lymphatic tissues is returned to the circulation at the right and left lymphatic ducts, which feed into veins in the upper portion of the chest.

166. **A is correct.** Type B negative blood carries B antigens and not the Rh factor. It does not carry A antigens. There are no O antigens. Thus type B negative blood makes antibodies that will only attach A antigens and Rh antigens.

167. **D is correct.** Old erythrocytes are destroyed in the spleen and the liver.

168. **C is correct.** The innate immune response does not involve humoral immunity (B-cell) or cell mediated immunity (T-cell). The innate immune system responds to any and every foreign invader with the white blood cells called granulocytes as well as with inflammation and other actions.

EXPLANATIONS TO QUESTIONS IN LECTURE 8

169. **D is correct.** Neither actin (the thin filament) nor myosin (the thick filament) changes its length during a muscular contraction; instead the proportion of actin and myosin overlap increases.

170. **C is correct.** Permanent sequestering of calcium in the sarcoplasmic reticulum would prevent calcium from binding to troponin, which is what causes the conformational change that moves tropomyosin away from the myosin binding sites on actin. Choice A would occur if calcium were present and ATP were not. Loss of ATP would prevent the myosin from releasing from actin. B is incorrect because a resorption of calcium from bone would result in a decrease in bone density, not an increase. D is incorrect because loss of calcium would not cause depolymerization of actin filaments. If this actually occurred, it would pose a serious problem every time calcium was re-sequestered into the SR after the completion of a contraction.

171. **C is correct.** Shivering results from the increase of muscle tone. When muscle tone increases beyond a certain critical point, it creates the familiar indiscriminate muscle activity typical of shivering. While shivering may serve as a warning, it does more than that, and that is not its primary purpose; A is out. B is simply not the case, nor is D.

172. **D is correct.** Muscles cause movement by contracting (eliminating B and C), bringing origin and insertion closer together, usually by moving the insertion. Neurons cause contractions in muscles, not tendons, and the neural signals are not initiated by the muscle. This eliminates A.

173. **B is correct.** Antagonistic muscles move bones in opposite directions relative to a joint. In order to produce movement, one must relax while the other contracts.

174. **C is correct.** All muscles will need more energy and protein if they are being used rigorously, but in humans, mature skeletal muscle cells do not divide.

175. **A is correct.** MCAT doesn't test anatomy. This question is asking if you know that tendons connect muscle to bone. Ligaments connect bone-to-bone. Tendons are not cartilage.

176. **D is correct.** Peristalsis is a function of smooth muscle only. Shivering is an example of temperature regulation by skeletal muscle. Skeletal muscle may assist in venous blood movement and lymph fluid movement.

177. **B is correct.** Gap junctions allow for the spread of the action potential throughout the heart.

178. **C is correct.** Peristalsis is a smooth-muscle activity, and smooth muscle is innervated by the autonomic nervous system. The skeletal muscles are innervated by the somatic nervous system, so interruption of the autonomic nervous system would not affect the knee-jerk reflex or the diaphragm. An action potential in cardiac muscle is conducted from cell to cell by gap junctions.

179. **A is correct.** The heart requires long steady contractions in order to pump blood. We know we want adjacent heart cells to contract at the same time; B is out. We're not concerned about a neuron here; C is out. Sodium flows into the cell, not out. D is out as well.

180. **C is correct.** Smooth muscle contains thick and thin filaments, so it requires calcium to contract.

181. **A is correct.** This question asks you to recognize that smooth muscle and cardiac muscle are involuntary, and then to recognize muscle types of the different structures. Of the muscles listed, only the diaphragm is skeletal muscle.

182. **D is correct.** All muscles contract in response to increased cytosolic calcium concentration.

183. **C is correct.** You should know that the vagus nerve is a parasympathetic nerve. Heart rate is set by the SA node, and the SA node is innervated by the parasympathetic vagus nerve. The pace of the SA node is faster than the normal heart beats, but the parasympathetic vagus nerve tonically slows the contractions of the heart to its resting pace. Without tonic inhibition from the vagus, the heart would normally beat at 100 – 120 beats per minute.

184. **A is correct.** The word dilation here is the give away. Smooth muscle must be relaxing in order to dilate the vessels.

---

185. **C is correct.** Parathyroid hormone increases blood calcium by increasing osteocyte activity, and increasing osteoclast number.

186. **A is correct.** Synovial fluid acts as a lubricant, decreasing friction between the ends of the bones as they move. Bone cells receive circulation to keep them adequately hydrated; they do not need synovial fluid for this purpose. B is out. Synovial fluid persists in adults, after bones have stopped growing, so we know C is wrong. Synovial fluid is found in synovial joints, which allow movement. Its purpose is not to prevent movement. D is out.

187. **C is correct.** Of the tissues listed, only cartilage does not contain nerves.

188. **A is correct.** Ligaments attach bone to bone. Tendons attach bone to muscle.

189. **C is correct.** Bone does not regulate the body or blood temperature. It does store calcium and phosphate, support and protect the body, produce blood cells, and store fat.

190. **A is correct.** Yellow bone marrow is usually found in the medullary cavity of long bones.

191. **B is correct.** Hydroxyapatite is made up of calcium and phosphate in a compound that includes hydroxyl groups as well. You should know that bone acts as a storage place for phosphate and calcium. The hydroxyl you can get from the name.

192. **B is correct.** Spongy bone contains the blood stem cells important for blood cell synthesis. Red blood cell storage is the job of the liver and spleen. Fat storage is in long bones.

EXPLANATIONS TO QUESTIONS IN LECTURE 9

193. **B is correct.** According to the Punnett square, each time they have a boy, there is a 50% chance that he will be color-blind. The chance that both boys are colorblind is the chance of this happening twice, or 0.5^2.

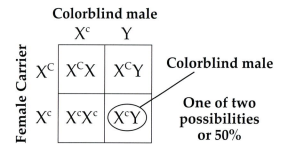

194. **A is correct.** D is a male. Since the mother had the disease, the mother must have been homozygous recessive. The father could not have been a carrier. Thus, female A received an X^h from her mother and an X^H from her father. Since she is only a carrier, it is possible that she passed on only her good X and that all her children are healthy.

195. **C is correct.** From the dihybrid cross Punnett square we have the phenotype ratio 9:3:3:1. Thus, 9 of 16 individuals display both dominant phenotypes.

196. **B is correct.** A dihybrid is heterozygous for both traits. See the dihybrid Punnett square in this chapter and count the dihybrid offspring.

197. **A is correct.** Both girls must carry the trait because they receive their father's recessive chromosome.

198. **B is correct.** A dihybrid cross is when two individuals that are hybrids at two different genes are crossed. i.e. AaBb x AaBb

199. **D is correct.** Sex-linked traits occur on the X or Y chromosomes. The X and Y chromosomes are not homologous.

200. **C is correct.** The woman has two recessive Xs. Her mother gave her one X and her father gave her one X. Her father only had one X to start with, so it must have been recessive and he must have been colorblind. Her son will receive a recessive X from her and a Y from his father so he will be colorblind.

---

201. **B is correct.** First you must recognize that 10% of the gene pool is represented by the recessive allele. Then you must realize that only the homozygous recessives display the recessive phenotype. Now use the binomial theorem to see that homozygous recessives are represented by 0.1^2 of the population.

202. **C is correct.** Catastrophic events will not cause significant genetic drift to a large homogeneous (well-mixed) population. Emigration, selection, and mutation all affect the HW equilibrium.

203. **D is correct.** Only D does not represent a type of isolation (geographic, seasonal, behavioral) which leads to speciation. The birds are migratory, and thus are not geographically isolated. This makes D the best answer choice.

204. **C is correct.** *r*-strategy is more efficient when the prevailing conditions are governed by density independent factors such as harsh environment, short seasons, etc. Commercial predation methods have a similar effect. *r*-strategists are able to better withstand massive predation from man or others, because they produce so many more offspring than they need to continue the species. *K*-strategists fare better when the prevailing conditions are governed by density dependent factors such as limited resources. *K*-strategists can better exploit limited resources by specializing.

205. **C is correct.** A single class encompasses several orders; two different families can belong to the same order.

206. **A is correct.** Remember: Darn King Phillip Came Over For Good Soup. Domain, kingdom, phylum, class, order, family, genus, species. The epithet *Canus lupus* indicates that the wolf is genus *Canus* and species *lupus*. Family is a broader category than either genus or species, so any organism that is of the species lupus, must also be *Canus* and canidae.

207. **D is correct.** This is an application of the binomial theorem under Hardy Weinberg Equilibrium. $p^2 + 2pq + q^2 = 1$ and $p + q = 1$. So, $p^2 = 0.36$. $p = 0.6$. $q = 0.4$. $2pq = 0.48$.

208. **D is correct.** Corn depends upon humans for its survival as a species. Humans ensure its survival. In return, humans are provided with food. Both species benefit from the relationship.

---

209. **B is correct.** Prokaryotes arose at least 3.6 billion years ago.

210. **C is correct.** Vertebrates, a subphylum of Chordata, contain backbones.

211. **D is correct.** Ants are not chordates. Tunicates, sponge-like creatures that attach to the sea floor, are actually chordates.

212. **C is correct.** Life originated in an atmosphere with little or no oxygen.

213. **B is correct.** The taxonomy of *Homo sapiens* is Domain: Eukarya; Kingdom: Animalia; Phylum: Chordata; Subphylum: Vertabrata; Class: Mammalia; Order: Primata; Family: Homididae; Genus: *Homo*; Species: *Sapiens*.

214. **C is correct.** You need to memorize that the taxonomy of humans is animalia, chordata, mammalian, primata, homididae, *Homo sapiens*.

215.	**C is correct.** Urey and Miller demonstrated that organic molecules may be created from inorganic molecules under the primordial earth conditions. Urey-Miller did not prove the existence of life on earth, (you do), nor did they prove that humans have evolved from bacteria, photosynthetic or not.

216.	**B is correct.** If you answered A, you would have to explain what the first living organism ate because he had no one else to eat. Even when there were millions of living organisms, they could not survive off each other because one organism would have to eat many others, and there just wouldn't be enough to go around initially.

Photo Credits

Covers

Front cover, Anatomical Overlays - Man Running Front View: © LindaMarieB/iStockphoto.com

Back cover, T4 bacteriophage infecting bacterium: © Russell Kightley / Science Source

Chapter 1

Pg. 1, Body Builder: © Steve Williams Photo/Getty Images

Pg. 2, Dolomedes fimbriatus on black Water: © Alasdair James /iStockphoto.com

Pg. 3, Overhead shot of food containing Omega 3: © Tooga/Getty Images

Pg. 4, Fat cells, TEM: © Steve Gschmeissner/Photo Researchers, Inc.

Pg. 5, Peanuts: © RedHelga/iStockphoto.com

Pg. 6, Picture symbolizing high-protein diet (meat, fish, vegetables): © Ulrich Kerth/Getty Images

Pg. 9, False-color TEM of collagen fibrils: © J. Gross/ Biozentrum, University of Basel/Photo Researchers, Inc.

Pg. 10, Glucose level blood test: © Alexander Raths/ iStockphoto.com

Pg. 10, Plant Cells (SEM): © Alice J. Belling, colorization by Meredith Carlson/Photo Researchers, Inc.

Pg. 10, Colored SEM of a liver cell (hepatocyte): © Professors Pietro M. Motta & Tomonori Naguro/ Photo Researchers, Inc.

Pg. 11, Front view of Holstein cow, 5 years old, standing: © Alexander Raths/iStockphoto.com

Pg. 11, Insect termite white ant: © defun/iStockphoto.com

Pg. 12, DNA molecule: © Sci-Comm Studios/Science Source

Pg. 13, Human Bone, Microscopic View: © David Scharf/ Getty Images

Pg. 13, Small child: © Jaroslaw Wojcik/iStockphoto.com

Pg. 15, Hexokinase enzyme with glucose: © Kenneth Eward/Biografx/Science Photo Library

Pg. 16, Vitamin supplements: © Sarah Lee/iStockphoto.com

Pg. 18, Pancreas cell, SEM: © Steve Gschmeissner/Photo Researchers, Inc.

Pg. 21, Sugar in a silver spoon: © Homiel/iStockphoto.com

Pg. 22, Streptococcus viridans: © Biophoto Associates/Photo Researchers, Inc.

Pg. 24, Mitochondrion, SEM: © Dr. David Furness, Keele University/Photo Researchers, Inc.

Pg. 26, Hans Krebs, German-Anglo Biochemist: © Science Source/Photo Researchers, Inc.

Chapter 2

Pg. 31, Conceptual computer illustration of the DNA double helix: © David Parker/Science Photo Library

Pg. 43, Southern blotting: © James King-Holmes/Photo Researchers, Inc.

Pg. 47, Col TEM of structural gene operon from E. coli: © Professor Oscar Mille/Photo Researchers, Inc.

Pg. 49, Male white lion: © Tony Camacho/Photo Researchers, Inc.

Pg. 52, Cancer cell: © Quest/Photo Researchers, Inc.

Pg. 54, Chromosomes, computer artwork: © SCIEPRO/ Science Photo Library

Chapter 3

Pg. 65, Bacteriophages: © Science Picture Co/Getty Images

Pg. 66, T4 bacteriophages: © Russell Kightley/Science Source

Pg. 68, Close-up of a cold sore on the lower lip: © Dr. P. Marazzi/Photo Researchers, Inc.

Pg. 70, Gut bacterium: © Hazel Appleton, Health Protection Agency Centre for Infections/Photo Researchers, Inc.

Pg. 71, E. coli bacteria: © Andrew Syred/Photo Researchers, Inc.

Pg. 71, MRSA: resistant Staphylococcus bacteria: © K. Lounatmaa/Photo Researchers, Inc.

Pg. 71, SEM of Treponema pallidum on cultures of cotton-tail rabbit epithelium cells: © Science Source/Photo Researchers, Inc.

Pg. 71, Bacterial capsule: © Dr. Kari Lounatmaa/Photo Researchers, Inc.

Pg. 75, Diffusion: © Charles D. Winters/Photo Researchers, Inc.

Pg. 77, Cholera bacteria: © Juergen Berger/Photo Researchers, Inc.

Pg. 79, Gut bacterium reproducing: © Hazel Appleton, Health Protection Agency Centre for Infections/ Photo Researchers, Inc.

Pg. 82, False color transmission electron micrograph of a thin section through an endospore of Bacillus subtilis: © Dr. Tony Brain/Photo Researchers, Inc.

Pg. 84, Strawberry Gray Mold disease: © Craftvision/ iStockphoto.com

Pg. 84, Trichophyton Interdigitale, or athlete's foot: © Biophoto Associates/Photo Researchers, Inc.

Pg. 84, Athlete's foot: © carroteater/iStockphoto.com

Pg. 84, SEM of Saccharomyces cerevisiae: © SciMAT/Photo Researchers, Inc.

Chapter 4

Pg. 108, Hand and thumb tack: © Dana Kelley Photos

Chapter 5

Pg. 127, Atopic dermatitis: © bravo1954/iStockphoto.com

Pg. 128, Third world healthcare: © Scott Camazine/Photo Researchers, Inc.

Pg. 128, Light micrograph of pancreatic islets of Langerhans, stained with H&E (Hematoxylin and Eosin): © Biophoto Associates/Photo Researchers, Inc.

Pg. 133, Sperm production: © Susumu Nishinaga/Photo Researchers, Inc.

Pg. 135, Sperm cell fertilizing an egg cell, colored ESEM (environmental scanning electron micrograph): © Thierry Berrod, Mona Lisa Production/Photo Researchers, Inc.

Pg. 135, Ovulation: © Profs. P.M. Motta & J. Van Blerkom/ Photo Researchers, Inc.

Pg. 137, An illustration depicting the initial stages of fetal development: © Jim Dowdalls/Photo Researchers, Inc.

Chapter 6

Pg. 143, Stomach lining: © Steve Gschmeissner/Photo Researchers, Inc.

Pg. 144, False-color scanning electron micrograph of a section through the wall of the human duodenum: © David Scharf/Photo Researchers, Inc.

Pg. 145, Pancreas surface: © Susumu Nishinaga/Photo Researchers, Inc.

Pg. 146, Radiologic evaluation of the large intestine often involves performing a barium enema: © Medical Body Scans/Photo Researchers, Inc.

Pg. 149, Baker's Dozen Chocolate Chip Cookies Isolated on White: © Frances Twitty/iStockphoto.com

Pg. 151, The normal surface pattern of the jejunal (small intestine) mucosa: © Biophoto Associates/Photo Researchers, Inc.

Pg. 151, Celiac Disease: © Biophoto Associates/Photo Researchers, Inc.

Pg. 152, Nutrition label: © Brandon Laufenberg/ iStockphoto.com

Chapter 7

Pg. 165, Color enhanced scanning electron micrograph (SEM) of a single red blood cell in a capillary: © Dr. Cecil H. Fox/Photo Researchers, Inc.

Pg. 167, Doctor checking patient blood pressure: © Rudyanto Wijaya/iStockphoto.com

Pg. 170, X-ray of a lung cancer: © muratseyit/iStockphoto.com

Pg. 176, Blood clot: © Steve Gschmeissner/Photo Researchers, Inc.

Pg. 179, Histology - Immune System: Human macrophage ingesting pseudomonas: © David M. Phillips/Photo Researchers, Inc.

Pg. 182, Blood storage: © Tek Image/Photo Researchers, Inc.

Chapter 8

Pg. 185, Michelangelo's David: © irakite/iStockphoto.com

Pg. 186, Light micrograph of a section through human skeletal (striated) muscle: © Manfred Kage/Photo Researchers, Inc.

Pg. 190, Muscular back of a man kneeling: © Max Delson Martins Santos/iStockphoto.com

Pg. 192, Heart muscle: © Manfred Kage/Photo Researchers, Inc.

Pg. 193, Light micrograph of a section of human smooth muscle: © SPL/Photo Researchers, Inc.

Pg. 195, Osteoblast: © CNRI/Photo Researchers, Inc.

Pg. 195, Transmission electron micrograph (TEM) of human bone showing an osteocyte: © Biophoto Associates/ Photo Researchers, Inc.

Pg. 195, Computer-generated image of multi-nucleated osteoclasts etching away trabecular bone in a process called bone resorption: © Gary Carlson/ Photo Researchers, Inc.

Pg. 196, Human skeleton, front view: © red_frog/ iStockphoto.com

Pg. 198, Hyaline cartilage: © Innerspace Imaging/Photo Researchers, Inc.

Pg. 199, Sample of skin from the back of a human hand: © Eye of Science/Photo Researchers, Inc.

INDEX

Symbols

1,25 dihydroxycholecalciferol (DOHCC) 130
2,3-DPG 170
3 to 1 ratio 203, 213
5'→3' 12, 290
5′→ 3′ directionality 33
5′ cap 38
5 stage cycle 188, 189
9+2 93
α-amylase 142
α-cells 129
α-subunit 105
β-cells 129
β-pleated sheet 6, 7

A

absolute refractory period 103
accommodation 103
acetylcholine 19, 107, 108, 110, 118, 124, 143, 168, 189, 236, 277, 293, 294
acetyl CoA 21, 24, 26, 27
acid hydrolases 92
acidosis 155, 172, 297
acinar cells 145
acquired immunity 178, 179
acrosome 133, 135
ACTH 121, 125, 126, 131, 138, 139, 234
actin 93, 177, 186, 188, 191, 258, 299
action potential 95, 100-107, 113, 124, 165, 168, 189, 192-194, 236, 237, 258, 259, 277, 283, 289, 293, 297, 299
activated 18, 36, 37, 82, 85, 91, 104, 129, 133, 143, 145, 148, 176, 189, 199, 237, 240, 242, 279, 293, 296
activators 19, 36, 37
active site 15-17, 20, 188, 272
active transport 10, 76, 92, 98, 104, 149, 157, 249, 258, 292, 296
Adaptive radiation 208
adenine (A) 33
adenohypophysis 125
adenosine phosphate 33, 223
ADH 14, 121, 123, 126, 127, 138, 139, 158, 240, 249, 281, 297
Adipocytes 3, 28, 92, 116
adrenal cortex 122-124, 126, 127, 138, 140, 158, 235, 249, 278, 281, 294, 295, 297
adrenal glands 125, 127
adrenaline 110, 128
adrenal medulla 122, 127, 128, 140, 234
adrenergic 110, 234, 241, 249
Adrenocorticotropic hormone (ACTH) 126
Aerobic respiration 24, 272
agglutinate 179
agonist 185, 186

agranular 92
Agranular leukocytes 176
albumin 8, 152, 155, 176, 248, 280
aldosterone 122-124, 127, 158, 162, 241, 242, 249, 294, 297
Aldosterone 124, 127, 131, 138, 139, 156, 158, 278, 281, 294, 297
alimentary tract 22
allele 204-206, 209, 210, 301
alleles 54, 182, 204, 210, 266, 285
all-or-nothing 103, 107
allosteric activators 19
allosteric inhibitors 19
Allosteric interactions 18
allosteric regulation 19
almost universal 45, 290
alveolar sacs 170
alveoli 169, 170, 173, 255, 298
amino acids 5, 6, 18, 21, 27, 31, 35, 45, 46, 53, 62, 77, 91, 126, 127, 129, 145, 147, 151, 155, 156, 162, 176, 216, 260, 271, 280, 290, 296
aminoacyl site 46
amphipathic 4, 73, 152
amplified 42, 44
ampulla of Vater 145
amylase 142, 145, 147, 148, 156, 247, 279, 296
amylopectin 10
amylose 10
anabolism 21
Anaerobic respiration 22
analogous 97, 137, 208
Anaphase 56, 60, 61, 273
anaphase II 58, 60
androgen 132, 134, 243, 244
aneuploidy 50
angiotensin 158, 249, 250, 280, 294, 297
Animalia 207, 301
anneal 42
anomers 10
ANS 108-110, 116
antagonist 118, 185
anterior cavity 112
anterior pituitary 121, 125, 126, 128, 131, 134, 140, 243, 294
anterior pituitary hormones 121, 125, 126
antibiotics 157, 235, 277, 292
antibodies 8, 43, 77, 90, 176, 179, 180, 182, 184, 228, 244, 252, 253, 265, 275, 281, 282, 285, 298
antibody 43, 179, 180, 242, 252, 253, 281, 282, 298
anticodon 46, 53
Antidiuretic hormone (ADH) 127
antigen 179, 180, 252, 253, 265, 281, 282, 285, 298
antigenic determinant 179
antiparallel 6, 33
antiport 157, 162

antisense 36, 274
antisense strand 36, 274
anus 141
aorta 163, 167, 168, 172, 297
apical 95, 144, 157
apoenzyme 16
apoferritin 155
apoproteins 4, 152
apoptosis 137
aqueous humor 112
Archaea 70, 71, 77, 86, 207
arrector pili 200
arteries 118, 152, 163-168, 170, 172, 173, 193, 234, 279, 297, 298
Arteries 165, 166, 183
arterioles 163-166, 168, 249, 280
Arterioles 165, 183
arthropods 84
ascending colon 141, 146
Ascomycota 84, 85
A site 46, 53
aster 56
astrocytes 106
a tail 69, 133, 211, 214, 291
ATP 10, 12, 16, 19, 21-24, 26-29, 34, 65, 76, 79, 98, 162, 188, 190, 218, 219, 240, 258-261, 271-273, 282, 283, 290, 292, 299
atria 164, 165, 168, 297
atrioventricular node (AV node) 165
auricle 114
autolysis 92
autonomic nervous system 108, 110, 164, 193, 194, 234, 249, 276, 281, 299
autonomic nervous system (ANS) 108
autorhythmic 164
Autotrophs 86
axon 100, 102, 106, 107, 237, 293
axoneme 93
axon hillock 100, 102, 237

B

B12 143, 146, 155
bacilli 71
backward mutation 51
Bacteria 71, 79, 83, 86, 228, 232, 275, 276, 292
bacterial envelope 77
bacteriophage 81, 230, 232, 275, 276, 291
Barr body 205
basal ganglia 111
basal lamina 97
basal metabolic rate 128, 138
base-pair 36, 48, 49
base-pairing 33
Basidiomycota 84, 85
basidiospores 85
basolateral 95, 149, 246
basophils 176
Basophils 177, 179
B-cell immunity 179
behavioral isolation 208
bicarbonate 143, 145, 147, 171

bicarbonate ion 143, 145, 171
bidirectional 34
Big Bang Theory 211
bile 145, 148, 152, 155, 157, 246, 247, 250, 279
bilirubin 145
binary fission 79, 83, 230, 231, 272, 292
binomial theorem 301
bladder 95, 145, 157, 162, 193
blastocyst 136, 137, 242
blastopore 137, 211
blood cell production 195
blood cot 177
Blood filtration 155
Blood reservoir 199
Blood storage 155
Blood types 182
B lymphocytes 179, 180, 183, 184, 252
Bohr shift 170
bolus 142
Bone 185, 195, 197, 201, 259, 260, 262, 283, 284, 300
Bowman's capsule 157, 162, 248, 250, 279, 280, 296, 297
brain 100, 106, 107, 108, 110-114, 118, 125, 129, 137, 234, 249, 294
Braun's lipoprotein 77
bronchi 169, 170
bronchioles 169, 170, 173, 193, 298
Brownian motion 75, 104, 107
brush border 144, 145, 152, 162, 246, 297
budding 85, 87, 92
bulbourethral glands 133
bulk flow 91
bundle of His 165

C

calcitonin 125, 128, 131, 260
Calcitonin 128, 138, 139, 295
calcium 16, 104, 107, 128, 130, 131, 138, 188, 191, 192, 194, 195, 197, 202, 258, 262, 277, 282, 283, 295, 299, 300
callus 200
Calmodulin 18
Calvin cycle 70
cAMP 36, 37, 120, 122, 126, 129, 240, 278
canaliculi 195, 202
canal of Schlemm 112
cancer 52, 137, 170, 180, 225, 273, 274
capacitation 133
capillaries 144, 156, 163, 164, 166-168, 170, 171, 174, 176, 178, 194, 199, 255, 282, 297, 298
Capillaries 165, 166, 170, 173, 183
Capillary 165
capsid 65, 81, 229, 231, 232, 291
CAP site 37
capsule 71, 79, 157, 162, 198, 248, 250, 279, 280, 296, 297
carbamino hemoglobin 172, 173
Carbohydrate metabolism 155, 250
carbohydrates 1, 4, 8-10, 14, 18, 77, 92, 120, 129, 138, 142, 145, 147-149, 151, 155, 271, 289, 296
carbon dioxide pressure 170-173
carbonic anhydrase 171, 173, 298
carboxypolypeptidase 145

carcinogens 52
cardiac muscle 95, 103, 109, 164, 165, 168, 185, 192-194, 234-236, 276, 297, 299
cardiac sphincter 142
carrier 29, 68, 76, 121, 129, 152, 205, 206, 242, 254, 267, 282, 289, 300
carrier proteins 76, 121
Cartilage 185, 198, 201, 259, 298
Cartilaginous joints 198
catabolic 14, 21, 28
catabolism 21, 289
catabolite activator protein 37
catalase 16
catalyst 15, 281, 296, 298
catecholamines 122, 125, 128, 240
cDNA 40, 61
cell fission 85
cell-mediated 179
cell membrane 90-92, 95, 97, 107, 120, 121, 126, 182, 231, 235, 236, 240, 242, 258, 291, 294
cell membranes 4, 135, 165
cellulose 10, 14, 77, 84, 146, 222, 276
central chemoreceptors 172
Central Dogma 31
central nervous system (CNS) 108
centrioles 56, 84, 93, 222, 232, 272, 276
Centrioles 56, 291
centromeres 56, 291
centrosome 93
cerebellum 111, 118, 293
cerebral cortex 111, 118, 294
cerebrum 111, 118, 294
Ceruminous glands 200
chaperones 46, 216, 271
chemical concentration gradient 75
chemical synapse 104, 105
chemoreceptors 112, 114, 172
chemotaxis 178
chemotrophs 70
chiasma 58
Chief cells 143, 296
chief (peptic) cells 143
chitin 14, 84, 87, 276
chloride shift 171
cholecystokinin 147
cholesterol 4, 73, 74, 83, 92, 121, 152, 155, 235, 246, 279, 292, 294
cholinergic receptors 110
Chordata 207, 211, 213, 214, 293, 301
chromatids 55, 56, 58, 60, 63, 274
chromatin 54, 56, 71
chromosomal mutation 48
chromosome 34, 48, 50, 54-56, 58, 60, 63, 79, 80, 206, 222, 226, 229, 230, 265, 285, 291, 300
chylomicrons 4, 152, 156, 246, 296
chyme 142, 145, 148, 152, 289
chymotrypsin 145, 148, 279
cilia 8, 93, 98, 106, 135, 140, 162, 169, 170, 295, 297
Cilia 116, 170
ciliary muscle 112
ciliary processes 112

cisterna 91
cisternal space 91
citric acid cycle 21, 26, 29
Class 207, 213, 269, 301
clathrin coated pit 90
Cleavage 136, 139
clone 40-42, 79
cloned 42
clone library 40
cloning 41, 42
closed circulatory system 164, 255
CNS 108, 110, 111, 116
coacervates 211
Coarse hair 170
coated vesicle 90
cocci 71, 292
cochlea 114, 118, 293
coding 36, 37, 48, 266
co-dominant 182
codon 45, 46, 49, 53, 290
coelom 211
Coenzymes 16, 28
cofactor 16, 170, 249
Collagen 8, 9, 97, 195, 200
collecting duct 127, 158, 162, 250, 281
collecting tubule 127, 158
commensalism 209, 210
common bile duct 145
Compact bone 195
Competitive inhibitors 17
complement 80, 90, 179, 275, 290
complementary DNA 40
complementary strands 33, 42, 290
complement proteins 90
complete dominance 204, 266, 285
cones 112, 113
conformation 6, 8, 17, 46, 73, 190
conjugated proteins 9
conjugation 79, 80, 83, 85, 230, 292
conjugative plasmid 80
connective tissue 97, 144, 145, 176, 198, 200, 202, 259, 260, 283
consensus sequence 36
constitutive heterochromatin 54
constitutive secretion 91
Control proteins 18
convergent evolution 208
cooperativity 19, 170
cornea 112
corpus albicans 134, 278
corpus luteum 134, 136, 138, 140, 242, 278, 289, 295
cortex 82, 111, 118, 122-127, 138, 140, 157, 158, 235, 249, 278, 281, 294, 295, 297
cortical reaction 135
cortisol 122, 127, 234, 242
Cortisol 127, 138, 139, 156, 240, 277
cosubstrates 16
Cowper's glands 133
cranial nerves 108, 234
cristae 96
crossing over 50, 58, 222, 272, 291

crypts of Lieberkuhn 144
C-terminus 46
cyclic AMP 12, 105, 124, 240, 241, 277, 278
cystic duct 145
cytochromes 9, 13, 27, 258, 282
Cytochromes 9
Cytokinesis 56, 61
cytoplasmic streaming 84, 93
Cytoplasmic streaming 84
cytosine (C) 33
cytoskeleton 93, 95, 293
cytosol 22, 29, 35, 37, 46, 47, 53, 73, 77, 91, 92, 100, 121-223, 235,
    240, 252, 253, 258, 259, 272, 276-278
cytotoxic 180, 184, 298
cytotoxic T cells 180, 298

D

degenerative 45
dehydration 2, 14, 46, 199, 297
deletion 38, 49, 274
deletions 49, 50, 224
denatured 8, 39, 218
dendrites 100, 105, 107, 108, 293
dense bodies 193
density dependent 208, 267, 301
density independent factors 301
deoxyribonuclease 145
deoxyribonucleic acid 33, 43
depolarization 101, 192, 236, 277
dermis 137, 199, 200
descending colon 141, 146
Desmosomes 95, 116
determination 137
determined 83, 93, 137, 216, 271
Detoxification 155
deuterostomes 211
developmental isolation 207
dextrinase 144
diapedesis 178, 179
diaphragm 169, 194, 234, 276, 299
diaphysis 195, 259
diastole 164
differentiate 134, 137, 176, 179, 180, 195, 243, 259, 262, 283
differentiation 136, 137
diffusion 10, 22, 24, 75, 76, 92, 104, 149, 151, 156, 157, 165, 199,
    296
digest 10, 39, 87, 92, 109-111, 135, 145, 147, 148, 293, 296
dihybrid cross 204, 206, 266, 300, 301
dihydroxycholecalciferol 130
diploid 54, 58, 85, 87, 226, 243, 292
disjunction 56, 60
distal tubule 158, 162, 249, 250, 280
disulfide bonds 6, 216, 217, 253, 272, 281
Divergent evolution 208
DNA 12, 14, 28, 31, 33-37, 39-45, 47, 48, 50, 53-56, 58, 61, 62, 65,
    67, 71, 79-81, 83, 86, 89, 96, 124, 216, 223-226, 229-232,
    252, 273-276, 289-292
DNA ligase 35, 39, 44, 230
DNA polymerase 34-36, 40, 44, 290
DOHCC 130

domains 70, 207
dominant 182, 203-206, 209, 210, 265, 266, 285, 300
dormant 82, 85
dorsal 107, 108, 211, 293
dorsal root ganglion 108
double helix 12, 14, 33-36, 39, 44, 224, 225
double stranded 33, 35, 36, 39, 44, 54, 71, 79, 223, 225, 229-232,
    274, 289
duodenum 141, 144, 145, 147, 148, 279, 296
Duplications 50
dynein 93

E

E. coli 37, 71, 83, 146, 231
ectoderm 137, 140, 295
effector 108, 118, 120, 121, 123, 128, 130, 234, 279, 293, 294
eicosanoids 4
elastic 77, 165, 198, 200
electrical gradient 75
electrical synapses 104, 164, 192
Electrical synapses 104
electrochemical gradient 75, 76, 98, 236, 292
electrolyte absorption 146
electromagnetic receptors 112
electron transport chain (ETC) 27
elements 41, 50, 202, 211, 229
elongation 36, 46
embryonic stem cells 136
emigration 209, 210
Endocrine 119, 123, 138, 139
Endocrine glands 119, 123
endocytosis 90, 92, 232, 291
endocytotic 91, 92
endoderm 137, 140, 295
endoplasmic reticulum (ER) 91, 235
endospores 82, 228
endosymbiont theory 96
endothelium 97, 157, 177
energy storage 14, 92, 149, 152, 195, 289
Enhancers 37
enslavement 209
enterocytes 10, 22, 55, 144-146, 149, 151, 156
enterokinase 145, 279
envelope 47, 65, 77, 84, 89-91, 222, 232, 271
Environmental sensory input 199
enzymes 8, 10, 15-20, 23, 37, 39, 74, 77, 84, 91, 92, 105, 119, 120,
    122, 124, 129, 133, 135, 142, 144, 145, 148, 151, 177, 178,
    184, 199, 216, 217, 224, 235, 240, 246, 247, 271, 272, 277,
    279, 289, 292, 293, 296
enzyme specificity 15
enzyme-substrate complex 15
eosinophils 176
Eosinophils 177, 179
ependymal cells 106, 169
epidermis 199, 200
epididymus 133, 140, 295
epiglottis 170
epinephrine 110, 122, 128, 234, 240, 241, 244, 260, 276-278
Epinephrine 116, 122, 128, 131, 138, 139, 165, 240, 241, 276, 278,
    294

gluconeogenesis 21, 127, 129, 131, 138, 151, 155, 156

glucose 10, 11, 14, 15, 19, 22, 23, 27, 29, 37, 76, 77, 92, 98, 100, 123, 126, 127, 129, 131, 138, 147, 149, 151, 155-157, 162, 218, 219, 234, 235, 240, 248, 260, 261, 272, 277, 280, 283, 294, 295, 296

glucose 6-phosphate 22, 92, 235

glyceraldehyde 3-phosphate 22, 219, 272

glycerol 3, 4, 21, 26, 73, 127, 147

glycocalyx 71, 97

glycogen 10, 14, 22, 92, 100, 127, 129, 149, 155, 156, 190, 219, 235, 240, 241, 260, 261, 277, 283, 289, 296

glycogenesis 149, 155

glycogenolysis 14, 21, 129, 149, 156, 276, 296

Glycolipids 4

Glycolysis 1, 22, 28, 29, 218, 219, 290, 296

glycoproteins 74, 217, 232

Glycoproteins 8

glycosaminoglycans 96

glycosidases 92

glycosylated 91

glycosylation 91

goblet cells 144, 156, 170

Golgi 47, 91, 92, 98, 116, 120, 152, 177, 235, 242, 293

Golgi apparatus 91, 98, 120, 152, 235, 242

Golgi complex 91, 235

gonadal hormones 122

gonads 122, 132, 137

G-proteins 18, 105, 240, 278

gradient 10, 22, 27, 29, 75-77, 79, 98, 149, 151, 157, 217, 236, 249, 259, 281, 282, 283, 292, 298

Gram-negative bacteria 77

gram-positive bacteria 77, 82

Gram staining 77

granular 91, 158, 176

granular cells 158

granular leukocytes 176

granulosa cells 134, 135, 243, 278

gray matter 106

guanine (G) 33

gustatory 114

H

habitat isolation 207

hair cells 114, 118, 293

Haldane effect 172

haploid 54, 58, 60, 84, 85, 87, 226, 243, 292

hapten 179

Hardy-Weinberg 203, 209, 210, 213, 265, 266

Hardy-Weinberg equilibrium 209, 210

Haversian (central) canals 195

HCG 20, 132, 136-139, 242, 278

head 114, 133, 135, 175, 184, 188

heart 97, 105, 109-111, 118, 131, 137, 140, 163, 164, 166-168, 192, 194, 234, 236, 246, 255, 277, 278, 290, 295, 297, 299, 300

helicase 34

helix 6-8, 12, 14, 28, 33-36, 39, 44, 83, 224, 225, 259

helix destabilizer proteins 35

helper T cell 179

helper T cells 180

hematocrit 176, 248, 280

heme 9, 13, 16, 27, 170, 254, 255, 282

heme cofactor 170

hemoglobin 8, 9, 19, 48, 145, 170-173, 176, 190, 254, 255, 260, 261, 282, 283, 298

hemopoiesis 195

hepatic cells 10

hepatic portal vein 154

hepatic sinusoids 154

hepatic vein 154

heterochromatin 31, 54, 55

heterogeneous nuclear RNA [hnRNA] 38

Heterotrophs 86

heterozygous 182, 204, 209, 265, 266, 285, 300

Hexokinase 15, 19, 22

hGH 121, 125, 126, 138, 139

high-density lipoproteins 152

high density lipoproteins (HDL) 4, 246

histamine 143, 179, 252

Histamine 178

histones 54, 71, 96

hollow nerve cord 211

holoenzyme 16

homeostasis 37, 157, 158, 235

homologous 50, 54, 58, 204, 206, 208, 285, 301

homologues 54

homoplastic 208

homozygous 182, 204, 205, 210, 265-267, 285, 300, 301

hopanoids 73, 74

Hormonal communication 99

hormones 4, 8, 14, 75, 90, 99, 119-128, 130-132, 138, 143, 145, 147, 156, 176, 193, 240-244, 260, 276, 278, 279, 281, 289, 294, 295

host 65, 71, 79, 81, 83, 229-232, 252, 275, 276, 291

human chorionic gonadotropin (HCG) 136

Human growth hormone (hGH) 126

humoral 179, 281, 298, 299

hyaline 198

hybrid inviability 208

hydrochloric acid (HCl) 143

hydrogen bond 2, 217

hydrogen ion concentration 20, 170, 289

hydrolases 19, 92

hydrolysis 2, 14, 18, 24, 34, 141, 148, 219, 246, 296

Hydrophilic 2, 28

hydrophobic 2- 4, 6, 44, 156, 235

hydrostatic pressure 77, 158, 164, 166, 248, 279, 280, 296

Hydrostatic pressure 157

hydroxyapatite 13, 195

hyperplasia 190

hyperpolarization 101, 112

hypertonic 77

hypertrophy 190, 192

hyphae 84, 85

hypodermis 200

hypothalamus 110, 111, 118, 125, 126, 131, 186, 191, 234, 242, 243, 295

hypotonic 77

I

ileum 141, 143, 144, 148
immigration 209
Immunity 181, 183, 199
immunoglobulin 179
immunoglobulins 176, 184, 253
implantation 136, 140, 295
inclusion bodies 71
incus 114
induced 15, 48, 152, 216
induced fit 15
induces 7, 137
inflammation 99, 178-180, 299
inhibin 132, 243
inhibitors 17, 19, 258
inhibitory postsynaptic potential (IPSP) 105
initiation 36, 37, 44, 46, 194
initiation complex 36, 37, 46
initiation factors 36, 46
innate immunity 178, 298
inner ear 114
inner membrane 27, 96
inner mitochondrial membrane 24, 29, 290
insensible fluid loss 199
insertion 49, 185, 186, 191, 225, 229, 274, 299
insertion or deletion 49
insulin 8, 10, 91, 100, 119, 121, 123, 128, 129, 131, 145, 147, 219, 240, 294, 295
Insulin 10, 129, 131, 138, 139, 156, 296
integral 74, 76, 91, 98, 101, 235, 259, 283
integumentary system 200
interbridge 77
intercalated disc 192
Intercostal muscles 169
intermediate 29, 94, 99, 152, 193, 272
intermediate-density lipoproteins 152
intermediate filaments 94, 193
intermembrane space 27, 29, 96
Interneurons 108, 116
interphase 55, 58, 63, 226, 274, 291
interstitial fluid 99, 143, 152, 174
intracellular second messenger 120
intrinsic factor 143
intrinsic proteins 74
introns 38, 40, 44, 223, 273, 290
inversion 50
involuntary 109, 110, 142, 192, 193, 194, 299
ionophores 104
IPSP 105
iris 113, 193, 194, 234
iron 13, 16, 155, 170, 235, 254, 277
irreversible covalent modification 18
irreversible inhibitors 17
isomerases 19
isotonic 77
isozyme 22

J

jejunum 141, 144
juxtaglomerular apparatus 158, 249

K

karyotype 205
Keratinocytes 200
ketone bodies 155
ketosis 155
kidney 10, 22, 95, 124, 127, 131, 138, 151, 155, 157, 158, 247, 248, 250, 261, 280, 294, 295, 296, 297
killer T cells 180
kinase 18, 19, 219, 235, 240-242, 258, 259, 277, 283
kinetochore 56
kinetochore microtubules 56
Kingdom 87, 207, 213, 301
Krebs cycle 12, 26, 27, 29, 96, 149, 272, 289, 290
K-selection 208, 210, 213
Kupffer cells 155

L

lac operon 37
lactase 144, 247
lacteal 144, 246
Lactose 22, 37, 147
lacZ 40, 41
lagging strand 35
lamellae 71, 195
Langerhans cells 200
large intestine 141, 146, 148, 184, 250, 296
large population 42, 209, 210
large subunit 46, 53
larynx 128, 132, 169, 170
latent 122, 232
latent period 122
Law of Independent Assortment 204, 213, 265, 266, 285
Law of Segregation 204, 213, 265, 266
leading strand 35
leaflets 73, 74
leakage channels 76
left atrium 164, 297
left ventricle 163, 164, 168, 297
lens 112
Leukocytes 176, 183
leukotrienes 4, 173
Leydig cells 132, 243, 279
LH 121, 125, 126, 132, 134, 138, 139, 140, 242-244, 278, 295
Lieberkuhn 144
ligament 185, 191
ligases 19
lipase 145, 148, 152, 246, 247, 279, 296
Lipase 145, 147, 161
lipases 92, 246, 296
lipid 3, 4, 65, 73-76, 92, 121, 122, 124, 128, 156, 222, 229, 235, 241, 294
lipid anchored proteins 74
lipopolysaccharides 77
lipoprotein lipase 152
lipoproteins 4, 152, 155, 246

Lipoproteins 4, 74, 152

liposomes 73

liver 10, 14, 22, 26, 29, 55, 92, 98, 127, 129, 131, 137, 145, 149, 151, 152, 154-156, 176, 179, 199, 235, 240, 241, 250, 261, 278, 283, 290, 293, 296, 298, 300

Local mediators 99

lock and key theory 15

locus 204, 265, 285

loop of Henle 157, 158, 162, 250, 281, 296, 297

low-density lipoproteins 152

low density lipoproteins (LDL) 4, 246

lower esophageal sphincter 142

luteal surge 134, 140, 295

lyases 19

lymph 97, 144, 152, 156, 174-176, 178, 180, 184, 186, 195, 202, 279, 296, 298, 299

Lymphatic System 163, 174

Lymphocytes 176, 177, 183

lymphokines 99, 178

lysogenic 230, 231, 275, 291

lysosomes 47, 91, 92, 261, 293

Lysosomes 92, 116

lysozyme 144

lytic 230, 275

M

macrophages 90, 176, 178, 179, 252

malignant 52

malleus 114

maltase 144

Mammalia 207, 213, 214, 301

mast cells 179

matrix 8, 13, 24, 26, 27, 29, 35, 36, 53, 91, 96, 97, 104, 176, 195, 198, 217, 259, 260, 283, 289

matrix of a mitochondrion 24

mechanical isolation 207

mechanoreceptors 112

medulla 111, 118, 122, 125, 127, 128, 138, 140, 157, 158, 169, 172, 234, 281, 293, 297

medullary cavity 195, 300

Megakaryocytes 177

Meiosis 31, 58, 60, 61, 292

meiosis I 58, 60, 278

melanin 200

Melanocytes 200

melted 39

melting 39

melting temperature (Tm) 39

membrane 4, 8, 9, 22, 24, 26, 27, 29, 42-44, 47, 56, 60, 65, 71, 73-77, 83, 89-93, 95-98, 100-105, 107, 112, 114, 118-122, 124, 126, 129, 144, 151, 152, 157, 158, 177, 179, 182, 186, 189, 192, 198, 222, 231, 232, 235-237, 240-242, 247, 250, 252, 253, 258, 272, 273, 276-278, 280, 281, 289-294

memory B cells 179, 252, 281

memory T cells 180

Mendelian ratio 203, 285

Mendel's First Law of Heredity 204

Mendel's Second Law of Heredity 204

menstrual cycle 126, 134, 140, 243, 278

Merkel cells 200

mesencephalon 111

mesoderm 137, 140, 295

mesosome 71

Metabolism 1, 14, 21, 26, 28

metaphase 56, 58, 60, 63, 222, 226, 291

metaphase II 58, 60

metaphysis 195

methylation 39

micelle 73, 152

Michaelis constant 16

microfilaments 56, 93, 94

microglia 106

microtubule-organizing center 93

microtubules 55, 56, 79, 93, 94, 98, 169

microvilli 93, 114, 118, 144

middle ear 114

midpiece 133

mineral corticoids 122, 127

Minerals 1, 13, 28

mineral storage 195, 202

Minus-strand RNA 86

missense 48

missense mutation 48

Mitochondria 89, 96, 116, 273

mitosis 55, 56, 58, 60, 63, 79, 84, 87, 126, 136, 176, 190, 191, 195, 222, 226, 231, 259, 272, 273, 275, 278, 283, 291, 292

Mitosis 31, 56, 61, 222

mitotic spindle 93

moistens 170

Monera 70, 207

monocistronic 37

Monocytes 177, 179

monosaccharides 10, 21, 22, 145

morula 136, 137, 140

Motor (efferent) neurons 108

motor unit 189, 258

mouth 22, 141, 142, 148, 246, 250

movement 75, 93, 95, 114, 133, 140, 142, 165, 173, 185, 186, 191, 195, 197, 198, 217, 271, 297, 299, 300

mRNA (messenger RNA) 35

mucous 119, 143, 170

mucous cells 143, 170

Mucus 143, 170

multinucleate 186

multiunit 193

Multiunit smooth muscle 193

murein 77

muscarinic 110, 277

Muscle 97, 185-188, 190, 192, 193, 201, 259, 260, 277, 282, 284

muscle tissue 97, 185, 254, 261

mutagens 48, 52, 224, 225, 274

mutation 48-51, 224, 225, 240, 271, 274, 301

mutational equilibrium 209

mutualism 209, 210

mycelium 84

myelin 106

myofibril 186

myoglobin 8, 190, 254, 255, 258, 259, 260, 282

myosin 177, 186, 188, 191, 258, 299

N

NADH 12, 21-24, 26-29, 219, 272, 290
Nails 200
nasal cavity 170
natural killer cells 179, 252
Negative cooperativity 19
negative feedback 18-20, 123, 126, 140, 180, 219, 279, 294
nephron 157, 158, 162, 237, 247, 249, 250, 280, 297
nerves 97, 100, 108, 111, 198, 200, 234, 259, 283, 295, 300
nervous tissue 97, 106
neural plate 137
neural tube 137
neuroglia 106
neurohypophysis 126
neurolemmocytes 106
neuromuscular synapse 189
neuron 8, 100-102, 104, 105, 107, 108, 118, 189, 193, 194, 231, 236, 237, 258, 259, 289, 293, 299
Neuronal communication 99
neurotransmitter 99, 104, 105, 108, 110, 234, 249, 276, 277, 293, 294
neurotransmitters 99, 105, 110, 119
neurula 137
neurulation 137
neutral mutation 48, 271
neutrophils 90, 176, 178, 179, 180
Neutrophils 177, 178
niche 208
nicked 80
nicotinic 8, 110
nitrocellulose 42, 43
nociceptors 112
nodes of Ranvier 106
non-coding 37
Noncompetitive inhibitors 17
nondisjunction 60, 226, 274
nonsense 46, 49
nonsense mutation 49
noradrenaline 110, 128
norepinephrine 110, 122, 128, 240, 276
Northern blot 43
notochord 137, 211, 214
N-terminus 46
nuclear envelope 47, 84, 89, 91, 222
nuclear pores 35, 46, 89
nucleases 92
nucleic acid hybridization 39, 43
nucleic acids 9, 12, 15, 148, 218, 232, 271, 293
nucleoid 71, 292
nucleolus 35, 46, 56, 89, 98, 290, 292
nucleosidases 144
nucleosome 54
nucleotide derivatives 1
Nucleotides 1, 12, 26, 28, 147
nucleus 4, 35, 36, 38, 44-47, 53-56, 71, 83, 84, 89, 91, 98, 110, 121, 122, 128, 135, 176, 177, 192, 193, 200, 222, 223, 230, 231, 240, 242, 243, 250, 255, 259, 273, 276-278, 282, 290, 292-294

O

O_2 molecule 170
oblongata 169, 293
oils 3
Okazaki fragments 35, 44, 274
olfactory 114
oligodendrocytes 106
oncogenes 52
one nucleus 71, 192, 193, 259
oocyte 58, 60, 134, 135, 243, 278, 291
oogonium 58
Oomycota 84
open system 174, 175, 282
operator 37
operon 37
order 9, 16, 19, 33, 35, 40, 53, 55, 60, 65, 76, 80, 83, 92, 95, 103, 105, 109, 114, 121, 148, 185, 210, 214, 231, 240, 244, 261, 264, 267, 271, 272, 275, 276, 278, 280, 286, 291, 294, 297-301
Order 207, 213, 301
organ of Corti 114, 293
organs 92, 95, 97, 100, 109, 128, 137, 138, 184, 195, 234, 242, 295
origin 34, 39, 47, 79, 185, 186, 191, 214, 299
origin of replication 34, 79
osmotic pressure 77, 149, 157, 166, 174, 176, 279
Osteoblasts 195, 201, 259
Osteoclasts 195, 201, 259, 278
Osteocytes 195, 201
osteon (Haversian system) 195
Osteoprogenitor (or osteogenic) cells 195
outer ear 114
outer membrane 24, 26, 77, 96, 292
oval window 114
oviduct 134
ovulation 58, 134-136, 138, 140, 243, 244, 278, 295
ovum 60, 135, 137, 140, 242
oxidative phosphorylation 22, 27, 219
oxidoreductases 19
oxygen dissociation curve 19, 170, 254, 282
oxyhemoglobin 170
oxyhemoglobin (HbO_2) dissociation curve 170
oxytocin 121, 124, 126
Oxytocin 126, 138, 294

P

palindromic 39
pancreas 22, 98, 119, 124, 128, 129, 137, 140, 145, 147, 154, 237, 246, 247, 296
pancreatic 18, 119, 121, 145, 147, 148, 156, 184, 246, 279
pancreatic amylase 145, 148
pancreatic hormones 121
Paracrine System 4, 89, 99
parallel 6, 110
parasitism 209, 210
parasympathetic 109, 110, 111, 113, 118, 164, 168, 194, 234, 276, 277, 300
parathyroid glands 130
parathyroid hormone 121, 124, 130, 131, 202, 260, 294
Parietal cells 143, 296

parietal (oxyntic) cells 143
partial 170, 173, 249
passive diffusion 76
PCR 38, 39, 42, 44-47, 62, 290
penicillin 77, 87, 292
Penicillin 17, 83
pepsin 8, 18, 142, 143, 148, 296
pepsinogen 18, 143, 296
peptide 5, 6, 18, 47, 53, 119-122, 125-130, 132, 136, 143, 147, 217, 243, 271, 272, 278, 281, 291
peptide bonds 5, 18, 271
peptide hormones 119-122, 125, 128, 147
Peptide hormones 120, 240, 281
peptidoglycan 17, 77, 83, 87, 229, 276, 292
peptidyl site 46, 291
peptidyl transferase 46
perforin 180
perichondrium 198, 259
perilymph 114
Peripheral 74, 116
peripheral chemoreceptors 172
peripheral nervous system (PNS) 108
periplasmic space 77
peristalsis 142, 145, 191, 194
peristaltic action 142
Peroxisomes 92, 116
phagocytosis 79, 90, 92, 93, 179
phagocytotic cells 178, 198, 252
phagosome 90
pharyngeal slits 211
pharynx 169, 170
phenotype 204, 205, 210, 273, 285, 300, 301
pH of 6 145, 279
phosphatases 19, 235
phosphate 4, 12, 22, 33, 34, 44, 46, 73, 92, 130, 157, 188, 197, 202, 218, 219, 223, 228, 235, 240, 258, 260, 272, 283, 289, 290, 300
phosphate group 4, 12, 22, 33, 73, 258, 272, 283, 289, 290
phosphodiester bond 33
phosphodiester bonds 12, 14, 39, 44, 289
phospholipid 4, 53, 73, 75-77, 89, 92, 96, 98, 235, 276, 277, 289, 292, 293
Phospholipids 28, 235
photosynthetic bacteria 211
phototrophs 70
Phylum 207, 213, 301
pigments 112, 113, 157
pili 79, 193, 200
piloerection 199, 234
pinna 114
pinocytosis 90, 92, 165
placenta 122, 132, 136, 137, 242, 278
Plantae 207
Plasma 86, 176, 179, 183, 252, 253, 282, 298
plasma cells 179, 180, 184, 252, 253, 298
plasma membrane 47, 71, 73, 74, 77, 83, 91, 277, 292
plasmid 40, 80, 230, 275
platelet plug 177
Platelets 162, 177, 183, 282, 297
pleated sheet 6, 7

PNS 108, 116
point mutation 48
polar body 60, 135
polarity 75, 101
polyadenylated 38
poly A tail 38
polycistronic 37
polymerase chain reaction (PCR) 42
polymorphic 43
polymorphism 208
polypeptides 5, 46, 47, 126, 145, 151, 156, 253, 282
polyploidy 50
pons 111
porin 24
positive cooperativity 19
Positive feedback 18
postabsorptive state 152
posterior pituitary 14, 121, 126, 131, 249, 281, 295
posterior pituitary hormones 121
postganglionic neurons 110, 234, 276
Post-transcriptional Processing 31, 38
post-translational modifications 46
$p + q = 1$ 209, 212, 301
pre-mRNA 38, 223
preprohormone 120
primary antibody 43
primary follicle 134
primary oocyte 58, 134, 243, 291
primary response 179
primary spermatocyte 58, 63, 226, 274, 291
primary structure 6, 8, 216, 235, 254, 271, 282
primary transcript 38, 44
primary urine 157
Primase 34
primer 34-36, 44, 290
primitive streak 137
prions 67
probe 40, 42, 43
proenzyme 18
progesterone 20, 122, 126, 134, 136-138, 140, 242, 278, 289, 295
prohormone 120
prokaryotes 31, 34-38, 53, 71, 73, 83, 96, 214, 222, 223, 225, 229, 230, 232, 273-275, 291, 292
Prokaryotes 36, 46, 65, 70, 71, 86, 225, 272, 276, 291, 292, 301
Prolactin 121, 126, 138, 139
proline 7, 14, 217, 271
promoter 36, 37
prophage 230, 232, 275
Prophase 56, 62
prophase I 58, 63, 226, 291
prophase II 60, 63, 226
prostaglandins 4, 178, 289
Prostaglandins 99
prostate 133
prosthetic 9, 13, 16, 112
proteases 92, 247, 296
protection 38, 97, 178, 195, 197, 295
Protection 199
protein kinase 18, 235, 240-242, 277
Protein metabolism 155

Proteins 1, 5, 9, 26, 28, 47, 74, 86, 91, 92, 141, 147, 151, 174, 216, 277, 282
proteoglycans 96, 198, 259
Proteolytic cleavage 18
prothrombin 155, 177
Protista 207
proton-motive force 27
proto-oncogenes 52
protoplast 77
protostomes 211
protrombin activator 177
provirus 66
proximal tubule 10, 157, 158, 162, 237, 249, 280, 296
P site 46, 53
PTH 121, 130, 139, 244
pulmonary circulation 164, 297
pulmonary veins 164, 173, 298
Punnett square 204, 205, 285, 300
pupil 113
purines 33
Purkinje fibers 164, 165
pus 179
pyloric sphincter 145
pyrimidines 33, 224, 274
pyrophosphate group 34
pyruvate 21, 22, 24, 26, 29, 219, 272, 290
pyruvic acid 22, 26

Q

quaternary structure 6- 8, 46, 216, 254, 272, 281, 282

R

rapid 84, 99, 100, 135, 186, 194, 252, 259, 264, 297
reabsorption 127, 130, 131, 157, 158, 162, 250, 294
receptor 47, 90, 104, 105, 108, 112, 119-121, 124, 129, 231, 240-242, 250, 252, 277, 278, 291, 294
receptor mediated endocytosis 90
receptors 8, 74, 90, 107, 110, 112, 119-122, 186, 199, 234, 240, 241, 244, 249, 276-278, 293, 294
recessive 182, 203-206, 209, 210, 226, 265-267, 285, 300, 301
recognition sequence 39
recombinant DNA 39, 40, 42
rectum 141
red bone marrow 195
reduction division 60
regulated secretion 91
relative refractory period 103
release factors 46
renal calyx 158
renal corpuscle 157, 158
renal pelvis 157, 158
renal pyramids 158
repetitive sequence DNA 31
replication 34-36, 44, 48, 58, 63, 79, 124, 222-226, 273, 274, 290
replication fork 34
replication units 34
replicons 34
replisome 34
repolarization 101

repressors 36, 37
residue 5
respiration 21, 22, 24, 27, 29, 96, 100, 219, 254, 258, 259, 272, 283, 290, 291, 298
resting potential 101, 107, 236, 237, 277, 293
restriction endonucleases 39
Restriction enzymes 39
Restriction fragment length polymorphisms 43
restriction site 39
reticulocytes 176
retina 112, 113, 118
retinal 112
retroviruses 229
reverse transcriptase 40, 276
Reversible covalent modification 18
RFLP 43
Rh factors 182
rhodopsin 112
riboflavin 146
ribonuclease 145
ribosome 44, 46, 47, 53, 71, 98, 290, 291
ribosomes 35, 46, 47, 53, 71, 83, 89, 91, 92, 96, 223, 232, 253, 261, 276, 277, 291, 292
right atrium 164, 168
right lymphatic duct 175, 246
right ventricle 164, 168, 297
RNA 12, 14, 28, 31, 33-36, 38, 39, 42-47, 54, 55, 62, 65, 67, 71, 86, 89, 96, 98, 223, 229, 231, 232, 273, 274, 276, 289-291, 293
RNA polymerase 34, 36
RNA primer 34
rods 112, 113
roughage 146, 149
rough ER 47, 91, 92, 120, 122, 235, 242, 277
round window 114
rRNA 35, 38, 46, 53, 89, 273, 290, 292
rRNA (ribosomal RNA) 35
r-selection 208, 210

S

saltatory conduction 106, 107, 237, 277
saltus 106
saprophytic 84, 87, 292
sarcolemma 186, 189, 258, 259, 283
sarcomere 186, 188, 191, 258, 259, 283
sarcomeres 190, 192, 193
sarcoplasmic reticulum 186, 188, 189, 191, 194, 258, 277, 283, 299
satellite cells 106
Saturated fatty acids 3
saturation kinetics 16
scala tympani 114
scala vestibuli 114
Schwann cells 106, 107
seasonal isolation 207
sebaceous 119, 200
sebaceous (oil) gland 200
secondary active transport 10, 76, 149, 249, 292
secondary antibody-enzyme 43
secondary follicle 134
secondary oocytes 60
secondary response 179, 180

secondary spermatocytes 60

secondary structure 6, 216

second messenger system 105, 124, 126, 129, 240, 250, 281, 294

secreted 47, 71, 124, 127, 134, 145, 147, 148, 157, 170, 240, 242, 243, 246-248, 280, 294-296

Secretin 147

secretory vesicles 91, 120

Secretory vesicles 91

sedimentation coefficients 46

segmentation 145

selective hybrid elimination 208

selectively permeable 76

self-antigens 180

Semen 133, 139

semicircular canals 114, 118

semiconservative 34, 290

semidiscontinuous 35

seminal vesicles 132, 133

seminiferous tubules 132, 133, 243, 295

semipermeable 75

sense strand 36, 44, 225, 274

Sensory (afferent) neurons 108

septa 84, 165

Sertoli cells 132, 133, 243, 279

serum 156, 176, 246, 279

sex chromosome 205, 226

sex-linked 205, 206, 226, 267

sex-linked disease 205

sex pilus 79, 80

side chains 5, 6

sigmoid colon 141, 146

signal peptide 47, 53, 291

signal-recognition particle (SRP) 47

signal sequence 91

silent mutation 48

Simple epithelium 97

single nucleotide polymorphisms (SNPs) 43

single stranded 35, 39, 42, 225, 230, 231, 289

single-unit 193

Single unit smooth muscle 193

sinoatrial node (SA node) 164

sinusoids 154

size 42-44, 52, 55, 65, 75-77, 103, 126, 136, 140, 145, 165, 186, 189, 217, 218, 242, 248, 264, 271, 272, 274, 280, 293

Skeletal muscle 185, 186, 191, 192, 259, 299

skin 4, 8, 49, 94, 95, 114, 118, 128, 137, 178, 186, 194, 199, 200, 202, 224, 265, 281, 295

slime layer 71, 79

slow oxidative (type I) fibers 190

small intestine 16, 20, 22, 77, 118, 131, 141, 144, 145, 147-149, 162, 246, 281, 289, 296

small subunit 46

smooth endoplasmic reticulum 92, 98, 121, 152, 156, 235, 240, 246, 253, 261, 293, 294

smooth muscle 4, 99, 104, 105, 109, 110, 142, 144, 165, 173, 174, 185, 193, 194, 200, 241, 249, 276, 278, 298, 299

SNPs 43

snRNPs 38, 223, 273

snurps 38

sodium 8, 10, 75, 98, 101, 107, 112, 124, 147, 149, 151, 157, 158, 168, 194, 217, 249, 250, 277, 280, 292, 293, 294, 296, 297

solenoids 54

somatic mutation 48

somatic nervous system 108, 110, 118, 234, 276, 299

Somatostatin 128

somatotropin 126

Southern blotting 43, 44

Speciation 208, 213

Species 207, 213, 301

specific 13, 15, 16, 18, 33, 37, 43, 65, 75, 76, 90, 99, 100, 105, 111, 119, 120, 126, 151, 164, 176, 178-180, 199, 216, 231, 253, 276, 281, 282, 290, 291, 298

sperm 58, 60, 98, 132, 133, 135, 140, 295

Spermatogonia 132, 139

spermatogonium 58, 274

spike proteins 68

spinal nerves 108

spindle apparatus 56, 63, 93, 98, 291

spindle microtubules 56

spirilla 71

spirochetes 71, 292

spliceosome 38

Spongy bone 195, 300

spontaneous 48

sporangiophores 85

spore coat 82

spores 85, 228, 292

SRP 47, 53

SSB tetramer 35

SSB tetramer proteins 35

stapes 114

starch 10, 14, 142, 149, 279, 289, 296

stem cell 176, 260

stem cells 136, 184, 200, 243, 252, 281, 298, 300

stereocilia 114

sterility 208

steroid hormones 75, 119, 122, 124, 127, 294

Steroid hormones 121, 240

Steroids 4, 28, 124, 278, 294

sticky ends 39

stomach 16, 20, 97, 141-145, 147, 148, 154, 156, 178, 184, 193, 237, 246, 250, 279, 289, 296, 298

stratified epithelium 97

striated 186, 192

structural 4, 6, 8, 47, 94, 96, 97, 202, 259, 260, 283

structural proteins 8, 96, 259, 283

substrate level phosphorylation 22, 27

substrate-level phosphorylation 23, 26, 219

substrates 15, 16, 19, 289

sucrase 144

sudoriferous 119

Sudoriferous (sweat) glands 200

Sulfanilamide 17

supercoils 54

superficial fascia 200

superior and inferior vena cava 163

support 96, 97, 100, 106, 108, 126, 195, 197, 202, 275, 300

support tissue 97, 108, 126

suppressor T cells 180

survival of the fittest 208

Svedberg units (S) 46

symbiosis 146, 209

sympathetic 109-111, 113, 118, 128, 138, 165, 199, 234, 249, 276, 278, 281, 293-295
synapse 100, 102, 104, 105, 107, 108, 110, 118, 124, 189, 234, 237, 293, 294
synaptic cleft 104, 189
synaptonemal complex 58
synergistic 186
synergistic muscles 186
synovial fluid 198, 202, 300
Synovial joints 198
synthase 19, 27, 29, 272, 290
synthetases 19
systemic circulation 164, 166, 297
systems 10, 12, 52, 97, 99, 105, 109-111, 132, 185, 224, 234, 249, 260, 261, 276, 283, 294
Systole 164

T

T$_3$ 122, 125, 126, 128, 138, 139, 294
T$_4$ 122, 125, 126, 128, 138, 139, 294, 295
tail 38, 65, 68, 69, 98, 133, 211, 214, 223, 273, 291, 303
target cell 120, 121, 240, 242, 250
taxonomical classification 207
T-cell immunity 179
Telomerase 35
Telomeres 35, 62
telophase 56, 60, 63, 222, 226, 273, 291
telophase II 60, 63
temperate virus 66
temperature 4, 16, 20, 39, 92, 112, 140, 170, 171, 173, 185, 191, 193, 199, 200, 202, 228, 234, 262, 289, 295, 299, 300
template strand 36
tendon 185, 191
termination 36, 45, 46, 53, 112
termination sequence 36, 53
Terpenes 4
tertiary structure 6, 7, 216, 218, 254, 255, 271, 281, 282, 289
testes 132, 140, 242, 243, 295
testosterone 122, 132, 138, 140, 242, 243, 244, 278, 279, 294, 295
tetrads 58, 60, 291
thalamus 111, 118
thermoreceptors 112
Thermoregulation 199, 201
thiamin 146
thick 71, 77, 91, 97, 165, 186, 191, 193, 200, 297, 299
thick and the thin filament 186
thick filament 186, 191, 299
thin filament 186, 191, 299
thoracic duct 152, 175, 246, 279
threshold stimulus 103, 107
thrombin 177
thromboxanes 4
thymine (T) 33
thyroid 122, 125, 126, 128, 130, 137, 138, 240, 244, 249, 279, 281, 294, 295
thyroid hormones 122, 128, 240
Thyroid-stimulating hormone (TSH) 126
thyroxine 122, 124, 128, 131, 294
Tight junctions 95

tissue 13, 14, 92, 95-97, 106, 108, 119, 122, 123, 126, 127, 129, 131, 132, 134, 136, 137, 144, 145, 152, 165, 176, 178-180, 184, 185, 195, 198, 200, 202, 240, 252, 254, 259, 260, 261, 265, 281, 283, 294
T-lymphocytes 180
totipotent 136
trachea 128, 169, 170, 173
traits 54, 206, 266, 285, 300, 301
transcription 35-38, 47, 48, 53, 63, 89, 105, 121, 122, 124, 126, 129, 137, 223, 229, 230, 240, 250, 273, 278, 290, 292, 294, 298
transduction 79, 81, 83, 230, 292
transferases 19
Transformation 80, 86
transition 48
transition mutation 48
Translation 31, 44, 46, 47, 53, 62, 91, 290, 291
translocation 46, 50
transport 8-10, 13, 24, 27, 29, 75, 76, 90-92, 96, 98, 100, 104, 121, 126, 149, 151, 152, 157, 158, 165, 176, 219, 240, 246, 247, 249, 250, 258, 272, 282, 290, 292-294, 296
transport maximum 157, 247
transposable elements 50
transposons 50, 229, 275
transverse colon 141, 146
transversion 48
transversion mutation 48
Triacylglycerols 3
triglycerides 3, 145, 149, 152, 156, 246, 260, 289, 296
triiodothyronine 122, 128
tRNA 35, 38, 46, 53, 273, 291
tRNA (transfer RNA) 35
tropomyosin 186, 188, 258, 299
troponin 186, 188, 258, 299
trypsin 16, 145, 148, 246, 279, 296
Trypsin 145, 147, 161
TSH 121, 124-126, 128, 138, 139, 244, 279, 294
T-tubules 188, 189, 194, 258, 259, 283
tubulin 8, 55, 93
tumbles 79
tumor 14, 52, 124, 240, 252, 273, 294
Turnover number 16
two fatty acid chains 73
tympanic membrane 114, 118
tyrosine 119, 121, 122, 125, 128
tyrosine derivatives 119, 122, 125, 128

U

unambiguous 45
unidirectional 104
Unsaturated fatty acids 3
uracil 12, 35, 223-225, 273, 289
urea 148, 151, 155-157, 162, 176
ureter 157, 162
urethra 133, 157, 162
Urey-Miller experiment 211, 214

V

vaccine 244, 253, 282
vagus nerve 164, 168, 194, 234, 277, 300

vasa recta 158

vas deferens 133

vasopressin 127, 162, 294

vector 40, 41, 81

veins 152, 163, 164, 166, 173-175, 193, 279, 298

vena cava 154, 163, 173

ventricle 163, 164, 168, 297

venules 163-166, 173

Venules 166, 183

Vertebrata 207, 211, 213, 214, 293

very low-density lipoproteins 152

very low density lipoproteins (VLDL) 4, 246

vestibular membrane 114

villi 144, 152, 297

virion 65, 230, 275

virus 40, 65, 67, 81, 230-232, 252, 275, 276, 291, 292

visceral 104, 105, 193, 234

vitamin A 4, 112

vitamin D 4, 130, 199

Vitamin D synthesis 199

vitamin K 146

vitamins 16, 155

Vitamin storage 155

vitreous humor 112

vocal cords 170

Volkmann's canals 195, 202

voltage gated potassium channels 101

voltage gated sodium channels 8, 101

voluntary muscle tissue 185

W

warms 170

Water 1, 2, 28, 77, 138, 157, 199

water absorption 146

Western blot 43

white matter 106

wild type 51

Y

yellow bone marrow 195

Z

zona pellucida 134, 135, 243, 278

Zygomycota 84, 85

zygospore 85

zygote 60, 85, 98, 135, 136, 140

zymogen 18, 91, 143, 148, 246

An Unedited Student Review of This Book

The following review of this book was written by Teri R—. from New York. Teri scored a 43 out of 45 possible points on the MCAT. She is currently attending UCSF medical school, one of the most selective medical schools in the country.

"The Examkrackers MCAT books are the best MCAT prep materials I've seen-and I looked at many before deciding. The worst part about studying for the MCAT is figuring out what you need to cover and getting the material organized. These books do all that for you so that you can spend your time learning. The books are well and carefully written, with great diagrams and really useful mnemonic tricks, so you don't waste time trying to figure out what the book is saying. They are concise enough that you can get through all of the subjects without cramming unnecessary details, and they really give you a strategy for the exam. The study questions in each section cover all the important concepts, and let you check your learning after each section. Alternating between reading and answering questions in MCAT format really helps make the material stick, and means there are no surprises on the day of the exam-the exam format seems really familiar and this helps enormously with the anxiety. Basically, these books make it clear what you need to do to be completely prepared for the MCAT and deliver it to you in a straightforward and easy-to-follow form. The mass of material you could study is overwhelming, so I decided to trust these books—I used nothing but the Examkrackers books in all subjects and got a 13-15 on Verbal, a 14 on Physical Sciences, and a 14 on Biological Sciences. Thanks to Jonathan Orsay and Examkrackers, I was admitted to all of my top-choice schools (Columbia, Cornell, Stanford, and UCSF). I will always be grateful. I could not recommend the Examkrackers books more strongly. Please contact me if you have any questions."

Sincerely,
Teri R—

About the Author

Jonathan Orsay is uniquely qualified to write an MCAT preparation book. He graduated on the Dean's list with a B.A. in History from Columbia University. While considering medical school, he sat for the real MCAT three times from 1989 to 1996. He scored in the 90 percentiles on all sections before becoming an MCAT instructor. He has lectured in MCAT test preparation for thousands of hours and across the country. He has taught premeds from such prestigious Universities as Harvard and Columbia. He was the editor of one of the best selling MCAT prep books in 1996 and again in 1997. He has written and published the following books and audio products in MCAT preparation: "Examkrackers MCAT Physics"; "Examkrackers MCAT Chemistry"; "Examkrackers MCAT Organic Chemistry"; "Examkrackers MCAT Biology"; "Examkrackers MCAT Verbal Reasoning & Math"; "Examkrackers 1001 questions in MCAT Physics", "Examkrackers MCAT Audio Osmosis with Jordan and Jon".

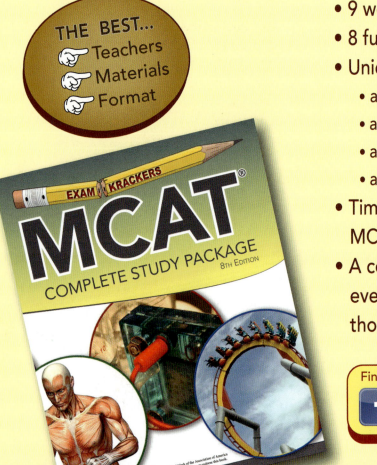